Regression & Linear Modeling

SAGE was founded in 1965 by Sara Miller McCune to support the dissemination of usable knowledge by publishing innovative and high-quality research and teaching content. Today, we publish over 900 journals, including those of more than 400 learned societies, more than 800 new books per year, and a growing range of library products including archives, data, case studies, reports, and video. SAGE remains majority-owned by our founder, and after Sara's lifetime will become owned by a charitable trust that secures our continued independence.

Los Angeles | London | New Delhi | Singapore | Washington DC | Melbourne

Regression & Linear Modeling

Best Practices and Modern Methods

Jason W. Osborne
Clemson University

Los Angeles | London | New Delhi
Singapore | Washington DC | Melbourne

FOR INFORMATION:

SAGE Publications, Inc.
2455 Teller Road
Thousand Oaks, California 91320
E-mail: order@sagepub.com

SAGE Publications Ltd.
1 Oliver's Yard
55 City Road
London EC1Y 1SP
United Kingdom

SAGE Publications India Pvt. Ltd.
B 1/I 1 Mohan Cooperative Industrial Area
Mathura Road, New Delhi 110 044
India

SAGE Publications Asia-Pacific Pte. Ltd.
3 Church Street
#10-04 Samsung Hub
Singapore 049483

Printed in the United States of America

Library of Congress Cataloging-in-Publication Data

Names: Osborne, Jason W., author.

Title: Regression & linear modeling : best practices and modern methods / Jason W. Osborne, Clemson University.

Other titles: Regression and linear modeling

Description: Los Angeles : SAGE, [2017] | Includes bibliographical references and index.

Identifiers: LCCN 2015042929 | ISBN 978-1-5063-0276-8 (hardcover : alk. paper)

Subjects: LCSH: Regression analysis. | Linear models (Statistics)

Classification: LCC QA278.2 .O833 2017 | DDC 519.5/36— dc23 LC record available at http://lccn.loc.gov/2015042929

This book is printed on acid-free paper.

Acquisitions Editor: Leah Fargotstein
eLearning Editor: Katie Ancheta
Editorial Assistant: Yvonne McDuffee
Production Editor: Kelly DeRosa
Copy Editor: Christina West
Typesetter: C&M Digitals (P) Ltd
Proofreader: Scott Oney
Indexer: Will Ragsdale
Cover Designer: Candice Harman
Marketing Manager: Susannah Goldes

16 17 18 19 20 10 9 8 7 6 5 4 3 2 1

BRIEF CONTENTS

DETAILED CONTENTS

PREFACE

When I was learning statistics in graduate school, we were taught analysis of variance (ANOVA) one semester, and the next semester, we learned multiple regression. As I was training in psychology, there was no mention of logistic, probit, or log-linear models, and so forth. There was no acknowledgment that ANOVA and regression would produce identical results if appropriately modeled. There were clearly faculty members who were "ANOVA" researchers or "regression" researchers (in other words, who analyzed their data primarily in either ANOVA or regression modalities, regardless of whether the data were most appropriately analyzed that way).

Latent variable modeling was in its infancy, and hierarchical linear modeling was being invented, but not widely disseminated yet, at least to graduate students. A series of events led me to step out of the psychology program and take a job as a statistician in a department of family medicine, where I found I was simultaneously well prepared and ill prepared for that task. I was strongly prepared in many areas, particularly in psychometrics and linear modeling, but I had never heard of logistic regression, survival analysis, or other common tools in the health sciences. Thus, I began taking epidemiology courses to try to obtain a little background in that area. This is where I was exposed to some basic logistic regression concepts. I will never forget a lecture where the professor was talking about blood pressure as a risk factor for heart attack and stroke—but every example was broken into "high versus low" (binary predictors) when the original measurement was continuous. I asked why we don't just keep the predictor continuous, and she looked at me incredulously and responded that we don't do that. But my research showed that not only could we do that, authors often did (although not commonly back then).

These scholars were stuck in methodological silos. Their students, barring fortuitous experiences outside the norm, would likewise be stuck in self-limiting silos. Once I started teaching graduate statistics courses, I committed to pushing my students outside the traditional norms, both because I thought it was interesting and because I thought it did them a valuable service. Back then, there were not good regression texts that presented anything other than ordinary least squares (OLS) regression, often not even including ANOVA as clearly part of the same family. The big names in statistical texts of the time, including Cohen and Cohen, Pedhazur, Stevens, and the classic multivariate text by Tabachnick and Fidell (all of which are still on my bookshelf), were amazing, but I found myself yearning for something that presented a more flexible, broad-ranging view of statistical modeling. Although revisions of these classic texts included expanded views of linear modeling (e.g., Cohen, Cohen, West,

& Aiken, 2003, and the fourth edition of Tabachnick & Fidell, 2001), in my opinion, they were far from integrated and inclusive enough.

Another aspect of statistical practice that I yearned for was to take the common examination of interaction (moderator) effects in ANOVA and to translate them to routine and common practice in linear models. Many texts failed to demonstrate how to incorporate these routinely into a variety of linear models. Although Aiken and West's 1991 masterpiece on interactions in regression attempted to bring those "regression" researchers into the brave new world of interaction effects (and also presented a brilliant tutorial on exploring nonlinear effects), these aspects of linear modeling were often not thoroughly integrated into texts and thus are often not included in research.

In my prior book on logistic regression, I attempted to bring logistic models to social scientists and to also share my enthusiasm for interactions and nonlinear effects (not to mention nonlinear interactions) in an accessible form. Although Menard and others had attempted to do this previously, I knew too many scholars and students who had a deep fear of (or misconceptions concerning) these types of models. Using my students as willing experimental participants, I wrote the book and used it to teach the class prior to publication, allowing me to see how students in a variety of disciplines reacted to the new tools and to adjust the pedagogical approach appropriately. In short, they seemed enthusiastic to have a new way to think about quantitative analyses where outcomes were not continuous.

As I reflected on that experience, I realized that I still felt dissatisfied—that my job was left only half accomplished. As I looked forward to teaching a regression course this past year, I decided to expand on what I had accomplished in the logistic regression book. My goal, my hubris, was so grand as to envision a text that presented a comprehensive introduction to generalized linear modeling, so that any researcher with any type of independent and dependent variable could appropriately model the data without kludgy fixes (like dichotomizing continuous variables). I also continued my grand experiment of pilot testing my text (or at least the first 10 chapters) with my own graduate students prior to publication. They seemed to really enjoy the way the material was presented, and I think your students will too.

In this book, I have attempted to coherently and accessibly present ANOVA, OLS regression, logistic models, Poisson models, and log-linear models that represent a broad introduction to the amazing spectrum of the generalized linear model. Furthermore, after introducing a diverse set of these models, I attempted to gently and coherently bring readers along with exploring curvilinear effects in a variety of models. After that, the book moves to exploring nonadditive (interaction or moderation) effects in a variety of models, and it finally integrates the last two ideas in a chapter on nonlinear interactions in a variety of models. I pushed these chapters into the book because I believe these types of effects are more common than we are led to believe by our literatures due to the fact that I believe many researchers fail to routinely examine their data for these types of effects. I wanted my students to think it was commonplace and sensible (not to mention really fun!) to examine their data in these ways. I hope you are persuaded to make these types of effects more commonplace in your courses and research.

The other ax I continuously grind in this book (and attempt to coherently model) is the idea of exploring data quality in your statistical models. Throughout the book, I use real data, often from large publicly available databases, which provides numerous opportunities to example inappropriately influential cases and how to deal with them to produce cleaner, more representative results.

In short, this book represents my attempt to create a 21st century text that will provide a broad, accessible introduction to modern statistical analysis. I mean no disrespect to those giants who came before, and on whose shoulders I have stood in creating this work. In fact, the classic texts I reference above were constant companions as I wrote this book. These authors deserve my gratitude for creating the foundation of this work (and many others). However, I hope you find this more integrative, applied approach to regression and linear modeling valuable. I think this is how modern statistics should approach data analysis, at least at the basic level. I chose not to delve deeply into the exciting world of latent variable modeling or multilevel modeling, because there are excellent, modern texts on these topics. Honestly, this book is what I wish I had access to when I first started teaching graduate statistics and what I will use for future classes. I hope you find it a worthwhile attempt at presenting a new vision of statistical modeling.

I have developed some materials to accompany the book. Everything (data sets, answer keys for selected enrichment exercises, and some guides to performing these analyses in IBM® SPSS® Statistics*—and perhaps other software packages) will be on the book's website at study.sagepub.com/osbornerlm by the time you read this. Any errata, comments, or suggestions are welcome at jason@jwosborne.com or jasonwosborne@gmail.com.

Jason Osborne
Clemson, South Carolina, 2015

---

*SPSS is a registered trademark of International Business Machines Corporation.

ACKNOWLEDGMENTS

A book like this and a career like this are developed over long periods of time, and neither is developed in a vacuum. I have been fortunate to have been surrounded by family, students, colleagues, staff, and superiors who supported, guided, and often cajoled me toward whatever success I have experienced.

As always, I acknowledge my family as the primary source of support and motivation for what I do. My parents have been lifelong cheerleaders for whatever harebrained odyssey I chose to undertake. My wife has been continually supportive of my career, despite the burden it may have placed on her. My children have inspired me to want to be better than I might otherwise have been.[1]

Throughout my career, my students have often inspired me to attempt translational communication of best practices to the applied setting. For this book, I continued the experiment piloted with my prior book. As I wrote the chapters, I used them as primary materials for a class I taught during Spring 2015 at the University of Louisville. My students at the time (Nida Ali, Billie Castle, Bridget Cauley, Kish Cumi, Shaun Digan, Patty Kuo, Shankar Naskar, Sarah Roane, Dominic Schmuck, and Katie Watterson) and especially my volunteer teaching assistants (Joy Cox and Holley Pitts) helped troubleshoot and shape the book in interesting and important ways. They deserve many thanks for their patience, careful reading of the chapters, and efforts to keep me from creating problems or confusion.

The anonymous reviewers of the book proposal and the early drafts of the book also deserve much thanks for eagerly and thoroughly engaging in that thankless task. The comments were insightful, helpful, and often inspiring.

Special thanks are due to my colleagues at several former institutions (University of Oklahoma, North Carolina State University, Old Dominion University, and University of Louisville) who challenged me to think differently and more intelligently about statistics, who asked interesting questions, or who offered to collaborate. With great colleagues, the work is easier and more enjoyable. Thanks to each of you—you know who you are.

Larry Rudner and Bill Schafer (editors of *Practical Assessment, Research & Evaluation*, which can be found online at http://pareonline.net) deserve heartfelt appreciation. When a young and very green assistant professor reached out to them and asked to join their editorial board, they kindly and generously supported me. During the intervening 15 years, they have provided suggestions and recommendations on everything from submitted articles to career aspirations. Without their prodding, many of my

---

1   The cat is specifically excluded from these acknowledgments. He knows what he did.

publications would have been far less useful, and my books might not have happened at all. Larry was particularly excellent in gently prodding me to think bigger and more broadly about some key projects. I cannot remember a suggestion from you two that was not, in retrospect, spot on. Thank you both.

Along those lines, I have benefitted from over a decade of collaboration with editor and publisher Vicki Knight. She has made the work of putting books together enjoyable and easy, and she has always made sure the final product was better than it otherwise would have been. Her insightful suggestions, patience, and extensive experience in the field were invaluable to me. Thank you, Vicki, and thanks to your excellent team.

I would also like to acknowledge the following reviewers:

Jocelyn H. Bolin, Ball State University

Clint Bowers, University of Central Florida

Martin Dempster, Queen's University Belfast

Birol Emir, Columbia University

Jeffrey D. Kromrey, University of South Florida

Bruce McCollaum, University of North Carolina at Greensboro

Craig Parks, Washington State University

Cort W. Rudolph, Saint Louis University

Michael D. Toland, University of Kentucky

Timothy W. Victor, University of Pennsylvania

ABOUT THE AUTHOR

Jason W. Osborne is currently Associate Provost and Dean of the Graduate School at Clemson University in Clemson, South Carolina, where he is also Professor of Applied Statistics in the Department of Mathematical Sciences. He is author of more than 70 peer-reviewed articles and seven books, many of which focus on best practices in statistical methods. He has also been active in research related to educational psychology and evaluation. His work has been cited in scholarly publications more than 8,900 times according to Google Scholar, and he is also an Accredited Professional Statistician (awarded by the American Statistical Association). His books are supported by a website (study.sagepub.com/osbornerlm) where ancillary materials, updates, data sets, and occasional errata can be found. Comments and questions can be directed to his personal e-mail at jasonwosborne@gmail.com or jason@jwosborne.com.

Jason is the proud father of Collin, Andrew, and Olivia (in order of appearance), each of whom he considers an outlier in the positive tail of the distribution of life. He considers himself fortunate to be married to Sherri, who has learned to accept and celebrate[1] his nerdly obsessions and tendencies.

---

1   Or at least she gives the appearance of accepting and celebrating . . . which is all one can really ask.

1

A NERDLY MANIFESTO

When I was in graduate school, my adviser and I were analyzing data from an experiment where we manipulated academic feedback to university students. Students were assigned to "succeed" or "fail" on an academic-type task, largely because we randomly assigned them (without their knowledge) to receive one of two versions of a common anagram-type task. One version contained a large percentage of anagrams that were impossible to solve, whereas the other contained many anagrams that should have been easy to solve. Thus, at the end of the timed experiment, about half of the students felt they had not done well, and about half felt they had done well.[1] We evaluated whether their self-esteem was influenced by these experiences, and as a function of several other factors that I will spare you from discussing at this time.

When it came time to analyze the data, my adviser directed me to perform an analysis of variance (ANOVA). I countered that I would rather perform a regression analysis. Her training had clearly led her to think about this as an "ANOVA" type of analysis, since the independent variable (IV; success versus failure condition) was categorical and the dependent variable (DV; change in self-esteem) was continuous. I was a graduate student who had just completed some great regression courses, and my position was that I could analyze the data in a regression framework despite the types of variables in the analysis. Who was right? It turns out we were both right, and also that the question was faulty. What was not widely taught back in the late 1980s and early 1990s in the graduate statistics courses I had taken was that both ANOVA and regression models were special cases of the generalized linear model (GLM), and that for a given set of data with appropriate variables, both will produce identical results.

The purpose of this book is nothing short of upending over a century of quantitative dogma and mischaracterization of our field. I will also grumble in various places throughout the book about anachronistic practices within our discipline (not only about ANOVA and regression still being taught as completely different courses in many graduate schools). For example, many books cover *t*-tests and simple ANOVA as completely different chapters, when in reality there is no fundamental difference.

---

1 Of course, following guidelines for ethical treatment of human subjects, they were thoroughly debriefed and reassured that they are each beautiful human beings.

One (the *t*-test) is easier to hand calculate and is also much more limited in terms of utility (it can only deal with a single binary IV). Under these conditions, both analyses produce identical results once you realize that a *t* is merely the square root of an *F*. Similarly, simple correlation and simple regression are often different chapters despite being fundamentally the same family of analyses. Simple correlation is easier to hand calculate and is much more limited (it can only evaluate simple relationships between two variables), but both produce identical outcomes for the same data set.

In the 21st century, do we really need to waste students' time learning techniques that no practicing researcher or statistician needs? I think not, and I will spend the next few hundred pages attempting to persuade you that we can streamline and improve how we teach and learn statistical methods by understanding that much of what we do, and much of what statisticians debate, leads to the same results because most of the techniques we commonly fret over are all part of the same underlying model, and they produce identical results once you know how to look at them. But don't take my word for it—I will demonstrate my points when we get to those chapters.

Fundamentally, we as a discipline need to acknowledge that (a) there is almost no utility to teaching statistics through hand calculations, (b) we should teach statistics the way that professional, practicing statisticians use the methods, and (c) our understanding of the field has fundamentally changed from just a few decades ago, yet we still teach the methods in a way little changed from decades ago.

Although there are mathematical statisticians who need to understand matrix algebra and calculus applications for statistics, few practitioners need to know these skills. And certainly, few students need to start off learning hand calculations. We no longer need techniques that provide computational shortcuts because we have virtually unlimited computer processing power on our desks and laps. I believe that what we need in the modern 21st century statistics class is a *practitioner-focused* text that mentors students in using best practices in applied statistics in an authentic way—the way actual researchers and statisticians use the methods, and the way they are likely to apply them. Just as people no longer need to understand the details of the internal combustion engine in order to be safe drivers, we do not need to spend semesters teaching and learning formulae and hand calculations. We no longer need to teach the Pearson correlation or Student's *t*-test,[2] and we no longer do anyone justice by treating ANOVA and regression as two completely separate and distinct processes.

What we do need is a cadre of statisticians who understand the best way to match the analysis technique with the research question, who understand the appropriate interpretations of the results of these analyses, and who are trained to follow modern guidelines (e.g., reporting effect sizes and confidence intervals [CIs]).

This textbook is designed to be different. It is designed to help professors teach in a different, and modern, way. It is designed to help students master concepts thoroughly so that they will be competent at data analysis when faced with their own data.

Through the journey we are about to undertake, we will explore a variety of models within the GLM, including ANOVA, ordinary least squares (OLS) regression, logistic regression, and other techniques, as manifestations of a single underlying process. We

---

2 Except as really impressive historical contributions upon which our entire field is based.

will discuss best practices in using these techniques, and we will focus on accurate and appropriate interpretation of outcomes. We will also delve into important aspects of linear modeling, such as appropriate modeling of curvilinear effects and interactions. My goal is that by the end of this text, you should have an excellent working knowledge of the most common analytic techniques. This book should be a start, not an end point, for those of you with desires to be competent, professional researchers.

I also want this to be a text that is accessible and helpful not only to those of you using it in the context of a course, but also for those of you looking to refresh your knowledge of these techniques. I want this book to stay on your shelf for the long term, not to be one of the books that gets sold back to the bookstore at the end of the semester. Toward that end, by the time this book is published, I will have developed online resources for you to use, including chapter-by-chapter guides to performing these analyses in SPSS (via point-and-click as well as via syntax) and other statistical programs, answer keys to select problems, links to data sets, and so on. I prefer to put this material on the web both to reduce the cost of the book and to also allow me to update, expand, and edit the material more easily (and in color!). Thus, if you go to study.sagepub.com/osbornerlm, you will find resources such as these for this book, which will enhance your experience.

The Variables Lead the Way

Most of you will be aware of some basic aspects of scientific measurement. Measurement is the act of assigning numbers to characteristics of an individual (or phenomenon). There are different types of measurement that we all learn about in our basic methodology courses (from most desirable to least): ratio, interval, ordinal, and nominal. Throughout this book, we will often discuss two different classes of variables as we discuss what analytic techniques are most appropriate in a given situation: categorical and continuous. Before we go much further, it is appropriate to briefly digress and discuss what types of variables you will see in the wild. What differentiates different types of measurement, briefly, is whether they contain three different ingredients:

1. Ordinality,
2. Equal intervals, and
3. A "true" zero point.

Ordinality

In all measurement, we assume that higher numbers relate to increasing amounts of something.

Equal Intervals

This is the assumption that the distance between numbers is the same across the range of measurement. This might seem a silly issue—how can the difference between

4 and 5 be different than between 10 and 11? They are both a difference of 1.0, right? In many sciences, this is something we don't have to consider. Inches and pounds and degrees and velocity are all measured on scales that have equal intervals. One inch and one pound and one degree difference is the same regardless of where we are in the range of measurement. But what about IQ? Some authors have argued that the difference between an IQ of 100 and 110 is not the same as between 40 and 50—both are increments of 10, but both have profoundly different implications for the expectations and lives of those two individuals. It is questionable whether the difference between student projects that receive an F versus a D− would be the same as between A− and A. There has certainly been debate in the measurement literature about whether differences between points on Likert-type scales meet this criterion (e.g., the difference between 1 ["strongly agree"] and 2 ["agree"] is the same as between 2 ["agree"] and 3 ["neutral"]). To meet this condition, it must be reasonable to assert that increments are identical across the entire range of a scale.

True Zero Point

This characteristic of measurement refers to whether a scale or construct has a meaningful, legitimate zero point that truly means the complete absence of what is being measured. In theory, weight has a meaningful zero point—the complete absence of weight. This is, of course, a theoretical construct, as everything on the planet has some weight (perhaps even the exotic things that particle physicists talk about). Distance is another example of something with a conceptually meaningful zero point. Constructs including zero in their range of possible numbers do not always meet this characteristic. For example, temperature measured on Fahrenheit and Celsius scales does *not* meet this goal. These scales' zero points are important, but they are not true zeros in the sense of a complete absence of something. The Kelvin scale, on the other hand, has a theoretically meaningful zero point that means the complete absence of molecular motion. Social science constructs can meet this criterion as well. Income can be 0, meaning you can have no income. The score on a final exam can be 0, or the number of days absent can be 0. However, constructs such as GRE scores, IQ, and self-esteem might not have a true zero point that is meaningful.

Different Classifications of Measurement

Ratio Measurement

Ratio measurement incorporates all three of these characteristics and is generally considered the most desirable type of measurement: Higher numbers indicate more of something, there are equal increments between numbers, and zero is a true absence of something, even if it is not theoretically possible to find an example of an individual with an observed 0. Weight is a good example—even air has weight, but the construct of weight has a meaningful true zero. Thus, weight is an example of ratio measurement, which allows us to interpret ratios of numbers, such as the fact that 5 kg is exactly twice as heavy as 2.5 kg and half as heavy as 10 kg.[3] The number of cigarettes smoked is another example of ratio measurement, because it has a true zero point; thus, we can say that smoking 12 cigarettes a day is twice as much as smoking six a day. However, because IQ does not have a true zero point, we cannot conclude that someone with an IQ of 150 is twice as smart as someone with an IQ of 75 and only half as smart as someone with an IQ of 300.

As we will see in later chapters, count data are sometimes tricky to analyze, particularly when they are *zero-inflated* (many cases with zero counts) or highly skewed. Stay tuned for a more thorough discussion of this.

Interval Measurement

Interval measurement incorporates ordinality and equal intervals—but not a true zero. For example, temperature measured in Celsius is interval. One degree is one degree, regardless of whether it is at −200 on the scale or +200. However, we cannot say that 100 degrees Celsius is twice as hot as 50. In general, it is nice to have ratio measurement, but most statistical techniques are perfectly fine with interval measurement.

Ordinal Measurement

Ordinal measurement is merely a rank ordering, containing only the characteristic that higher numbers mean more of something. For example, if you take four runners in a race, such as the three medalists from the last Olympics in the 10,000-meter race and me, the gold medal winner could be assigned a 4 (fastest), the silver medal winner a 3 (second fastest), the bronze medalist a 2 (third fastest), and I would get a 1 (fourth fastest). However, there is no way to know how far apart each individual is. We only know their relative ranking. Odds are that I am humorously far behind the bronze medal winner and the other three are relatively closely matched. Many statistical techniques are robust enough to deal effectively with this level of measurement, but this type of variable needs careful handling to avoid problematic interpretation of results.

Nominal Measurement

We will consider any of the previous three types of measurement as "continuous" for our initial discussions in this book (which will allow us to fully occupy measurement nerds writing me hate mail). Variables that I will refer to as "categorical" will fall into the fourth category of measurement: *nominal measurement.* Here, numbers are assigned to groups in an ad hoc fashion as labels only. In this way, I might assign racial/ethnicity categories in my data numeric codes[4] as follows: 1 for Black/African

---

3   We should note that the perception and effects of weight are not necessarily a ratio. Dropping a 5-kg weight on your foot is not necessarily twice as painful as dropping a 2.5-kg weight on your foot, and it is not necessarily half as painful as the 10-kg weight. Don't ask me how I know this.

4   I am aware that in many disciplines, there is great controversy over the validity of the concept of racial groupings. I share this concern and am merely using this as a convenient (and common) example, whether you agree or disagree with the process or details of grouping individuals into race-based groups.

American, 2 for White/Caucasian, 3 for Asian/Pacific Islander, 4 for Hispanic/Latino individuals, and so forth. Individuals assigned a 2 do not have more "race" than those assigned a 1, and those assigned a 3 do not have more than those assigned a 1 or 2. In this type of categorical variable, the number is merely a label, a placeholder. Thus, one cannot simply place variables with this sort of measurement in a regression equation and expect sensible results, as it violates the expectations of measurement and contains none of the characteristics mentioned above.

In most early chapters, we will deal with "continuous" and binary variables (categorical variables with only two groups; e.g., yes and no or 1 and 0). In later chapters, we will explore how to appropriately use categorical variables with more than two groups. End of digression. Thank you for your patience.

It's All About Relationships!

If you are reading (or taking a class with) this book, I assume that you have already had a basic statistics course that exposed you to simple descriptive analyses and perhaps some simple inferential statistics (e.g., simple correlations or Pearson product moment correlations, t-tests, ANOVA, simple regression, and maybe other statistics such as chi-square or odds ratios). In many classes and in many textbooks, these are taught as different analyses, with different assumptions, formulae, and interpretations.[5] Indeed, for much of the last century since these statistics were invented, this has been the case. However, more recently,[6] we (as a field) have begun talking about the GLM, which is a simple model for thinking about relationships between variables, with many of the aforementioned statistical techniques as *special cases* of this model. Thus, we can talk about t-tests as a special case of ANOVA, and ANOVA as a special case of the GLM, in which the IV is categorical and the DV is continuous. We can talk about OLS regression as a special case of GLM in which both the IV and DV are continuous (and simple correlation as essentially the same thing as regression, albeit in a more limited package). We can talk about logistic and probit regression as special cases of GLM in which the DV is binary and the IV is continuous or categorical. In fact, we can talk about all of these analyses as examining the simple relationship between two variables. In reality, what we want to know is this: Does one variable relate to, or affect, another? The type of analysis we choose should be driven not by outdated ideas or training, but by what model most appropriately fits the data being analyzed. As I will show you later in this book, the choice of analysis (e.g., ANOVA versus regression) is largely moot and, if performed properly, will yield identical results. The difference is that some analyses are *easier* to perform using one framework or another.

In Table 1.1, I have attempted to simplify the path we will take for much of this book. The general framework for most basic inferential statistics developed over the

---

5 ANOVA is from Mars, regression is from Venus? I see a best-selling novel on my horizon (and perhaps a movie deal too!).

6 By "recently," I mean for several long decades . . .

Table 1.1 Examples of the Generalized Linear Model as a Function of Independent Variable and Dependent Variable Type

	Continuous DV	*Binary DV*	*Unordered Multicategory DV*	*Ordered Categorical DV*	*Count DV*
Continuous IV	OLS regression	Binary logistic regression	Multinomial logistic regression	Ordinal logistic regression	OLS, Poisson regression
Mixed continuous and categorical IV	OLS regression	Binary logistic regression	Multinomial logistic regression	Ordinal logistic regression	OLS, Poisson regression
Binary/ categorical IV only	ANOVA	Log-linear models	Log-linear models	Ordinal logistic regression	Log-linear models

ANOVA, analysis of variance; DV, dependent variable; IV, independent variable; OLS, ordinary least squares.

past century depends on what type of variable the IV and DV are. It is most likely that in a previous statistics class, you may have learned that if you have two continuous variables, it is most appropriate for you to perform simple regression or simple correlation. If you have a categorical IV and a continuous DV, you probably learned that a *t*-test or ANOVA would be appropriate. Many, particularly those of us trained in the social sciences, would be left a bit flummoxed if presented with a categorical DV; however, in many other areas of science (especially health sciences or biological sciences), the quick answer is that logistic or probit regression would be appropriate and desirable in that situation.

A Brief Review of Basic Algebra and Linear Equations

Before we go much further, I thought it might be good to review some basic algebra around the type of linear equations we are going to use throughout this book. You might remember learning about line equations sometime in secondary school, specifically that a line

- Can always be described by exactly two points,
- Extends infinitely in both directions, and
- Can be completely described as a function of two pieces of information (which can be derived from the two points mentioned above) as follows:

  o The intercept (b_0), where the line crosses the Y axis and X = 0; and
  o The slope (b_1), which quantifies how much the line changes in Y for each increase or decrease of 1.0 in X.

We usually write the equation in the form of $Y = b_0 + b_1 X_1$.

Because lines extend infinitely in each direction, we can use lines to predict values that might not be observed or part of the data. Let us take a completely invented example of eating vegetables and health[7] as measured by diastolic blood pressure (that's the second number in the blood pressure measurement). In my fictitious example, presented in Figure 1.1, you can see some data points plotted on a line. Because we can see the intercept (where the line crosses the Y axis and X = 0), we know that the intercept (b_0) is 100. We can calculate the slope as the change in Y divided by the change in X (Equation 1.1):

$$\text{Slope} = \frac{(Y_2 - Y_1)}{(X_2 - X_1)} \text{ or } \frac{\Delta Y}{\Delta X} \tag{1.1}$$

Therefore, when the line moves from X = 0 to X = 1, the Y values move from 100 to 90, leaving us with the following:

$\text{Slope} = \dfrac{90 - 100}{1 - 0}$, which leaves us a slope of –10. Thus, we would write the line equation for this example as follows:

$$Y = 100 + (-10)X_1$$

This equation completely describes everything we know about this line, and everything we need to know. From it, we can predict values for Y given any value of X.

Figure 1.1   Fictitious Example of a Line

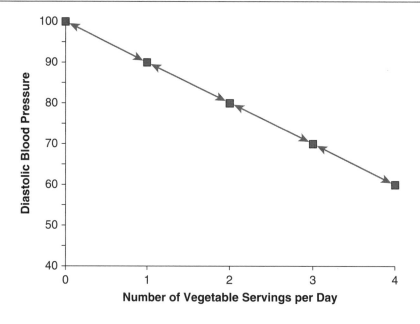

We know that when X = 0, Y = 100. We know that when X = 1, Y = 90. What if that was all we knew? What if someone eats five servings of vegetables a day?

$$Y = 100 + (-10)5$$
$$Y = 50$$

As we move into regression analyses, we must be careful not to extrapolate beyond our data, or beyond reason. What if we think someone eats 20 servings a day? Enter that into our equation and we can estimate a predicted value of Y as −100, which is nonsensical (blood pressure cannot be negative . . . even if you are deceased). What if we enter −5 for X? Blood pressure would be 150, which is not dangerously high but is also impossible, because nobody can eat negative servings of something. All of this is to say that line equations are valuable tools, and they can also be used inappropriately. We will try to avoid the latter.

The GLM in One Paragraph[8]

Simple regression uses the principles of a line, discussed above. Specifically, in simple linear regression, we use a "line of best fit" to summarize a distribution of data. Of course, because we are starting with a bunch of data points that do not necessarily fit exactly on the line, we need to add another concept not present in algebra, the error term. This is the difference between an actual data point and the line of best fit that summarizes the data (imperfectly). I have generated an example (again, entirely fictitious) in Figure 1.2, to represent this new dimension to our line.

In Figure 1.2, you can see that there are data points that appear to show a general relationship between two variables, and that we can use a line to summarize the nature of this relationship (more on how that is estimated in Chapter 2). Because not every data point falls exactly on the line, we also have errors, or residuals, in the equation. These are designated as A, B, and C in Figure 1.2. A residual or error is merely the difference between what would be predicted (the prediction is the line itself, incidentally) and the actual data point. Therefore, you can see that A is relatively large and represents a data point that is underpredicted (i.e., the predicted score is less than the actual score), B is smaller but related to a data point that is overpredicted, and C is almost on the line.

We still have a regression line, with the same characteristics as before, giving us a predicted value for Y, which we label $\hat{Y}$ (to distinguish it from the actual value for Y, the observed data; Equation 1.2):[9]

$$\hat{Y} = b_0 + b_1X_1 \tag{1.2}$$

---

8 Give or take a little . . .

9 I realize the irony of having just finished grumbling about how we don't need more textbooks with equations and then presenting equations. They will be used only when critical and will be useful later in the book.

Figure 1.2   A Fictitious Example of a Regression Line With Data

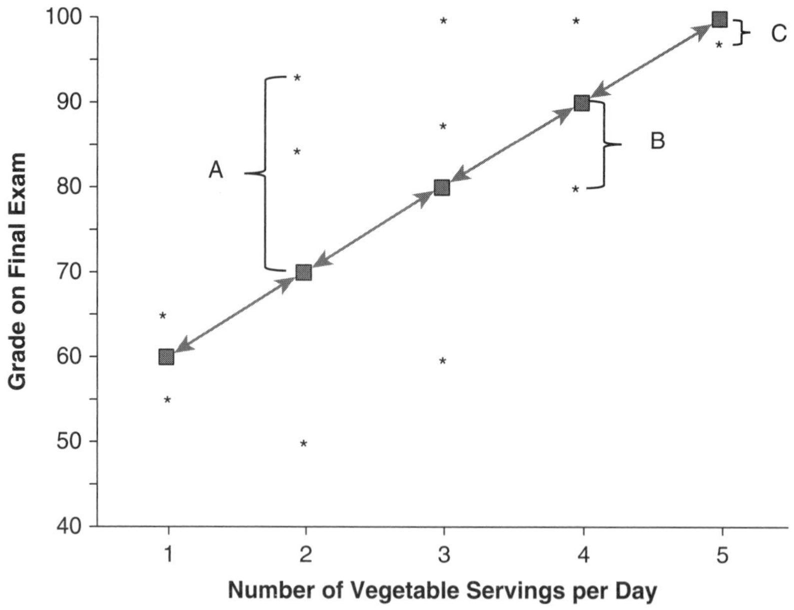

and we can also model the actual value for Y as a function of three pieces of information (Equation 1.3):

$$Y = b_0 + b_1 X_1 + e \qquad (1.3)$$

This equation simply means that we can decompose any person's score on a variable (Y) into several ingredients:

1. A starting point (b_0)—usually the intercept, or the place where a line intercepts an axis.

2. Slope—the expected change in the variable (b_1) as the result of where the person is on another variable (X_1).

3. The leftover stuff that is not explained by either of the other two ingredients (e)—often referred to as the residual, the unexplained variance, or error. These are all synonymous terms.

Typically, you will see equations like this in statistics textbooks related to regression (OLS or logistic or probit), but rarely will you see these in discussions of ANOVA (or *t*-test, or even correlation). This equation simply indicates that any individual score on one variable (i.e., the DV) is the function of (or can be decomposed into) a starting point (the intercept), the effect of another variable (the slope of the IV), and an unexplained portion (the residual). Intuitively, this view fits our traditional understanding of regression, but how does it relate to *t*-test and ANOVA types of analyses? If, instead of "intercept," we call b_0 (the starting point) the "grand mean" and b_1 the "effect of group membership," then we have a simple (one-way) ANOVA (ANOVA-type analyses also have residuals, or within-group error). In Chapter 4, I will demonstrate this and show that you can exactly reproduce the results of a simple ANOVA with an appropriately modeled regression analysis.

Furthermore, if we change the DV to categorical and add a logit or probit link function (more on link functions later), we have data appropriate for logistic or probit regression. We will explore this concept much more deeply (with empirical examples) in Chapter 4.

A Brief Consideration of Prediction Versus Explanation in Linear Modeling

There are two basic and general uses to which one can apply generalized linear modeling (usually regression): explanation and prediction. Take the silly and fictitious example in Figure 1.2, which compares the relationship between how many servings of vegetables a day a student eats and how well that student does on an exam in school. This is a simple algebra problem—the slope of the line is 10 (Y increases 10 for each 1 increase in X). We don't see an intercept on the graph (value of Y when X is 0), but we could either impute the answer or graphically determine it, as all lines extend infinitely in both directions. Knowing that the slope is 10, the intercept should be 50. This is an example of prediction, and of uses of lines. Therefore, we can create a line equation for this example as follows (Equation 1.4):

$$\hat{Y} = 50 + 10X \tag{1.4}$$

Of course, knowing that lines extend infinitely in all directions (even outside the data you currently have or used to create a line), we can have infinite fun with this line. If we know that Student A had three servings of vegetables, we can predict that this student should score an 80 on the exam in question. What if the student is vegetarian and ate 10 servings? That student should score 150 on the exam, even if the theoretical upper limit is 100. What about if the student has −3 servings of vegetables? That student should score a 20.

This example highlights the power of prediction and also warrants some caution. It allows us to predict, with greater or lesser accuracy, what someone should score on a variable as a function of another variable. The downsides, however, include the possibility that

1. Your variable is not a good predictor of the outcome,

2. You can predict irrational or impossible values,

3. You can run into difficulty if you predict data in ranges outside the original data, and

4. The prediction equation may be based on a sample that is not representative of the population, and thus may not produce valid predictions for cases not in the original sample.

In many areas of social and behavioral sciences, prediction is probably not used as often as explanation. When we use regression for explanatory purposes, we are trying not to predict outcomes for specific individuals; rather, we aim to understand phenomena by understanding the variables that (at least partly) explain outcomes related to those phenomena. The latter use, which in my experience is the more common use of linear modeling, leads us to a discussion of hypothesis testing.

A Brief Primer on Null Hypothesis Statistical Testing

Much of this book will be devoted to making good evidence-based decisions, which is a large part of what statistical procedures are used for. This entire book is devoted to inferential statistics, in which we are using probabilities and other statistical information to inform our decision making. For most of the last century, this has been operationalized as null hypothesis statistical testing (NHST). This is but a brief peek into some of the basics of NHST as well as some of the discussions around how to make more informed decisions outside of NHST.

NHST has been reviled in the modern literature by many as counterproductive and misunderstood (for an excellent overview of the issues, see Fidler & Cumming, 2008; Killeen, 2008; Schmidt, 1996). Many authors have acknowledged the significant issues with NHST, and some have proposed alternatives (Killeen, 2008; Schmidt, 1996). To understand the issue, we must delve into NHST a little deeper. The basis for NHST has roots in the very earliest aspects of statistical practice, where it was a source of bitter debate and disagreement. Despite this, it has become the de facto norm in most of the quantitative methods, and we must simultaneously work with it and seek to augment it in the best way possible.

In NHST, a scientist proposes two hypotheses. The first, a null hypothesis (H_0), is generally a hypothesis that there is no effect (no relationship, no difference between groups, etc.) in the population, and this can be stated in very simple forms for zero order correlation or simple regression (respectively) as follows:

$$H_0: r_{xy} = 0; \text{ or}$$
$$H_0: b_1 = 0; \text{ or}$$
$$H_0: \bar{x}_1 = \bar{x}_2$$

Conversely, alternative hypotheses are generally what the researcher expects to be the case in the population and is often misstated as the presence of a "significant effect" (i.e., a relationship between two or more variables, a significant odds ratio, significant mean differences between groups, etc.).

$$H_a: r_{xy} \neq 0; \text{ or}$$
$$H_a: b_1 \neq 0; \text{ or}$$
$$H_a: \bar{x}_1 \neq \bar{x}_2$$

In brief, how we typically and traditionally perform hypothesis testing is that we gather data, hoping that these data will accurately represent the nature of the population of interest. We will then calculate a statistic (e.g., F, t, or χ^2) that is related to a distribution. Specifically, what we are asking when we do this is "What is the probability that we would get the observed data in the sample if the null hypothesis was true in the population?" The larger the statistic is, the lower the probability that the observed data came from a population with the characteristics of the null hypothesis (H_0).

A Trivial and Silly Example of Hypothesis Testing

To begin exploring hypothesis testing, we will use some simple data from the Centers for Disease Control and Prevention 2001–2002 National Health and Nutrition Examination Survey.[10] Let us imagine that we are curious whether height and weight are related in some way—in other words, we suspect perhaps that taller people tend to weigh more, and shorter people tend to weigh less. Our null hypothesis would be that there is no relationship between these two variables in the population of interest, and our alternative hypothesis would be that there is a relationship (in other words, that the slope or correlation would be different from zero in the population). Without access to inferential statistics, we could have made anecdotal observations and argued about whether there is a strong relationship or not. Better yet, we could graph those observations, as shown in Figure 1.3.

It certainly looks as though taller people tend to weigh more, but descriptive statistics leave much open to debate. For example, there is a lot of variability in weight at any given height. For individuals who are 180 cm tall (just under 5 feet, 11 inches), the range in weight is about 50–170 kg (around 110–375 lb). Looking at individuals who are 100 cm tall, there is much less variability. Also, imagine a much smaller sample, which could show a very different picture, depending on a variety of factors. Therefore, depending on many details, a researcher could argue that height has a great deal to do with weight or not so much at all.

In inferential statistics and NHST, we try to make objective decisions about what we are going to conclude. We could calculate a simple correlation or regression coefficient, and test our hypotheses. Calculating either a correlation or simple regression will tell us that there is a reasonably strong relationship between the two variables within the sample. In this case, we see a correlation or standardized regression coefficient of $\beta = 0.80$. That seems large, but how do we know whether we should reject the null hypothesis and what we should conclude about the nature of the relationship between height and weight in the population?

We calculate another statistic, t, with a known distribution at a given sample size, which tells us the *probability that a sample drawn from a population with zero correlation between the two variables could produce the observed statistic.* If the t is sufficiently large (more on the actual calculations later in this book), the probability (p) drops

---

10 Data and more information are available at http://www.cdc.gov/nchs/nhanes/search/nhanes01_02.aspx. All data sets used in this book will be available on the book website at study.sagepub.com/osbornerlm.

Figure 1.3   Relationship Between Height (in Centimeters) and Weight (in Kilograms)

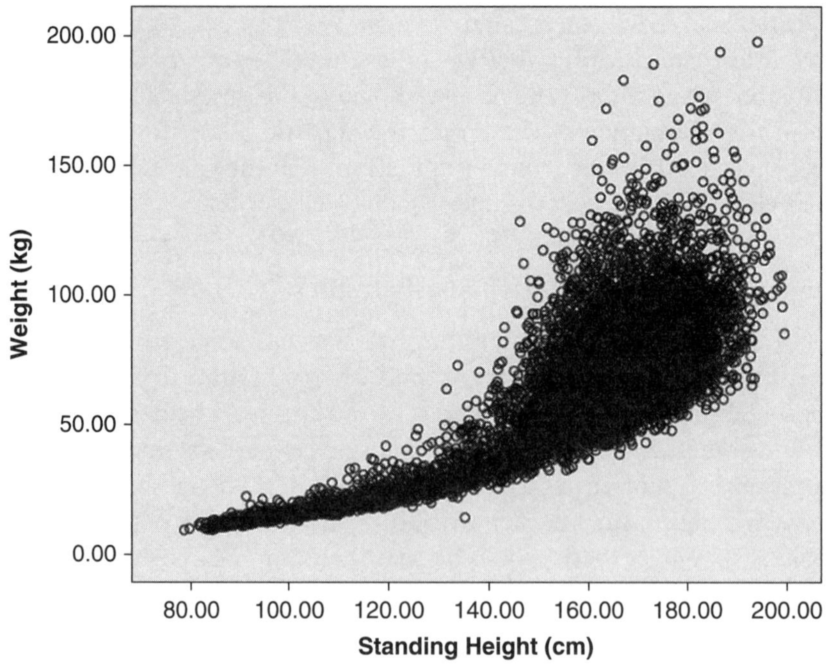

DATA SOURCE: 2001–2002 National Health and Nutrition Examination Survey, courtesy of the Centers for Disease Control and Prevention (http://www.cdc.gov/nchs/nhanes.htm).

below a target value of 0.05. When $p < .05$, it is standard practice to conclude that it is unlikely that the data were drawn from a population where the null hypothesis is true. Thus, we reject H_0. This leaves us with the alternative hypothesis, H_a, which concludes that the relationship is not equal to zero. This might sound weird, because we *know* the relationship is not zero—in fact, we calculated it to be 0.80! Herein lies an important concept—that *we are attempting to draw conclusions about the population of interest from our sample.* Therefore, it is all well and good that the sample has a relationship of $\beta = 0.80$, but the real question is whether it is reasonable to conclude that the relationship is not zero in the population. We can theoretically obtain a relationship of 0.80 if we have a small, unlucky (or biased) sample from a population in which there really is no relationship. Thus, in NHST, we can use several pieces of information to calculate how likely it is that we are incorrectly drawing a conclusion about the population of interest from our sample. In this case, with more than 9,000 individuals in our sample and a large effect observed, we can be fairly confident that the population relationship is not zero. In fact, based on the data I just summarized, the probability that the population relationship between weight and height is zero was calculated to be less than 0.0000001.

Furthermore, we should attempt to draw conclusions about the likely magnitude of the relationship within the population, rather than merely stop with the conclusion that it is not zero. More on that in a moment.

A Tale of Two Errors

We will continue to utilize NHST in this book, knowing that it is tradition (albeit a flawed and misunderstood one), and we will also seek to incorporate modern practices to supplement the information we obtain from NHST. Before we get into that discussion, we should briefly touch on Type I and Type II errors, two errors of inference commonly discussed and used occasionally later in this book.

In NHST, we face two different possible decisions regarding our hypotheses and two different (yet unknowable) states of "reality" in the population. Broadly, we can decide either to reject or not to reject the null hypothesis. In the unknowable reality of the population we are trying to investigate, we can also have a relationship that is or is not different from zero. As Table 1.2 shows, this gives us four possible outcomes, two of which are potential errors.

Thus, we hope that our data lead to a correct inference concerning the population, but it is possible that we make either a Type I (concluding that there is a relationship in the population when in reality there is none) or Type II error (failing to conclude there is a relationship in the population when there is one). As quantitative methods evolved early in the 20th century, a primary focus was on minimizing the probability of making a Type I error (i.e., concluding there are effects when in fact there are not). For example, if I am testing a new drug on patients and comparing them with placebo/control groups, I want to be very sure that the new drug is actually producing significant differences before I recommend doctors prescribe that drug. Likewise, we want to be relatively certain that a psychological or educational intervention will produce the desired differences over existing interventions before we recommend implementation. In the earlier decades of the 20th century, this decision rule ($\alpha = 0.05$) was more flexible (which is why you can still find reference to "setting alpha" in many other statistics texts—but you will not find it in this one!); although at this point, it seems that we routinely expect that alpha is fixed at 0.05, meaning that we give ourselves only a 5% chance of making a Type I error in our decision making.

Why not set the bar at 1% or 0.01% so that we are very certain of not making an error of this type? We could do that (and historically, many scholars have), but in

Table 1.2   Hypothesis Testing and Errors of Inference

| | | Population or Unknowable "Reality" | |
		There Is No Effect	There Is an Effect
Decision Based on Data Gathered	There is no effect	Correct decision	Type II error
	There is an effect	Type I error	Correct decision

doing so, we would drastically increase the odds of making the other type of error: a Type II error (concluding there is no effect when in fact there is an effect). Thus, the community of statistical practice informally settled on 5% as a compromise.

What Conclusions Can We Draw Based on NHST Results?

Significance tests *do not* tell us several critical things about our results. First, p values do not tell us the probability that the results would be replicated in a subsequent sample or study. In fact, it is power that gives us insight into the probability of replication given identical circumstances (Schmidt, 1996). Second, and most important, significance tests *do not* tell us the importance of a particular effect. We often see researchers in some disciplines use terms like "marginally significant," "significant," and "highly significant" to indicate ever-smaller p values. Yet p values are determined by multiple factors, including sample size, alpha, and effect size. As we discussed above, a very small effect in a very large sample can be statistically "significant" (i.e., have a very small p value, leading to rejection of H_0) but practically unimportant. Likewise, a large effect in a small sample might have a p value that does not cross the .05 threshold but might be otherwise important. In neither of these scenarios do we know anything about the probability of replicating the result unless we know the power of each test. Finally, p values do not tell us anything about the probability of making a Type II error (failing to reject a null hypothesis when there is a significant effect in the population). Only power can tell us the probability of making this type of error.

One common criticism of NHST is that in an absolute sense, the null hypothesis is almost always wrong. No relationship will be exactly zero if taken to enough decimal places (Cohen, 1988), so in an absolute sense, the basic function of NHST is a bit absurd. We can imagine a correlation of 0.00001 which is not equal to 0.00000 in the strictest mathematical sense; however, for practical purposes, the two are functionally the same. Given a large enough sample, an observed $r = 0.00001$ could lead to rejection of the null hypothesis, allowing us to conclude that the relationship in the population is probably different from zero. However, this does not mean a relationship that small is important. Conversely, it is possible to have a correlation, like 0.40, that looks very different from 0.00, but we might not be able to reject the null hypothesis at $p < .05$ because of low power or sample size.

Another criticism of NHST is that it is answering a question few researchers are interested in asking. As stated already, NHST as it is usually implemented asks the probability that we could get the observed data given a population where the null hypothesis is true. We are usually more interested in determining the probability of whether our alternative hypothesis is true. This more useful question is the one we do not answer directly with NHST (although CIs, effect sizes, and CIs around effect sizes can help). Thus, some scholars have argued that NHST is (a) misused broadly, (b) usually misinterpreted, and (c) mostly irrelevant to the research questions we really care about, leaving little rationale for continuing its use beyond that of blindly following tradition.

We seem, as a field, to be unwilling to part with tradition, so the next best thing to do is to encourage researchers to examine multiple pieces of information when reporting research and drawing conclusions. Inferential tests such as NHST were developed to *guide* researchers in determining whether they could conclude that there was a significant effect or not (Fisher, 1925), but these tests are not sufficient by themselves (decades of arguments of this nature are nicely summarized in Wilkinson, 1999).

In summary, NHST gives us one piece of information: how likely it is that the observed effects could have come from a population where the null hypothesis is true. Once we reject H_0, we are left with large and important questions, such as "If the population effect is not zero, what is it likely to be?" More on that in a moment. First, let us briefly review errors of inference.

So What Does Failure to Reject the Null Hypothesis Mean?

We must distinguish between *failure to reject the null hypothesis* and *acceptance of the null hypothesis*. This might seem like semantic tomfoolery, but the difference is conceptually critical. If you fail to reject the null hypothesis (i.e., if you do not get a *p* value of less than .05 when performing NHST), two different possibilities exist: (a) that you have no clear information about the nature of the effect (failure to reject the null hypothesis), or (b) that you have sufficient information about the nature of the effect, and you can conclude that the null is accurate (acceptance of the null hypothesis). It is very different to state that we cannot draw any conclusions than to state that we know there is no relationship. Unfortunately, these two statements often become conflated in traditional implementations of NHST. In the seminal works by Fisher (1925), it is not clear that he intended failure to reject the null hypothesis to mean the *acceptance* of the null (Schmidt, 1996).

This is an important distinction. Imagine the situation where you are evaluating two educational interventions, one that is very simple, traditional, and inexpensive, and one that uses millions of dollars of instructional technology in an attempt to improve student outcomes. Failure to reject the null could mean that you have insufficient information to draw any inferences, or it could mean that the two interventions are not producing identical outcomes. The ability to make a strong statement that two interventions produce the same effect is important from a policy (and in this case, financial) perspective. It means school districts could save taxpayers millions of dollars every year by implementing the "traditional" intervention in lieu of the high-technology intervention, as outcomes are identical. By contrast, not having enough information means just that: there is no conclusion possible.

The difference between being unable to draw no conclusions and being able to conclude that the null hypothesis is valid is related to the power of the study. If the study had sufficient power to detect appropriately sized effects and we did not detect any effect, we can be more confident in concluding that the null is supported. If the study did not have sufficient power to reliably detect appropriately sized effects, then no conclusion is possible. This is a common misconception in the scientific literature (Schmidt, 1996) and yet another reason to ensure you have the appropriate power in your research.

Moving Beyond NHST

Other Pieces of Information Necessary to Draw Proper Conclusions

As mentioned above, the p value and rejection of the null hypothesis (or not) is one part of the traditional equation, but it is not sufficient for statistical inference in the modern era. A rejected null hypothesis tells us what the population effect is likely *not to be* (i.e., likely not to be a null effect) but doesn't inform us as to what the effect is likely to be. Many recently developed guidelines (e.g., American Psychological Association; many journal guidelines) recommend or mandate that authors also report estimated effect sizes, CIs, and, where possible, CIs for effect sizes. I hope these make sense in context of the previous discussion.

Where NHST tells us what is likely not to be the case in the population, CIs help us estimate what the population parameter is likely to be (or, at least, the range it is likely to be found within). Taking our previous example of height and weight, we reject H_0 and have a *point estimate of the population parameter* of $\beta = 0.80$. But how precise or accurate is that point estimate? CIs can tell us the likely range of the true population parameter.

The most common CI is a 95% CI, which estimates a range within which we can be 95% confident that the true population parameter resides. Thus, if we had a small sample and poor precision, we might have a 95% CI of [0.20, 0.95], meaning that we are 95% confident that the true population relationship between height and weight is somewhere between $\beta = 0.20$ and $\beta = 0.95$. Given that β has a maximum range of -1.00 to 1.00, this tells us that we are not estimating the population parameter with a great deal of precision. On the other hand, if the 95% CI is [0.79, 0.81], then we can be very pleased with the precision of our estimate, and we also have a good sense of how strong the relationship is in the population.

Effect sizes are also important to understanding how strong these significant effects can be. We will talk about the details of various effect sizes as we move through various analyses in chapters to come, but let us start here with a simple dichotomy. One type of effect size attempts to quantify the importance of the effect by describing what proportion of the variance in the DV is attributable to the IV. Is a correlation of 0.10 important or unimportant? It translates to 1% of the variance accounted for, leaving 99% of the variance in the DV unaccounted for. By contrast, a correlation of 0.80 translates to 64% of the variance in the DV accounted for by the IV. That is more impressive.

Another type of effect size is an attempt to standardize and quantify the magnitude of the effect in a different metric altogether. A standardized regression coefficient or correlation is already an effect size, as it is standardized and substantively interpretable. But what if you are performing an ANOVA and you have two group means 10 points apart. Is that a large difference or small? In part, it depends on the standard deviation (SD).[11] Imagine that 10-point difference is observed where there is a SD of 100 points—dividing the difference by the SD leaves us with a small effect size of 0.10. Alternatively, if the SD were 1.0, then the effect size would be 10. We can use this type of effect size to compare effects across variables with different ranges and SDs, or even across different studies.

---

11   I am simplifying here for some conceptual clarity. Don't flame me yet!

Ideally, we could combine the two in some meaningful way, to estimate the likely magnitude of the effect in the population. Thanks to Bruce Thompson and other scholars in the area, we are slowly moving the field toward acknowledgment that we need to use our data for more than NHST. CIs, effect sizes, and CIs for effect sizes can help us estimate what the population parameters are likely to be once we conclude that it is likely not zero.

The Importance of Replication and Generalizability

I will never forget a class in graduate school when a guest speaker was talking about her research on resiliency in students at risk for adverse educational outcomes, using ethnographic and other qualitative research methods. I thought her results were fascinating but limited, in that she only studied three girls attending a single high school located in a particularly impoverished area. I naively asked her whether she thought we could generalize any of her findings to understand the plight of other students facing particular challenges. Her response to me was that she did not care to generalize beyond those three particular girls, and she did not particularly care if anyone else was interested in her research.

What I did not say to her at the time, but what I thought, was "Why should I (or anyone) care about your research if it only applies to those three girls and has no implications for anyone outside of them?" We often find important information in case studies such as this, primarily because case studies can reveal larger truths that apply to people outside the case study. I actually thought the researcher short-changed herself, and that her research did have important implications for understanding resilience in people. However, I continue to ponder her response. If someone conducts inquiry into something that truly has no implications for anyone else, and that nobody else could ever possibly care about, is it research? Or perhaps more specifically, is it worthwhile research?

Let me be clear—in the decades since that class, I have worked with many qualitative researchers on important and fun projects. I believe *good* research—qualitative, quantitative, or mixed methods—should tell us something about the world and the creatures that inhabit it. No one particular study can provide all of the answers to a question, but good research can lead the way. Single case studies are common in some fields (e.g., law, business, medicine) because they point the way to an issue that has broader implications beyond that one single individual. This is not a criticism of qualitative research. It is, however, a question that applies to any research, as raised recently by Richard Horton (2015), editor in chief of *The Lancet*. In his brief editorial, Horton cogently frames the highly sensitive issue scientists have been dealing with for more than a century: whether our results are reliable and reproducible. He writes:

> The case against science is straightforward: much of the scientific literature, perhaps half, may simply be untrue. Afflicted by studies with small sample sizes, tiny effects, invalid exploratory analyses, and flagrant conflicts of interest, together with an obsession for pursuing fashionable trends of dubious importance, science has taken a turn toward darkness. (Horton, 2015, p. 1380)

Dr. Horton goes on to hold culpable researchers seeking to create a cult of personality for themselves and who resist any call to improve scientific practice, as well as journals competing for ever-stronger impact factors and who resist any change that could reduce the problem (he does not mention the issue of predatory journals, but they are also culpable). He also points a finger at universities themselves for rewarding "productivity" and impact metrics, rather than true contribution. According to Dr. Horton (2015), "Part of the problem is that no-one is incentivized to be right" (p. 1380).[12] Other authors in our field, such as Bruce Thompson and Geoff Cumming (among a list of many notable scholars on the topic), have been calling for reform for years.

Nobody thinks the problem is intractable. The problem is that few are willing to change their practice to fix the problem. In that spirit, I seek to create a book that guides the reader toward more defensible practices. We will not solve a centuries-old problem with one book. However, by the end of this book, I hope your practices will be at least partially focused on creating generalizable, replicable, transparent science.

Where We Go From Here

I hope this introduction of some general concepts around inferential statistics and the GLM was helpful. If you feel a bit fuzzy on the concepts, have no fear! We will revisit these concepts repeatedly as we move through simple and increasingly complex analyses within the coming chapters. As we move through each type of statistical analysis, we will talk about these best practices, and we will give examples of how to incorporate them into your research. Don't forget to refer to my website (http://jwosborne .com) for access to data sets, answer keys, help with statistical software, and other resources keyed to enriching each chapter. For example, answers to this chapter's enrichment are available on the book website so that you can verify your mastery of the material before moving to the next chapter.

ENRICHMENT

1. Describe whether the following are examples of nominal, ordinal, interval, or ratio measurement.

   a. How fast an individual can run 100 meters
   b. Political party affiliation
   c. Ranking of top restaurants in your city
   d. Professor rating of teacher effectiveness on a scale of 1–5
   e. Number of children in a family

---

12 We leave aside, for the moment, situations in which scientists act indefensibly and unethically, such as recent highly publicized cases of falsification of data. The issue of external verification, which Horton highlights as an option that some fields have embraced, would help police this issue as well as some other undesirable behaviors.

  f. How happy you were today on a scale of 0–10

  g. Day of the month (first, third, or seventh day of the month)

2. Identify each of the following as an appropriate conclusion, Type I error, or Type II error.

  a. A researcher concluded that 12-year-old girls are taller than boys at the same age; in the population, this is accurate.

  b. A researcher concluded that there is no relationship between gender and how well someone learns statistics; in the population, there is no relationship between gender and statistical acumen.

  c. A researcher concluded that there is no relationship between height and weight; in the population, there actually is a strong relationship between height and weight.

  d. A researcher concluded there were significant differences between Republicans and Democrats in terms of attitudes toward deep-dish pizza; in fact, there were no differences in the population.

3. Without looking back at the chapter, can you describe the following?

  a. The difference between a null and an alternative hypothesis

  b. What a p value means or represents

  c. Some criticisms of null hypothesis statistical testing (NHST)

  d. What effect sizes can tell us that NHST cannot

  e. What CIs can tell us that NHST cannot

4. Identify the slope and intercept for each equation below, and graph them.

  a. $Y = 10 + 2X$

  b. $Y = 2 - 3X$

  c. $Y = -5 + 0.5X$

REFERENCES

Cohen, J. (1988). *Statistical power analysis for the behavioral sciences* (2nd ed.). Hillsdale, NJ: Lawrence Erlbaum.

Fidler, F., & Cumming, G. (2008). The new stats: Attitudes for the 21st century. In J. W. Osborne (Ed.), *Best practices in quantitative methods* (pp. 1–14). Thousand Oaks, CA: SAGE.

Fisher, R. A. (1925). *Statistical methods for research workers.* Edinburgh, Scotland: Oliver & Boyd.

Horton, R. (2015). Offline: What is medicine's 5 sigma? *The Lancet, 385*(9976), 1380. doi: 10.1016/S0140-6736(15)60696-1

Killeen, P. R. (2008). Replication statistics. In J. W. Osborne (Ed.), *Best practices in quantitative methods* (pp. 103–124). Thousand Oaks, CA: SAGE.

Schmidt, F. L. (1996). Statistical significance testing and cumulative knowledge in psychology: Implications for training of researchers. *Psychological Methods, 1*(2), 115–129. doi: 10.1037/1082-989X.1.2.115

Wilkinson, L., & Task Force on Statistical Inference, APA Board of Scientific Affairs. (1999). Statistical methods in psychology journals: Guidelines and explanations. *American Psychologist, 54*(8), 594–604. doi: 10.1037/0003-066X.54.8.594

BASIC ESTIMATION AND ASSUMPTIONS

2

Two types of estimation will be presented in this book: ordinary least squares (OLS) and maximum likelihood (ML) estimation. As we discussed in Chapter 1, there are a variety of statistical procedures that have been traditionally treated as separate, such as regression, analysis of variance (ANOVA), and logistic regression. However, the argument we will continue to develop is the idea that these are all different and complementary aspects of the generalized linear model (GLM). Although it may seem odd to have just finished Chapter 1 with an admonition that we should train researchers and statisticians in a modern and applied way and then launch into a chapter on estimation, it is critical for anyone performing statistical analyses to understand the *assumptions* and limitations of any procedure. This is intended to be a gentle, conceptual introduction to estimation so that you can understand the implications of different estimation procedures on the assumptions you must meet in order to have a defensible analysis.

Estimation and the GLM

As introduced in Chapter 1, we have a broad range of analysis options that can often be linked to the type of variables in the analysis (Table 2.1, below, is reproduced from Chapter 1).

Although all of these analyses are part of the GLM, some of them use different estimation procedures to perform the analyses. Procedures in the left column (e.g., regression and ANOVA) typically use OLS estimation, and those in the other columns tend to use ML estimation. This is not inherently important by itself; it is important to you, however, because each of these estimation procedures relies on different assumptions to reach valid and reliable conclusions. If we do not meet these assumptions when performing a particular analysis, we risk fundamental errors of inference and substantial misestimation of the model.

In this chapter, we will briefly discuss what OLS and ML estimation are, the assumptions related to each, and how to test some of the more important assumptions. By the end of this chapter, you should have reasonable conceptual knowledge of the important assumptions underlying each of these estimation procedures.

Table 2.1 Examples of the Generalized Linear Model as a Function of Independent Variable and Dependant Variable Type

	Continuous DV	Binary DV	Unordered Multicategory DV	Ordered Categorical DV	Count DV
Continuous IV	OLS regression	Binary logistic regression	Multinomial logistic regression	Ordinal logistic regression	OLS, Poisson regression
Mixed continuous and categorical IV					
Binary/ categorical IV only	ANOVA	Log-linear models	Log-linear models		Log-linear models

ANOVA, analysis of variance; DV, dependent variable; IV, independent variable; OLS, ordinary least squares.

In the coming chapters when we actually implement them to analyze data, we will also reinforce the importance of testing assumptions prior to interpreting results. Because this is primarily a regression textbook, most of the discussion will be contrasting OLS regression with logistic regression. When we discuss ANOVA in a separate chapter, we will discuss how OLS estimation plays out in that context.

What Is OLS Estimation?

Why do we call regression "ordinary least squares regression"? The "ordinary least squares" part refers to both the goal of the procedure and how it is calculated—the estimation method. The goal for OLS estimation is to create statistical models that minimize the distance from the observed value to the estimated value. This is represented by the *line of best fit*, which is exactly what it sounds like: the line that best fits the distribution of data. If you examine our fictitious example from Chapter 1 relating student grades to vegetable consumption (Figure 2.1), you can see that a line through the bivariate distribution of data points produces some relatively large residuals (e.g., A, where the actual data point is relatively far from the proposed line of best fit). Other residuals will be smaller (e.g., B and C, which are closer to the proposed line). If you subtract the observed values from the predicted values to quantify these residuals, some will be positive (e.g., A, where we would subtract 93 − 70 for a residual of 23) and some will be negative (e.g., B, where we would subtract 80 − 90 for a residual of −10). We cannot simply sum all of the residuals to see whether we have achieved our goal of the best fitting line, as the negative and positive residuals would cancel out. The mathematically expedient way to deal with this process, particularly many decades ago when these procedures were developed, is to square each number (now all are positive, squared distances) and sum. The line with the lowest

Figure 2.1   Fictitious Example of Ordinary Least Squares Estimation

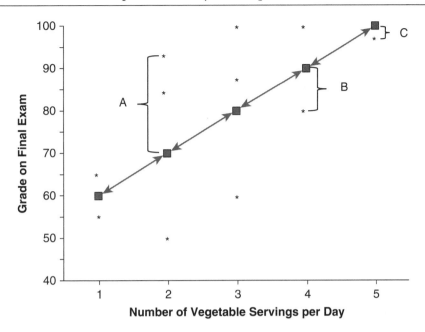

sum of squared distances is the *line of best fit*. This is where we get the phrase "least squares," and it also provides an added benefit of selectively punishing models with many large residuals as squared distances get larger much more quickly than simple distances (±2 squared is 4, ±3 squared is 9, ±4 squared is 16, etc.), which is desirable when the goal is to fit the data more closely.[1]

Research has shown that when assumptions are met in OLS regression, we can derive the following benefits: (a) the estimates produced are unbiased estimates of true regression properties within the population; (b) the standard errors decrease as sample size increases, meaning our estimates of population parameters are more precise; and (c) they are efficient, meaning that no other method of estimation will produce smaller standard errors (if you are interested in more on the technicalities of OLS estimation, an excellent introduction is provided in Cohen, Cohen, West, & Aiken, 2002).

Note particularly that the key phrase above is "when assumptions are met." It is often the case that when we read scholarly articles, even in the most respected journals, we do not know whether assumptions have been met because authors do not report having tested them. In fact, I wrote an entire book demonstrating why cleaning data and testing assumptions is so important (Osborne, 2013), and regression texts such as that by Cohen et al. (2002) clearly make the point that bad things can happen to analyses when assumptions are not met (e.g., the presence of even a single extreme outlier).

---

1   The "ordinary" part of OLS is not pejorative; rather, it correctly identifies this estimation procedure in contrast with other estimation procedures that use least squares estimation, such as weighted least squares or generalized least squares.

Later in this chapter, we will discuss the assumptions of OLS estimation and we will briefly discuss how to test some of the more important assumptions. We will also model this important step in the analysis process by discussing data cleaning and testing of assumptions as we progress through this book.

ML Estimation—A Gentle but Deeper Look

ML estimation is one of those developments in statistics that has spread primarily thanks to widespread access to statistical computing. Unlike OLS estimation, which is based on set equations that researchers or software can use to arrive at a calculated solution, ML estimation is an *iterative* procedure. In other words, the software uses starting values for coefficients, calculates a solution, and compares it with a criterion. If the solution and the criterion are farther apart than desired, new values are attempted and a new solution is found. The new solution is then examined and if found lacking, it is again adjusted and a third solution is attempted. It is hoped that with each iteration, the solution approaches the goals of the algorithm. At some point, the last iteration will be accepted as the final estimation of effects, and that is what the researcher will see in the output. You can imagine how computationally intensive this procedure is and why it was not widely used until computing power became widely available.

Without getting into too many technical details, the goal of ML estimation is to find a solution that provides intercepts and slopes for predictor variables that *maximizes* the *likelihood* of reproducing the individual's scores on the dependent variable (DV; Y) given their scores on the predictor variables (X_1, X_2, etc.). In other words, the algorithms are maximizing the likelihood that we would obtain the sample—the data, the observed scores—given the model and parameters being estimated. This might seem a bit backward in thinking—we already have the sample, don't we?

What we actually have are observed scores on variables for individuals within the sample that arose from some real dynamic or relationship within the population. The ML algorithm attempts to provide a model that maximizes the likelihood of producing the results observed. In essence, both OLS and ML are attempting to summarize the observed data to obtain inferences about the nature of the population or the dynamics that gave rise to the observed scores. The two procedures are merely using different mathematical techniques to get to that goal.

ML estimation is, in my mind, similar to the somewhat counterintuitive notion of hypothesis testing and *p* values we discussed briefly in Chapter 1. The actual interpretation of a *p* value is the probability of obtaining the observed data if in fact the null hypothesis (H_0) is true in the population. So conceptually, ML is trying to estimate the various parameters (slopes and intercepts) that best model (or re-create) the observed data. Thus, if we have a population wherein the height of women and their shoe sizes are strongly positively related (as evidenced by the observed data), ML will provide the coefficients and slopes that maximize the likelihood of obtaining the observed sample that contains the observed relationship between height and shoe size. ML will repeatedly attempt estimations based on slightly different coefficients until the fit with the observed data is as good as can be—in other words, that successive iterations fail to improve the fit by an appreciable amount.

Assumptions for OLS and ML Estimation

Model

The Model Is Complete

We assume that all important variables are in the model, and that unimportant variables are not present in the model. There is significant probability of misestimation of the model if important variables are left out. For example, if you are examining athletic performance and forget to include sex or age, you are likely not getting accurate estimates of effects. If you include variables that are not important, you not only reduce power to see effects and waste degrees of freedom, but you also risk misestimation through effects such as *suppression* or *collinearity*. A suppressor effect is a misestimation of the model by including a variable that is not related to the DV but is related to other IVs. This can cause the effects of the other IVs to be substantially distorted. Collinearity can happen when very highly correlated variables (or variables that are exactly redundant) are entered in the model. The analysis can produce irrational results if this is serious enough. We will cover diagnostics related to these effects in later chapters. In general, this assumption is satisfied by common sense and theory, ensuring that important background variables are included and unimportant ones are not.

The Model Is Linear

Another important assumption in linear modeling is the assumption of linearity.[2] The general assumption is that the correct form of the relationship is being modeled; however, in the case of OLS and logistic regression (and many other analyses), the assumption is that there is a linear relationship between the DV and the independent variable (IV). A similar generalization to planes and hyperdimensional relationships is in effect for multiple regression with two or more IVs, but thinking too deeply about hyperdimensional generalizations of linearity gives me a bit of a headache, so I tend to stick to the two- or three-dimensional examples.

This is a simple case of specifying a linear model, such as Equation 2.1a, when a line does not adequately capture the effect. We briefly reviewed line equations in Chapter 1. Curves can be represented similarly. For example, a quadratic equation (like in Equation 2.1b) has a squared term in it. Any equation with a squared term will produce a curve with one *inflection point*, or point where the slope changes direction (or where the slope equals zero). Cubic equations, with cubed terms, will have two inflection points, or changes in direction. We will discuss this in more detail later in this book. For now, you can see an example of a quadratic equation in Figure 2.2. If you have a curve in your data

---

2 Imagine the surprise . . . and then be surprised at how few researchers actually test for nonlinearity!

Figure 2.2   Curvilinear Relationship Between Student Age and Reading Achievement
Test Scores

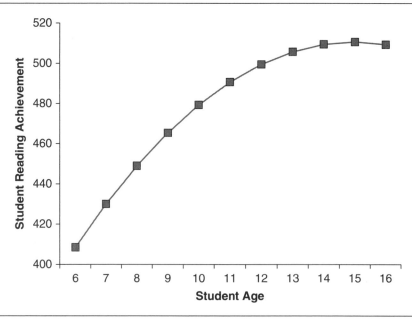

DATA SOURCE: Francis, D., Schatschneider, C., & Carlson, C. (2000). Introduction to individual growth curve analysis. In
D. Drotar (Ed.), *Handbook of research in pediatric and clinical child psychology* (pp. 51–73). New York, NY: Kluwer/Plenum.

such as this, and only model linear effects (which is common), you are asserting that b_2
is zero rather than testing whether it is zero or not. You will also be misestimating the
model, which usually has the effect of underestimating the size of the effect.

$$\hat{Y} = b_0 + b_1 X_1 \tag{2.1a}$$

$$\hat{Y} = b_0 + b_1 X_1 + b_2 X_1^2 \tag{2.1b}$$

I think it is easy to argue that most researchers ignore this important assumption
(as we almost never find discussions of curvilinearity in the research literature, and
we almost never see tests of this assumption). In my book on logistic regression, I
showed that curvilinear relationships are easy to find. Thus, I am skeptical that this
assumption is met routinely. Classic examples from the health and social sciences
include the relationship between arousal (i.e., stress) and performance (Loftus &
Ketcham, 1992; Sullivan & Bhagat, 1992; Yegiyan & Lang, 2010),[3] student achievement growth curves (Francis, Schatschneider, & Carlson, 2000; Rescorla & Rosenthal,
2004), grade point average and employment in high school students (Quirk, Keith,
& Quirk, 2001), dose-response relationships (Davis & Svendsgaard, 1990), and age
and life satisfaction (Mroczek & Spiro, 2005).

---

3   This is often attributed to Yerkes and Dodson (1908), or somewhat inaccurately referred to (sometimes by
me personally) as an anxiety-performance curve. See Teigen (1994) for a historical overview of this large
group of theories and studies.

For example, as Francis et al. (2000) showed, the general pattern for reaching achievement growth over time is curvilinear. In Figure 2.2, I present a growth curve modeled from their published data. Fortunately, there are simple ways to incorporate tests for curvilinear effects as statistical software packages begin to implement curvilinear regression options. We will explore these in detail later in this book.

Logistic regression is, by nature, nonlinear. Specifically, the way that logistic regression converts a dichotomous or categorical variable to a DV that can be predicted from other binary, categorical, or continuous variables involves a nonlinear transformation. For now, envision a DV that is an S-shaped curve representing the probability that an individual will be in one group or the other (like the one in Figure 2.3). Don't worry about the details of how the DV is created in logistic regression for now—we will have fun exploring that more thoroughly later. I find it interesting that although the basic character of logistic regression—the logit transformation—is curvilinear, there is a clear assumption of linearity as well. Specifically, there is an assumption that there is a linear relationship between any IVs and the DV—that IVs are "linear on the logit."[4] When using ML estimation in logistic regression, we can also easily explore and model relationships that are "nonlinear on the logit," something we will also explore in more depth in later chapters.

Figure 2.3   Standard Logistic Sigmoid Function

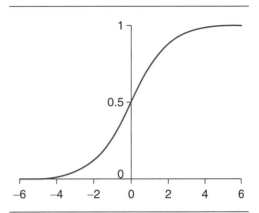

SOURCE: Wikipedia (http://upload.wikimedia.org/wikipedia/commons/thumb/8/88/Logistic-curve.svg/320px-Logistic-curve.svg.png).

The Model Is Additive

We also assume that the effects of different IVs in the analysis are *additive*. Look at a typical multiple regression equation, such as Equation 2.2a, below:

$$\hat{Y} = b_0 + b_1 X_1 + b_2 X_2 + \ldots + b_k X_k \tag{2.2a}$$

You can see that we explicitly propose the nature of these relationships between IVs and the DV as additive. This means that we assume the effect of one variable adds to the effect of another to help explain or understand the DV to some extent. However, it is not always the case that variables have additive relationships, or purely additive relationships. In curvilinear effects, those relationships are exponential, and when we talk about interactions between two (or more) variables, we are

---

4 This is also a great phrase to drop casually into conversations. Try it and watch your social capital climb!

Figure 2.4   An Example of an Interaction Between Two Variables, A Multiplicative Effect

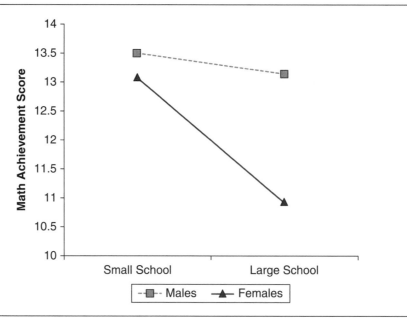

DATA SOURCE: National Center for Education Statistics High School and Beyond study data (http://nces.ed.gov/pub search/getpubcats.asp?sid=022).

talking about multiplicative effects. If you look at Figure 2.4, you can see an example of a multiplicative effect, where the effects of one variable depend on the effect of a second variable. For example, during recent decades, there has been extensive discussion about sex differences in math achievement test scores. This was a particular issue back in the 1980s, when the National Center for Education Statistics began the High School and Beyond (HS&B) study of high school students (information and data from the HS&B study are available at http://nces.ed.gov/pubsearch/getpubcats.asp?sid=022). In the 1980s, we knew that in general, girls underperformed on mathematics achievement tests compared with boys. As Figure 2.4 shows, that general pattern, however, is not the same in all types of schools. In this example, we see that there is a trend for much larger schools to have larger math achievement gaps and for smaller schools to have smaller achievement gaps. In other words, the effect of student sex differs depending on the size of the school (and probably many other variables). Again, there are many reasons for this that we will not get into, but the conclusion is interesting—that context matters. It is not simply a linear relationship; rather, there are a variety of relationships that depend on the context in which the students find themselves. Similarly, it is likely that when you look for them, you will find and model interaction effects when running logistic regressions. We will explore interaction effects in chapters to come.

As with curvilinear effects, if you fail to model these interaction effects and they are indeed present, the model is misspecified, and the results of the analysis are likely to be mistaken as well. To model a multiplicative (or nonadditive) effect, we need to include a multiplicative aspect to the analysis, such as the last term in Equation 2.2b, below:

$$\hat{Y} = b_0 + b_1 X_1 + b_2 X_2 + b_3 X_1 X_2 \qquad (2.2b)$$

The third term, which captures the effect of X_1 multiplied by X_2, was not present in Equation 2.1; thus, without examining whether there are multiplicative effects, we are *assuming* that b_3 is equal to zero without testing that assumption. Ironically, effects of this type are almost always examined in ANOVA analyses, because statistical software usually adds them to models by default. In regression, they are not automatically added, and many researchers thus do not examine them. By the end of this book, I hope to persuade you that these types of effects are some of the most interesting, and further that they are not difficult to test.

Variables

Variables Are Measured at Interval or Ratio Scale

In Chapter 1, we discussed some basic measurement language as well as some of the factors that separate interval or ratio variables from ordinal or categorical variables. When DVs are ordinal or categorical, we will have different regression analyses we can use to handle these appropriately (binary or multinomial logistic regression if categorical, or ordinal logistic regression for ordinal DVs). When IVs are ordinal or categorical, we can use dummy, effects, or other coding schemes to appropriately account for them. To the extent that this assumption is violated (and it is probably violated frequently when variables are assumed to be interval but are instead ordinal), the results from the analysis can be substantially misestimated.

Variables Are Measured Without Error

It is almost a dirty little secret in statistical science that we assume perfect measurement, yet rarely (if ever) achieve it. In most statistical procedures, we assume we are measuring the variables of interest well, and to the extent that we are not, biases and misestimations can occur.

In simple correlation and regression (and simple logistic regression and ANOVA), the effect is usually that of underestimating the effects in question. In my book on data cleaning (Osborne, 2013), I spend a good deal of time exploring the harmful effects of poor reliability on measurement. I will summarize the general arguments and evidence here. Let us take the example of simple linear regression. If there is a given effect in the population, and you have imperfect measurement of the IV and DV, then the observed relationship will be the relationship multiplied by the square root of the reliability of the IV multiplied by the reliability of the DV, as shown in Equation 2.3, which is adapted from Equation 2.10.5 in Cohen et al. (2002, p. 121):

$$r_{XY} = \rho_{xy} \sqrt{r_{XX}\, r_{YY}} \qquad (2.3)$$

where any correlation (or regression coefficient) is a function of the true correlation in the population (ρ_{xy}) multiplied by the square root of the reliabilities of X (r_{XX}) and Y (r_{YY}). Some examples of how this can lead to underestimation are presented in Table 2.2 and Figure 2.5.

As you can see in both Table 2.2 and Figure 2.5, a correlation will attenuate substantially as the assumption of perfect measurement is violated, even when reliability is in the "good" range of 0.80 or higher. For example, if you have a population correlation of 0.50 and reliability of 0.80, the correlation will be underestimated at 0.40. This does not

Table 2.2   Example Effects of Imperfect Reliability

	True Correlation Coefficient		
Reliability Estimate	*$\rho = 0.30$ (0.09)*	*$\rho = 0.50$ (0.25)*	*$\rho = 0.70$ (0.49)*
0.90	0.27 (0.07)	0.45 (0.20)	0.63 (0.39)
0.80	0.24 (0.06)	0.40 (0.16)	0.56 (0.31)
0.70	0.29 (0.04)	0.35 (0.12)	0.49 (0.24)
0.60	0.18 (0.03)	0.30 (0.09)	0.42 (0.18)

NOTE: Reliability estimates for this example assume the same reliability for both variables. Percent variance accounted for (shared variance, or coefficient of determination) is shown in parentheses.

Figure 2.5   The Effect of Imperfect Measurement on Effect Size

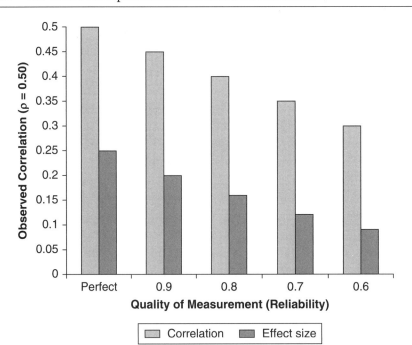

seem like a huge underestimation until you examine the variance accounted for, which is calculated as r^2. According to this metric, the true size of the effect dropped from 25% variance accounted for to 16%, a loss of about one-third of the effect. Of course, the underestimation rapidly becomes more serious when reliability drops below 0.80, which happens often, even in top research journals. With a reliability of 0.70, we have lost about one-half our effect.

In multiple regression and more complex procedures, the effects of poor reliability can become unpredictable and chaotic. For example, let us imagine that you are studying student achievement and attempting to control for the effects of a family's socioeconomic status (SES). Let us assume that SES and achievement are reasonably correlated.[5] We will also imagine that your measure of SES is not perfectly reliable. In this case, not only will your estimate of the effect of SES be attenuated (underestimated), but the effect of student achievement might be overestimated inappropriately.

This can lead other, related IVs to become overestimated if they are capturing variance that should have been removed by SES. I have dealt with this issue in more depth in other places and thus refer interested readers to those publications, rather than recapitulating the arguments here (Nimon, Zientek, & Henson, 2012; Osborne, 2013). Logistic regression also relies on reliable measurement of variables; therefore, the two are similar in that respect. This issue is largely avoidable in modern times. We can either choose instruments and measures that have a history of stronger reliability, or we can use structural equation modeling to more accurately model the true nature of our relationships.

Residuals and Distributions

The Residuals Are Normally Distributed, and Sum to Zero

Different estimation procedures have different assumptions. OLS is a parametric technique, meaning that it requires assumptions about the distribution of the population in order to be effective (these are discussed in most regression texts, but a particularly good reference is Cohen et al., 2002). Commonly used statistical tests such as ANOVA and OLS regression assume that the data come from populations with normally distributed characteristics. This is, of course, unknowable, but we can look at the distribution of the residuals; thus, an important assumption of OLS estimation is that the distribution of residuals (errors) is normal (Gaussian). We will examine whether this assumption is met by examining the distribution of residuals after a regression analysis is complete. We can even test whether that distribution significantly deviates from normal,[6] and if so, we have several remedies we can try, such as transformation of a variable or cleaning of data.

---

5 They usually correlated moderately in most research I have seen.

6 There are inferential tests to examine whether distributions of variables match standard distributions, such as the standard normal distribution. However, these tests should be used with caution, as larger samples have tremendous power to reject the hypothesis of similarity for minimally important deviations from a distribution.

An aspect of this assumption that is often not considered (but ought to be) is the concept of the *outlier*, or a data point that lies substantially outside the distribution of the rest of the sample. Another way to think about this is whether it is reasonable to assert that a particular data point is pulled from, or representative of, the population of interest. To the extent that outliers exist far outside the range of observed data points, the less likely it is that they represent the population of interest and thus the less legitimate it is to retain them in the sample for analysis. Consider the 6-year-old who reads at a college level. Certainly that individual is interesting, exceptional, and probably not representative of the population of 6-year-old children we wish to generalize to. Or consider data errors, such as the nurse in one survey I worked on many years ago who indicated her hourly salary was $45,000.00. Is this representative of the population of nurses? Probably not. She probably made an error and reported her annual salary. In this case, we used other information to estimate hourly wage from annual wage rather than leaving the data as is.

These data points tend to cause the distribution of residuals to be non-normal, and they also tend to be *inappropriately influential*, meaning that these data points tend to have a larger effect on the parameter estimates than other data points. They tend to add error variance and often lead to misestimation of the model. Thus, we will examine our data for them as well. In Figure 2.6, we have a distribution of residuals from a

Figure 2.6   Distribution of Residuals From a Regression of Weight and Height

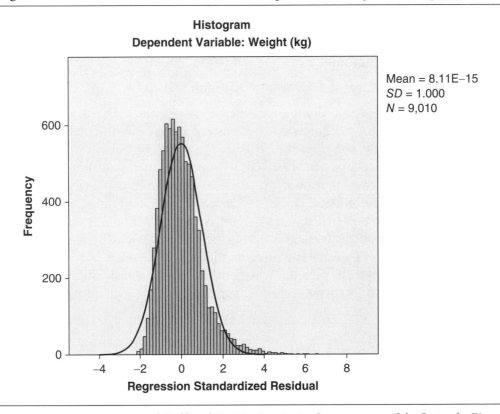

DATA SOURCE: 2001–2002 National Health and Nutrition Examination Survey, courtesy of the Centers for Disease Control and Prevention (http://www.cdc.gov/nchs/nhanes.htm).

regression of weight as a function of height. This distribution of residuals is relatively normally distributed *except for* the long tail to the right, where some residuals are up to 8 standard deviations (*SD*) from the mean. Trimming some of these data points would probably lead to more closely meeting this assumption.

By contrast, because of different mathematical assumptions used when the procedure was developed, ML estimation is a *nonparametric technique*, meaning it does not require any particular distributional assumptions of the population, and the residuals are free to be highly non-normal. However, inappropriately influential data points can also cause misestimation when using ML estimation, so when we perform logistic regression analyses, we will still examine residuals or indicators of influence to ensure our data represent the population as well as possible.

Homoscedasticity (or Constant Variance of the Residuals)

ANOVA has the assumption of equal variance across groups, which is a special case of this assumption: that the variance of the residuals is constant across the observed range of the DV. In other words, this means that if you plot the data points around the regression line (e.g., a zPRED versus zRESID plot), you should see a relatively homogeneous scattering of data points around the regression line at all points. In Figure 2.7, you see a plot that largely meets the assumption of constant variance across the observed range. The distribution of residuals is largely constant

Figure 2.7   Example Distribution of Standardized Residuals Plotted Across a Predicted Value of Y

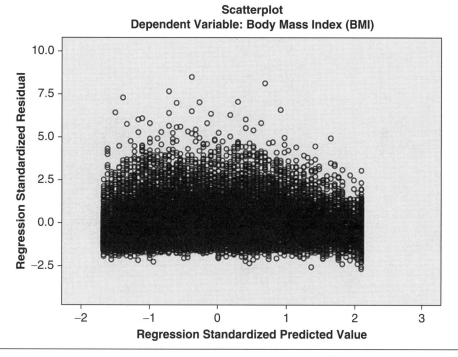

DATA SOURCE: National Health Interview Survey of 2010 (NHIS2010) from the National Center for Health Statistics (http://www.cdc.gov/nchs/nhis/nhis_2010_data_release.htm).

(within reason), and it would be more so if some simple data cleaning removed cases with residuals more than, say, exceeding 5 *SD*. Data transformations can also be utilized to help meet this assumption. By contrast, Figure 2.8 shows a distribution with marked heteroscedasticity evident; in other words, the variance of the residuals seems to vary broadly across the observed range of the DV. In this case, we would fail the assumption and produce a model that may be seriously misestimated. Data cleaning, transformation, or perhaps even curvilinear modeling might improve this situation.

Logistic regression is not a parametric procedure and has no assumption regarding the distribution of residuals or equality of variance. Of course, if you have problematic data points and have not cleaned your data, you could still experience substantial misestimation of the model. Reasonable cleaning of the data could remedy this issue in nonparametric analyses as well.

However, there are some interesting assumptions relating to sparseness that seem similar to me. Sparseness is a concept that can be understood by imagining lots of little boxes stacked together. In logistic regression, for example, each box represents a combination of the DV and IV. For example, if you are looking at blood pressure and odds of having a stroke, you have boxes for each range of blood pressure representing both people who have and who have not had strokes. In sampling from the

Figure 2.8   Example Distribution of Standardized Residuals Plotted Across a Predicted Value of Y

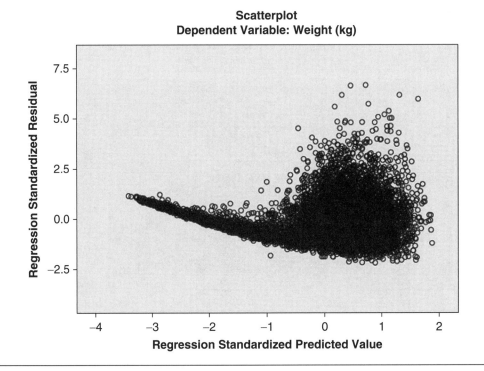

DATA SOURCE: 2001–2002 National Health and Nutrition Examination Survey, courtesy of the Centers for Disease Control and Prevention (http://www.cdc.gov/nchs/nhanes.htm).

population, you want to make sure you have your boxes filled as best you can. "Sparse" data refers to having too many of these boxes unfilled, which can prevent ML estimation to effectively form estimates.

Independence of Observations (Independence of Error Terms)

In most analyses, we assume that observations are independent unless we are specifically modeling nested data or repeated measures. Because much of our data in the world (especially in the social sciences, but also in many other sciences, such as health sciences) come from organisms that form hierarchies or groups, this assumption may be more or less tenable. For example, researchers sampling individuals from existing health centers or students from schools or classrooms are sampling individuals who are already more similar in many respects than individuals sampled at random from the entire population. This violates the assumption of independence of observation and may bias the results. For a brief primer on this concept, and the issues that can arise, you can refer to Osborne (2000) or Chapter 13 in this book, where we discuss hierarchical linear modeling applied to logistic regression.

Simple Univariate Data Cleaning and Data Transformations

In quantitative research methods, the standard normal distribution (or the *bell-shaped curve*) is a symmetrical distribution with known mathematical properties. Most relevant to our discussion, we know what percentage of a population falls at any given point of the normal distribution, which also gives us the probability that an individual with a given score (or above or below that score, as well) on the variable of interest would be drawn at random from a normally distributed population.

According to common statistics tables, we know that 34.13% of individuals fall between the mean and 1 *SD* above the mean. Thus, we know that in a perfectly normal distribution, 68.26% of the population will fall within 1 *SD* of the mean (34.13% between the mean and 1 *SD* above the mean, 34.13% between the mean and 1 *SD* below the mean). This means that most individuals randomly sampled from a normally distributed population should fall relatively close to the mean. It also means that we can calculate other interesting statistics, such as the percentage of individuals that should fall above or below a point. We also know that 50% of individuals fall below the mean (again, remember we are assuming a normal distribution—this does not apply to variables that do not conform to the standard normal distribution). An individual who is 1 *SD* above the mean is at the 84.13rd percentile (50% + 34.13%). An individual 1 *SD* below the mean is at the 15.87th percentile (50% − 34.13%). Likewise, 95.44% of the population should fall within ±2.0 *SD* from the mean,[7] and 99.74% of the population will fall within ±3.0 *SD* of the mean. In Figure 2.9, you can see an expectation that a particular percentage of the population will fall between any two points on a normally distributed variable.

---

7 That 95.0% fall within 1.96 *SD* of the mean gives rise to the hallowed $p < .05$ criterion for null hypothesis significance testing (for some historical background on null hypothesis statistical testing, see Fisher, 1925; Neyman & Pearson, 1936).

Figure 2.9   The Standard Normal Distribution

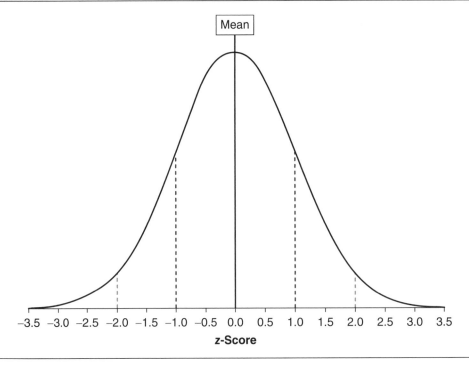

SOURCE: Adapted with permission from Osborne (2013).

We can use this information to our advantage in several ways. First, some scholars have argued that any data cleaning will distort the parameter estimates and make them *less* generalizable to the population. However, the truth is just the opposite when we are discussing judicious, defensible data cleaning following clear guidelines. What the normal distribution can tell us [as can any z table that you have lying around or that can be found in Appendix A of this book or by using the Excel function *normsdist(z)*] is the probability of selecting a case from a population with a given mean and *SD*. For example, for a population with a mean of 100 and a *SD* of 10, a score of 130 is 3 *SD* above the mean. This score has a $z = 3.00$—in other words, with a corresponding cumulative *one-tailed probability* of .99865. This one-tailed probability tells us the probability of getting a score less than 3 *SD* (between $-\infty$ and 3.00). Subtracting this value from 1.0 gives us the probability of getting a score above 3.00 (.00135) from a normally distributed population. The probability of getting a value more than 3 *SD* from the mean in either direction (*a two-tailed probability*) is twice that, or .0027. Therefore, I often focus on points at least 3 *SD* away from a mean ($z = \pm3$) because the probability of sampling a value that far from a given mean with a given *SD* is very small. We can then assert that the probability that a case that far from the mean is *not part of that target population* is relatively large. If our goal is to draw conclusions about a particular population, I think it is defensible to remove cases that have a low probability of being part of the population we seek to understand.

Univariate data cleaning can involve several different aspects, which I will only briefly discuss here.[8] In the context of GLM, I would recommend that at a minimum, you screen for data points far outside the expected distribution, that you examine missing data, and that you might want to consider transformations to improve normality. (Recall that even for OLS regression, we do not assume the variables are normally distributed, but rather that the residuals are normally distributed. We do assume normal distribution of variables in Pearson correlation, but I am going to argue that is an unnecessary technique we no longer need.)

Data Screening

There are several reasons why you might have a variable with some relatively extreme data points (by extreme I mean beyond 3 *SD*s from the mean). Some of them might be errors that could be fixed in some way, whereas some may need to be deleted or separated out for a different analysis. If you are using secondary data, it is even possible that they are codes for missingness or some other meaningful event but are not a valid entry on the variable in question. I think it is difficult to justify leaving these types of points in the data for analysis. More difficult are the data points that are "fringeliers"— cases near the edge of the distribution. For instance, the data in Figure 2.10 present a

Figure 2.10   Distribution of the Number of Faculty at U.S. Colleges and Universities

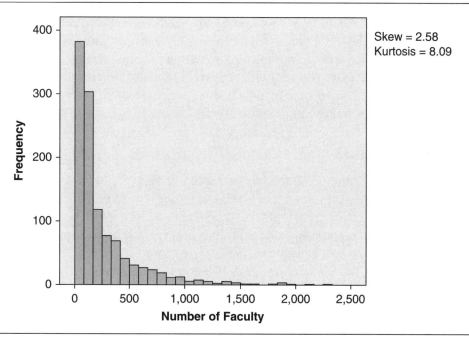

DATA SOURCE: Lock, R. (organizer). What's What Among American Colleges and Universities: American Association of University Professors (AAUP) faculty salary data (for the 1995 Data Analysis Exposition, sponsored by the Statistical Graphics Section of the American Statistical Association). These data are also available on the book website.

---

8   There is this brilliant book on the topic by some nerd named Osborne, if you want to get into more detail . . .

distribution of the number of faculty at institutions of higher education within the United States (data from the American Association of University Professors [AAUP]).[9] It is obviously a skewed distribution, but it is tough to know whether some of the higher scores are outliers (perhaps an extra zero accidentally turning a legitimate count of 200 into 2,000). If a distribution is merely skewed, removal of cases is not ideal, as it is less clear that they are not legitimate members of the target population. Instead, a transformation could be utilized to improve normality of the distribution.

Missing Data

The standard practice (and the default in most statistical packages) is to delete cases with missing data on any variable being analyzed. However, this can lead to substantial bias in the data if the data are not missing completely at random. This is called *complete case analysis*.

The second most common way for dealing with missing data is for researchers to substitute a value (often the mean) for the missing value. This retains the case in the analysis but is almost always an even worse option than complete case analysis, as this stacks many cases at the mean, artificially reducing the variance of the distribution. This can also introduce substantial bias into the estimates if the data are not missing completely at random.

So what is a nerd to do? There are other methods, such as imputation. This takes things we know about an individual and seeks to impute a reasonable value based on those existing data. For example, if we know your height and sex but are missing your shoe size, instead of just dropping you from the analysis or assigning you the average shoe size for your sex, we could impute a more reasonable shoe size by understanding where you are on the other two variables. If you are male and 6 feet tall, odds are that your shoe size is much larger than if you are female and 5 feet tall. Once we get further into regression, this will make more sense overall.

Transformation of Data

If you have a distribution like the one presented in Figure 2.10 (and if you feel the need to transform the data), there are many options. The traditional transformations are square root, log, or inverse transforms. These are special examples of power transformations (transformations that raise all values of a variable to an exponent/power). For example, a square root transformation can be characterized as $X^{1/2}$, and inverse transformations can be characterized as X^{-1}. Log transformations embody a class of power transformations in themselves. Although they have rarely discussed them in texts, some authors have talked about third and fourth roots (e.g., $X^{1/3}$, $X^{1/4}$) being useful in various circumstances. You might wonder why we should be limited to these options. Why would we not be able to use $X^{0.9}$, X^{-2}, or X^4 or any other possible exponent if it improves the quality of the data? In fact, for more than half a century, statisticians have

---

9   Lock, R. (organizer). What's What Among American Colleges and Universities: American Association of University Professors (AAUP) faculty salary data (for the 1995 Data Analysis Exposition, sponsored by the Statistical Graphics Section of the American Statistical Association).

been talking about this idea of using a continuum of transformations that provide a range of opportunities for closely calibrating a transformation to the needs of the data. Tukey (1957) is often credited with presenting the initial idea that transformations can be thought of as a class or family of similar mathematical functions. This idea was modified by Box and Cox (1964) to take the form of the Box-Cox series of transformations[10] (Equation 2.4):

$$X_i^{\lambda} = (X_i^{\lambda} - 1) / \lambda, \text{ where } \lambda \neq 0; \qquad (2.4)$$

$$X_i^{\lambda} = \log_e(X_i), \text{ where } \lambda = 0.$$

Although it is not implemented in all statistical packages,[11] there are ways to estimate the Box-Cox transformation coefficient, λ, through a variety of means; once an optimal λ is identified, the transformation is mathematically straightforward to implement in any software package. Implementing Box-Cox transformations within SPSS is discussed in detail at the end of this chapter, in Appendix A, and on the book website. Given that λ is potentially a continuum from $-\infty$ to ∞, we can theoretically calibrate this transformation to be maximally effective in moving a variable toward normality. In addition, as mentioned above, this family of transformations incorporates many traditional transformations, including the following:

$\lambda = 1.00$: no transformation needed; produces results identical to original data

$\lambda = 0.50$: square root transformation

$\lambda = 0.33$: cube root transformation

$\lambda = 0.25$: fourth root transformation

$\lambda = 2.00$: square transformation

$\lambda = 3.00$: cube transformation

$\lambda = 0.00$: natural log transformation

$\lambda = -0.50$: reciprocal square root transformation

$\lambda = -1.00$: reciprocal (inverse) transformation

$\lambda = -2.00$: reciprocal (inverse) square transformation

and so forth.

---

10 Since Box and Cox (1964), other authors have introduced modifications of these transformations for special circumstances (e.g., data with negative values, which should be addressed via anchoring at 1.0) or peculiar data types less common in the behavioral sciences. For example, John and Draper (1980) introduced their "modulus" transformation, which was designed to normalize distributions that are relatively symmetrical but not normal (i.e., removing kurtosis where skew is not an issue). In practice, most researchers will get good results from using the original Box-Cox family of transformations, which is preferable to those new to this idea, thanks to its computational simplicity.

11 SAS has a convenient and very well done implementation of Box-Cox within *proc transreg* that iteratively tests a variety of λ transformations and identifies several different good options for you. Many resources on the web provide guidance on how to use Box-Cox within SAS. Other software packages may also have this transform implemented.

University Size and Faculty Salary in the United States

In 1995, the AAUP collected data on the size of the institution (number of faculty) and average faculty salary from 1,161 US institutions.[12] As Figure 2.10 shows, the variable *number of faculty* deviates markedly from normal (skew = 2.58, kurtosis = 8.09). Figure 2.11 shows the result of a Box-Cox transformation where $\lambda = 0.00$ (after being anchored at 1.0). The macro used tested λ over a range of λ from −3.00 to 1.00. Because of the nature of these data (values ranging from 7 to more than 2,000 with a strong skew), this transformation attempt produced a wide range of outcomes across a variety of Box-Cox transformations, from extremely bad outcomes (skew < −30.0, where $\lambda < -1.20$) to very positive outcomes of $\lambda = 0.00$ (equivalent to a natural log transformation) that achieved the best result (skew = 0.139, kurtosis = −0.245 at $\lambda = 0.00$).

When minimum values of distributions deviate from 1.00, power transformations might become less effective. Thus, to explore this transformation, this variable was first anchored at 1.0 (because the minimum is 7, we can subtract 6 from each value in the distribution). Once this is accomplished, the distribution of the variable is much closer to the standard normal distribution (Figure 2.11). Recall once more that normality of the *variables* is an assumption in Pearson correlation (which we will not spend much time

Figure 2.11   Number of Faculty at Institutions in the United States After Transformation ($\lambda = 0.00$)

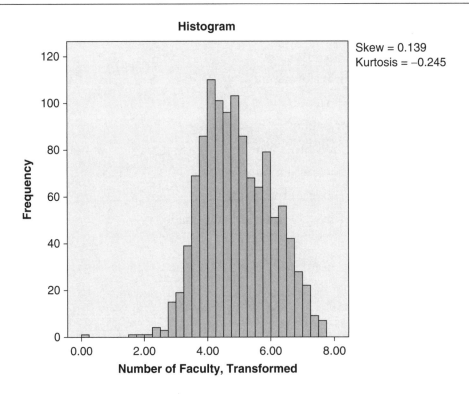

---

12   These data are available on the website for this book.

on); however, normal distribution of the residuals is important in OLS regression. Thus, this type of transformation would be desirable if the residuals from a regression analysis were non-normal. When we explore simple regression in the next chapter, we will explore the effect of this transformation.

What If We Cannot Meet the Assumptions?

We talked a lot about assumptions, particularly the importance of testing whether your data actually meet the assumptions of the statistical procedure you are performing. Too many researchers assume that they meet the assumptions and that the tests they are using are *robust* to violations of assumptions.[13] We must test whether we actually meet assumptions, and take actions as appropriate if not. What if your data are ill mannered and will not allow you to meet assumptions? There are other options aside from blindly driving forward. There is an entire class of analyses called "robust procedures" and nonparametric analyses that do not require the same assumptions, and Rand Wilcox (2008) and others have developed procedures that have different assumptions and will not misestimate parameters like OLS regression will (or ANOVA, or other parametric techniques). The tradeoff is often, however, lower power to detect effects. The point is that you must make ethical decisions when analyzing data. It is not an option to ignore the foundational underpinnings of the technique you are using and hope for the best.

Where We Go From Here

We have now spent two chapters getting oriented to some basics of the GLM, but we have not yet talked in detail about any particular analysis. It is time to change that. In the next chapter, we will begin to dig into regression, starting a journey that will end with strong mastery of best practices over a variety of complex and useful procedures.

ENRICHMENT

1. Describe each of the following assumptions of OLS regression (without referring back to the chapter):
   a. The model is complete.
   b. The model is linear.
   c. The model is additive.
   d. The variables are measured at an interval or ratio level.
   e. The variables are measured without error.

---

13   If you say this sentence several times quickly, it begins to sound like Dr. Seuss.

    f.   The residuals sum to 0.

    g. The residuals are normally distributed.

    h. The variance of the residuals is constant across the range observed.

    i.   The residuals are independent.

2. Which of the above-mentioned assumptions apply to ML estimation procedures?

3. You are reading an article in a journal that describes research based on data with poor reliability. The authors argue that reliability of 0.70 is common and acceptable. Do you agree with this position? If not, how would you counter that commonly held position?

4. If your residuals are not normally distributed, or fail to have constant variance, what are some things you can do to more completely meet these assumptions?

5. We will use OLS estimation for many procedures later in this book. In plain language, what does this estimation procedure seek to do?

6. Describe some ways to identify an extreme score (either visually or via z-score). Be able to discuss why it might be desirable to identify an extreme score in a distribution.

7. Download the AAUP data from the book website. Reproduce the transform of NUM_TOT to see if you can achieve similar results as those in Figure 2.11. Then explore the variable SAL_ALL, the average salary (in hundreds of dollars, for some reason), and evaluate whether that variable would benefit from transformation, and if so, whether Box-Cox transformation can produce a more normally distributed variable.

8. Download the GRADES data from the book website. This is a data file of student grades on a class exam from an undergraduate course I taught long ago. Typically, negatively skewed variables are tougher to transform than positively skewed variables—unless you use Box-Cox. See if you can transform this variable to be more normally distributed.

Log on to the book website at study.sagepub.com/osbornerlm to access files, data, and answer keys to evaluate your mastery.

Calculating Box-Cox λ by Hand

In current versions of SPSS, you have to use a script I adapted from prior scholars to provide a wide range of Box-Cox transforms. Many other statistical packages (SAS, R) have automatic routines to perform Box-Cox transforms.

    If you desire to estimate λ by hand, the general procedure is to

- Divide the variable into at least 10 regions or parts,
- Calculate the mean and *SD* for each region or part,

- Plot log(*SD*) versus log(mean) for the set of regions, and
- Estimate the slope of the plot, and use the slope (1 – *b*) as the initial estimate of λ.

As an example of this procedure, I revisit the second example, number of faculty at a university. After determining the 10 cutpoints that divide this variable into even parts, selecting each part and calculating the mean and *SD*, and then taking the $\log_{10}$ of each mean and *SD*, Figure 2.12 shows the plot of these data. I estimated the slope for each segment of the line, since there was a slight curve (segment slopes ranged from −1.61 for the first segment to 2.08 for the last) and averaged all, producing an average slope of 1.02. Interestingly, the estimated λ from this exercise would be −0.02, very close to the empirically derived 0.00 used in the example above.

Figure 2.12   Estimating λ Empirically in SPSS and Performing the Box-Cox Transformation

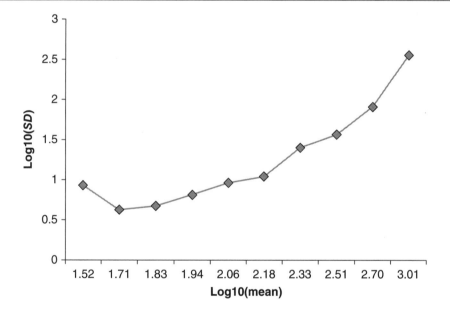

REFERENCES

Box, G. E. P., & Cox, D. R. (1964). An analysis of transformations. *Journal of the Royal Statistical Society, Series B, 26*(2), 211–234.

Cohen, J., Cohen, P., West, S., & Aiken, L. S. (2002). *Applied multiple regression/correlation analysis for the behavioral sciences*. Mahwah, NJ: Lawrence Erlbaum.

Davis, J. M., & Svendsgaard, D. J. (1990). U-shaped dose-response curves: Their occurrence and implications for risk assessment. *Journal of Toxicology and Environmental Health, Part A Current Issues, 30*(2), 71–83. doi: 10.1080/15287399009531412

Fisher, R. A. (1925). *Statistical methods for research workers.* Edinburgh, Scotland: Oliver & Boyd.

Francis, D., Schatschneider, C., & Carlson, C. (2000). Introduction to individual growth curve analysis. In D. Drotar (Ed.), *Handbook of research in pediatric and clinical child psychology* (pp. 51–73). New York, NY: Klewer/Plenum.

John, J. A., & Draper, N. R. (1980). An alternative family of transformations. *Applied Statistics, 29*(2), 190–197. doi: 10.2307/2986305

Loftus, E., & Ketcham, K. (1992). *Witness for the defense: The accused, the eyewitness, and the expert who puts memory on trial.* New York, NY: St. Martin's Griffin.

Mroczek, D. K., & Spiro, A., III. (2005). Change in life satisfaction during adulthood: Findings from the Veterans Affairs Normative Aging Study. *Journal of Personality and Social Psychology, 88*(1), 189–202. doi: 10.1037/0022-3514.88.1.189

Neyman, J., & Pearson, E. S. (1936). Contributions to the theory of testing statistical hypotheses. *Statistical Research Memoirs, 1,* 1–37.

Nimon, K., Zientek, L. R., & Henson, R. K. (2012). The assumption of a reliable instrument and other pitfalls to avoid when considering the reliability of data. *Frontiers in Psychology, 3,* 102. doi: 10.3389/fpsyg.2012.00102

Osborne, J. W. (2000). Advantages of hierarchical linear modeling. *Practical Assessment, Research & Evaluation, 7*(1).

Osborne, J. W. (2013). *Best practices in data cleaning: A complete guide to everything you need to do before and after collecting your data.* Thousand Oaks, CA: SAGE.

Quirk, K. J., Keith, T. Z., & Quirk, J. T. (2001). Employment during high school and student achievement: Longitudinal analysis of national data. *Journal of Educational Research, 95*(1), 4–10. doi: 10.1080/00220670109598778

Rescorla, L., & Rosenthal, A. S. (2004). Growth in standardized ability and achievement test scores from 3rd to 10th grade. *Journal of Educational Psychology, 96*(1), 85–96. doi: 10.1037/0022-0663.96.1.85

Sullivan, S. E., & Bhagat, R. S. (1992). Organizational stress, job satisfaction and job performance: Where do we go from here? *Journal of Management, 18*(2), 353–374. doi: 10.1177/014920639201800207

Teigen, K. H. (1994). Yerkes-Dodson: A law for all seasons. *Theory & Psychology, 4*(4), 525–547. doi: 10.1177/0959354394044004

Tukey, J. W. (1957). On the comparative anatomy of transformations. *Annals of Mathematical Statistics, 28*(3), 602–632.

Wilcox, R. (2008). Robust methods for detecting and describing associations. In J. W. Osborne (Ed.), *Best practices in quantitative methods.* Thousand Oaks, CA: SAGE.

Yegiyan, N. S., & Lang, A. (2010). Processing central and peripheral detail: How content arousal and emotional tone influence encoding. *Media Psychology, 13*(1), 77–99. doi: 10.1080/15213260903563014

Yerkes, R. M., & Dodson, J. D. (1908). The relation of strength of stimulus to rapidity of habit-formation. *Journal of Comparative Neurology and Psychology, 18*(5), 459–482. doi: 10.1002/cne.920180503

SIMPLE LINEAR MODELS WITH CONTINUOUS DEPENDENT VARIABLES

Simple Regression Analyses

3

Advance Organizer

In this chapter, we will cover some of the basic aspects of linear modeling: evaluating the simple relationship between two variables in which both the independent variable (IV) and the dependent variable (DV) are continuous. This is typically viewed as the "simple regression" chapter in any statistics textbook; however, in our world of the generalized linear model (GLM), this chapter will provide the foundation for just about all other analyses to come, including analysis of variance (ANOVA) (Chapter 4, in which we will demonstrate that you get the same results regardless of whether you analyze the data via ANOVA or regression), logistic regression (Chapter 5, wherein we model binary DVs), and most chapters that follow. If you master the concepts in these chapters, you will be well positioned to excel in all of the chapters to come (and indeed, much of the field of quantitative methods!). The concepts following in all of the other chapters are more or less simple generalizations or evolutions arising from the basic concepts learned here.

In this chapter, we want to cover

- Basic concepts for linear regression–type analyses,
- Why correlation analyses are unnecessary in the 21st century,
- The concept of the effect size, and how to interpret it,
- How to interpret confidence intervals and why they are valuable,
- Examples of testing, and
- Examples of American Psychological Association (APA)–compliant summaries of analyses.

Guidance on how to perform these analyses in various statistical packages will be available online at study.sagepub.com/osbornerlm.

It's All About Relationships!

As previously mentioned, I assume that you have already had a basic statistics course that exposed you to analyses such as simple correlations (e.g., Pearson product-moment correlations), *t*-tests, simple ANOVA, simple regression, and maybe things like chi-square, odds ratios, and so forth. In many classes and in many textbooks, these are taught as different analyses, with different assumptions, formulae, and interpretations. Indeed, for much of the last century since these statistics were invented, that has been the case. More recently, we (as a field) have begun talking about the GLM, which is a simple model for thinking about relationships between variables, with many of the aforementioned statistical techniques as *special cases* of the GLM. Thus, we can talk about *t*-tests as a special case of ANOVA, and ANOVA as a special case of GLM in which the IV is categorical and the DV is continuous. We can talk about ordinary least squares (OLS) regression as a special case of GLM in which both the IV and DV are continuous (and simple correlation as essentially the same thing as regression but much more limited). We can talk about logistic and probit regression as special cases of the GLM in which the DV is binary and the IV is continuous or categorical. In fact, we can talk about all of these analyses as examining the simple relationship between two variables. In reality, we want to know whether one variable relates to, or affects, another. We don't (or shouldn't) care whether the variables are categorical or continuous. Despite a century of "tradition," it really doesn't matter, as we will demonstrate through the course of this chapter. In previous chapters, we presented Table 3.1.

In this chapter, we are dealing with the top left quadrant of this table, as we will work with a continuous IV and DV. During the course of this chapter, I will make the argument that correlation is really a special (and more limited) case of OLS regression and, as such, should not be treated separately. Thus, by the end of this chapter, I will argue that we should remove correlation from this box.

We will also start with the question of whether student achievement scores in high school (zACH, a continuous variable reflecting standardized achievement scores, converted to *z*-scores, ranging from −2.01 to 2.36, centered at 0 with a standard deviation [*SD*] of 1.0) are related to affluence or poverty, as measured by family socioeconomic

Table 3.1   Examples of the Generalized Linear Model as a Function of Independent Variable and Dependant Variable Type

	Continuous DV	Binary DV	Unordered Multicategory DV	Ordered Categorical DV	Count DV
Continuous IV	Correlation/ OLS regression	Binary logistic regression	Multinomial logistic regression	Ordinal logistic regression	OLS, Poisson regression
Mixed continuous and categorical IV					
Binary/ categorical IV only	ANOVA	Log-linear models	Log-linear models		Log-linear models

ANOVA, analysis of variance; DV, dependent variable; IV, independent variable; OLS, ordinary least squares.

status (zSES, a continuous variable reflecting a composite of parent income and job status, converted to z-scores, ranging from −3.66 to 3.23, centered at 0 with a SD of 1.0) using data from the National Education Longitudinal Study of 1988 (NELS88).[1]

Basics of the Pearson Product-Moment Correlation Coefficient

The Pearson product-moment correlation (r) is one of the most common indicators of linear association in the social and behavioral sciences, and we will assume that you have already been exposed to it in a basic statistics class. It is also, in my opinion, largely irrelevant in a modern statistical world in which simple relationships are rarely of interest and regression analyses are easily accessible. Furthermore, it is redundant with, but inferior to, simple linear regression. More on that in a moment.

The Pearson product-moment correlation is designed to be a metric-free[2] indicator of the strength and linear relationship between two continuous variables, where $r = 0.00$ is an indicator of a perfect absence of a relationship, or shared variance, between two variables and ±1.00 is an indicator of perfect redundancy between two variables. Positive correlations between two variables indicate that one variable tends to increase as the other increases. Negative correlations indicate that one variable tends to decrease as the other increases (an inverse relationship, in other words).

Pearson correlations can be inferential statistics, so we have a null and alternative hypothesis when we perform this type of analysis:

$$H_0: r_{xy} = 0; \text{ or}$$

$$H_a: r_{xy} \neq 0.$$

In this case, the question is the probability of obtaining the observed results from a population in which the null hypothesis is true; in other words, what is the probability of obtaining the observed correlation if in reality the correlation in the population is 0.00?

---

1   These data were from the National Education Longitudinal Study of 1988 (NELS88) from the National Center for Education Statistics (NCES; http://nces.ed.gov/surveys/nels88/), a survey of students in eighth grade in the United States in 1988. Participants in this study were followed for many years on thousands of variables, similar to other studies from the NCES. This and all data sets used in this book will be available on the book's website so that you can replicate the results if desired. However, there is one note of caution. Because this is a book on statistics and not a book on student achievement, I purposefully refuse to appropriately weight the data sets we will use in this book in order to prevent people from drawing substantive conclusions based on these analyses. All analyses within this book are intended to illustrate particular techniques or topics within quantitative methods, and readers should not draw any conclusions based on their results. If you are interested in using these types of data sets for substantive research on things such as student success or public health, please refer to appropriate guides on using complex samples, such as Osborne (2011), which is also available at http://pareonline.net/pdf/v16n12.pdf.

2   This means that the interpretation is independent of the metric, range, or SD of the variables being correlated. Before this innovation (and the similar innovation in standardized regression), relationships between variables were evaluated as covariances, which could range from negative infinity to positive infinity, depending on the range of the variables, and were thus very difficult to interpret and compare without further context, just as unstandardized regression coefficients are.

If the correlation is strong enough, given the sample size, to reject the null hypothesis, we can interpret the correlation as being "significant," moving then to interpret the effect size. In Figure 3.1, we can see a scatterplot of a small random sample from NELS88 that shows, in general, what seems to be a modest positive relationship between zSES and zACH: those with lower socioeconomic status (SES) tend to have lower achievement, and those with higher SES tend to have higher achievement, but the relationship is by no means perfect. For example, there are students with very low SES (−2.0) who have relatively high achievement (ranging up to +2.0 and above), and there are students with high SES (+2.0) who have relatively low achievement (below −1.0). Thus, this is not a perfect correlation, but some variance in achievement is probably "explained" by SES; in other words, knowing something about a student's SES lets us understand something about what that student's achievement is likely to be.

Calculating r

The goal of this book is to focus on application and conceptualization, rather than formulae (which in the modern age of statistical computing tends to be less

Figure 3.1   The Relationship Between Student Achievement and Student Socioeconomic Status

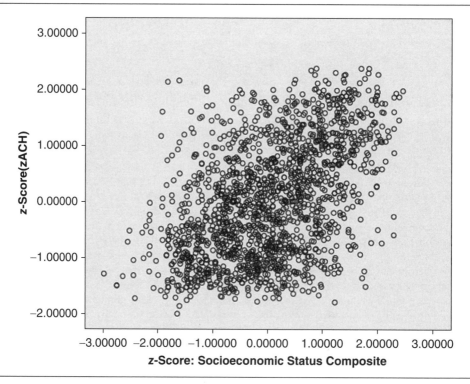

SOURCE: National Education Longitudinal Study of 1988 (NELS88) from the National Center for Education Statistics (http://nces.ed.gov/surveys/nels88/).

NOTE: This graph is from a reduced 10% sample of the data set, as the full sample tends to obscure the relationship in this scatterplot.

interesting to me). However, for the sake of comprehensiveness, I will occasionally present a formula for educational benefit. We have already started talking about variables that have been converted to z-scores, which makes the calculations simpler. In Equation 3.1 below, a simple z-score–based formula also provides some conceptual information about the calculations of r:

$$r_{xy} = 1 - \left(\frac{\Sigma \left(z_x - z_y \right)^2}{2(n-1)} \right) \qquad (3.1)$$

If you examine Equation 3.1 closely, it is intuitively telling us that where z-scores on two variables tend to correspond closely (the difference approaches 0), then the correlation coefficient will approach ±1.0. In other words, higher correspondence between the two variables will produce a correlation coefficient closer to ±1.0. The less consistent correspondence the two variables have, the closer to 0.00 the coefficient will be. An alternative formula for the correlation coefficient (Equation 3.2) again shows conceptually that it is a ratio of the correspondence (covariance) between two variables divided by the variance of the two variables.[3]

$$r_{xy} = \frac{\Sigma xy}{\Sigma x^2 \Sigma y^2} \qquad (3.2)$$

Effect Sizes and r

Rules of thumb for interpreting correlations have been proposed by many authors (e.g., Cohen, 1988). Typically these guidelines recommend that we interpret $r = 0.10$ as a "small" effect, 0.30 as a "medium" effect, and 0.50 as a "large" effect. However, the Pearson r is directly interpretable as a continuously variable effect size (e.g., 0.30 is larger than 0.20), and r^2 can also be directly interpreted as the percentage of the variance shared between the two variables. Thus, following Cohen's recommendation, variables with a "small" effect would share only 1% of their variance, variables with a "medium" effect would share only 9% variance, and those with a "large" effect would share 25%. This seems a bit uneven to me; however, because percent variance accounted for is also a continuous variable, ranging from 0% to 100%, my recommendation is that it be reported and interpreted appropriately.

A little mathematics leads us to Figure 3.2, where we show that the true effect size in a correlation, percent variance accounted for, is not a linear function of the magnitude of r. For example, a correlation of 0.10 represents 1% variance accounted for, a correlation of 0.20 represents 4% variance accounted for, a correlation of 0.30 represents 9% variance accounted for, and so forth. Thus, although one might think a correlation of 0.50 is relatively strong, being about halfway between 0.00 and 1.00, it is in reality representing only 25% variance accounted for and thus still leaves

---

3   Both equations are taken from Cohen, Cohen, West, & Aiken (2003).

Figure 3.2   The Relationship Between Correlation Coefficient and Percent Variance
Accounted For

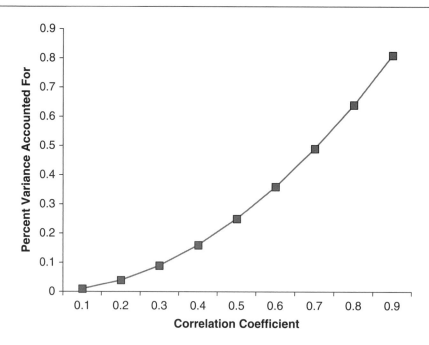

much variance unexplained (a correlation of just over $r = 0.70$ would equate to about
halfway if you think about percent variance accounted for). This concept of *variance
accounted for* will be one of the primary methods we will use to describe effect sizes
throughout much of the rest of this book.

A Real Data Example

The correlation between zACH and zSES within NELS88 was estimated to be $r_{(16,610)} =$
$0.51, p < .0001$, as you can see in Table 3.2. Given these results, we would reject the null
hypothesis and assert that there is a significant relationship between these two variables,
in the expected direction (as zACH increases, zSES also tends to increase), which should
be no surprise given the scatterplot in Figure 3.1, which shows overlap between the two
variables. But that is only half the battle! We also need to explain or quantify how strong
the relationship is, particularly in large samples where correlations close to zero can
lead to rejection of the null hypothesis but can be minuscule. This correlation of $r = 0.51$
might be viewed as relatively strong, because it is about halfway between 0.00 (no
relationship) and 1.00 (perfect relationship). However, this correlation converts to r^2
$= 0.26$, or about 26% of the variance in zACH accounted for by zSES (leaving almost
three-quarters, 74%, left unaccounted for). That is much more descriptive, to me, than
simply saying it is a "large" effect.

Thus, I encourage you to report percent variance accounted for and interpret it
substantively. Despite this example correlation being a "large" effect size according

Table 3.2   Correlation Between Socioeconomic Status and Achievement

Correlations

		zACH	zSES
zACH	Pearson correlation	1	0.505**
	p (2-tailed)		.000
	N	16,610	16,610
zSES	Pearson correlation	0.505**	1
	p (2-tailed)	.000	
	N	16,610	16,610

**Correlation is significant at the 0.01 level (two-tailed).

DATA SOURCE: National Education Longitudinal Study of 1988 (NELS88) from the National Center for Education Statistics (http://nces.ed.gov/surveys/nels88/).

to Cohen's rule of thumb, only about one-fourth of the variance is accounted for, or shared by the variables. This, to me, is not necessarily a "large" effect, although it is stronger than many we see in the social and behavioral sciences.

I have purposefully left this section on correlation short, as I believe it is not useful in the modern age of statistics. I also did not delve deeply into the assumptions of correlation, which are slightly different than regression. Again, I am not a fan of students learning simple correlation as an important mode of analysis. Let us quickly move to simple regression and see why.

The Basics of Simple Regression

Simple regression uses similar calculations to achieve the same goal as the Pearson product-moment correlation: to understand the extent to which we can understand one variable based on another variable. In regression, we create a *regression line* (and an associated *regression line equation*; also called the *line of best fit*) that summarizes the linear relationship between the two variables. Historically, these two procedures map onto two general goals in this type of research: explanation and prediction. These roughly correspond to two differing goals in research: being able to make valid projections concerning an outcome for a particular individual (prediction), or attempting to understand a phenomenon by examining a variable's correlates on a group level (explanation). There has been debate as to whether these two applications of regression are grossly different, as authors such as Scriven (1959) and Anderson and Shanteau (1977) assert, or are necessarily part and parcel of the same process (e.g., DeGroot, 1969; Kaplan, 1964; for an overview of this discussion, see Pedhazur, 1997, pp. 195–198). I believe both are necessarily part of the scientific process and that this dichotomy is an artificial and unnecessary one, particularly given that the numbers are identical regardless of the purpose. In my mind, the important, human aspect of statistical analysis is providing the context and interpretation of those numbers.

Regardless of your purpose, we usually seek to take our understanding of phenomena from our research and apply it to people who were not specifically involved in our research (a process called generalization).

This historical quirk separating correlation from regression can easily be dealt with in modern times by understanding that both analyses lead to identical conclusions, except that we get more information in simple linear regression and can do more things with the results. In Equation 3.3a, we can see that instead of one summary statistic for the relationship (the correlation coefficient, r, which is identical to the standardized regression coefficient, β, in simple OLS regression), we get three different pieces of information: an intercept (b_0; the value of Y when X is zero), a slope (b_1; the expected change in Y when X increases 1.0 in value), and the residual (e; the difference between the predicted value for the DV, Ŷ, and the actual observed value of the DV, Y). In Equation 3.3a, we decompose the observed values of Y into these three components: a starting point (intercept), effect of X on Y (slope), and the error term (which again is the difference between the predicted and observed scores). The equation is often presented in terms of predicted Y (which we denote as Ŷ), which is presented in Equation 3.3b. Because this second version of the regression line equation does not explicitly model the observed data, we do not have the error term in the equation.

$$Y = b_0 + b_1 X_1 + e \tag{3.3a}$$

$$\hat{Y} = b_0 + b_1 X_1 \tag{3.3b}$$

This is where the clear difference between correlation and regression is most obvious: in correlation, we take the same data and only estimate the standardized slope, leaving the other two pieces of information behind.[4] Let us return to the relationship between family socioeconomic status (zSES) and student achievement (zACH). Much has been written about the influence of the former on the latter, and we thus expect a reasonably strong effect (particularly if you have been paying attention in the chapter thus far).

Basic Calculations for Simple Regression

There are many different formulae for hand calculating (or using software like Excel to produce) estimates of the intercept (b_0) and slope (b_1). I present Equation 3.4a and 3.4b, both adapted from Cohen et al. (2002, p. 33), which hopefully will be intuitive and demonstrate some important concepts. Of course, nobody in the 21st century performs hand calculations,[5] and I do not expect you to either.

---

4   This will also explain why the assumptions for the two are slightly different despite being so closely related. There are no residuals in correlation; therefore, the assumption is that the variables are normally distributed, rather than the residuals, despite that being the more appropriate assumption.

5   Except in statistics classes or in the unlikely event that the alien overlords come and destroy our computers. If that happens, you will likely be using this book for other purposes—like lighting fires, rather than statistics—so I maintain my position that we do not need to teach hand calculation any more. If the alien overlords do come, we need to convince them that statisticians are vital to maintaining control over human society and—this is a critical point—are not at all tasty.

$$b_1 = r_{xy} \frac{sd_y}{sd_x} \tag{3.4a}$$

$$b_0 = \overline{Y} - b_1 \overline{X} \tag{3.4b}$$

As you can see, the unstandardized slope is a direct function of the correlation between the two variables and is scaled according to the ratio of the *SD*s of the two variables. The formula for intercept utilizes the slope and the mean of both variables to give us our regression equation.

Standardized Versus Unstandardized Regression Coefficients

Regression coefficients have two components, just like correlation coefficients: magnitude and direction. However, there is one complication. Unstandardized regression coefficients are influenced by the *SD* of the variables being analyzed. Thus, a b_1 of 100 might be very large, or very small, depending on the scale on which the variables are measured. If the scale is weekly income, that might be a strong effect. If the scale is population of a country, it might be a very week effect. Standardized regression coefficients (essentially, regression coefficients when variables are converted to *z*-scores, as ours have been, β_1) are interpreted the same way that correlation coefficients are. They range from 0.00 to ±1.00 in magnitude, and effect size is interpreted in the same way as a Pearson *r*. In fact, if we submit the same data to a simple linear regression analysis, we get identical results to the correlation analysis (as is obvious from Equation 3.4 if the *SD* for both variables is 1.0).

Hypothesis Testing in Simple Regression

We are explicitly testing several hypotheses, in classic null hypothesis statistical testing tradition for each parameter estimated, in addition to a hypothesis about the overall model.

H_0: $b_0 = 0$ (in other words, the intercept is zero)

H_a: $b_0 \neq 0$ (in other words, the intercept is not zero)

H_0: $b_1 = 0$ (in other words, there is no effect of zSES on zACH)

H_a: $b_1 \neq 0$ (in other words, there is a significant effect of zSES on zACH)

H_0: $R = 0$ (in other words, the multiple correlation is zero)

H_a: $R \neq 0$ (in other words, the multiple correlation is not zero)

Most regression texts focus on the slope, which is usually of more interest. In simple regression, this test and the results of the calculations will be identical to that of a simple correlation. One can have hypotheses about the intercept as well; this is less common, although it can be useful in some applications. Usually the intercept is non-sensical (e.g., when achievement is zero, when age is zero, when income is zero), which is why I think it is often ignored. However, when we center variables, like we have with achievement and SES, the intercept is not only meaningful, it is specifically the mean of the variable, so the intercept is the estimate of the DV when the IV is at the mean.

All of these values (slope, intercept, etc.) will be evaluated by calculating t, which is the ratio of b_1 (or whatever statistic we are testing) to the standard error of b_1. As t increases, the probability of observing the value from a population where the parameter value is zero drops, and once it drops below 0.05, we reject the null hypothesis and interpret the results. All statistical tests are in some way a ratio of signal to noise, so the larger the statistic is relative to the standard error, the more likely we are to reject H_0. Conversely, the more error in the data, the less likely we are to detect the effect.

A Real Data Example

Returning to our previous example of student achievement and family SES, our regression equation would look like this:

$$zACH = b_0 + b_1(zSES) + e \tag{3.5}$$

where b_0 is the intercept (or the average achievement score when zSES is 0, the average SES for students), and b_1 is the influence of SES on achievement. The other term in the equation, e, is the residual, or the difference between the predicted achievement score and the actual achievement score for each individual in the data set.

The Assumption That the Model Is Correctly Specified

As we reviewed in Chapter 2, there are several important assumptions around the model, some of which we will examine in later chapters. These include completeness, linearity, and additivity. It is almost certain that the model is not complete because we have only one variable predicting an outcome that we know is multiply determined. We will cover curvilinearity in later chapters, but one of the ways you can diagnose curvilinearity in data is by plotting standardized predicted values against standardized residuals (zPRED versus zRESID), as Figure 3.3 shows. If you saw significant curves in the scatterplot, this could indicate the need to model nonlinear relationships. There does not seem to be any indication of curvilinearity in this plot, and thus we can be relatively confident we meet this criterion.

We only have one variable in the equation, so additivity is not an issue.

Assumptions About the Variables

We also assume that the variables are measured without error and that they are measured at an interval or ratio level. Because we are using data that we did not collect or create, we do not have information about these aspects of the data; however, it is reasonable to at least assume interval measurement, given that the variables are continuous.

Figure 3.3   Scatterplot of Standardized Predicted Values (zPRED) Versus Standardized
Residuals (zRESID)

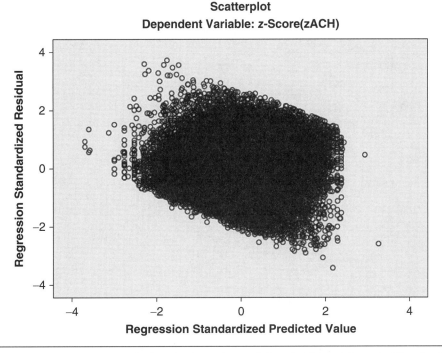

SOURCE: National Education Longitudinal Study of 1988 (NELS88) from the National Center for Education Statistics
(http://nces.ed.gov/surveys/nels88/).

Assumptions About Residuals

We assume that the residuals are normally distributed, and sum to zero. If you
examine the histogram of the residuals from this analysis in Figure 3.4, you will see
that they are largely normally distributed, and that the mean is about as close as you
can get to zero (0.000000000000000251, to be precise). Although the distribution is
not perfectly normal,[6] it is reasonably close and thus it is appropriate to say that these
assumptions are met. If you compute the actual skew and kurtosis (0.14 and −0.47,
respectively), these are also reasonably close to zero. There are some residuals that are
outside the −3 to +3 range, which might lead us to conclude that there are some inappropriately influential cases in the data. More on that in a moment.

Another assumption about residuals is that of *homoscedasticity*, or the assumption
that the variance of the residuals is relatively constant across the range of the variables.
If you examine the zPRED–zRESID plot (Figure 3.3), you can see that assumption is
largely met. According to those who study this topic, large differentials in variance (e.g.,
an area in which the variance is 10 times more than another area) is a serious violation
of this assumption.

---

6   Few things are in applied statistics, including the statisticians.

Figure 3.4 Distribution of Standardized Residuals

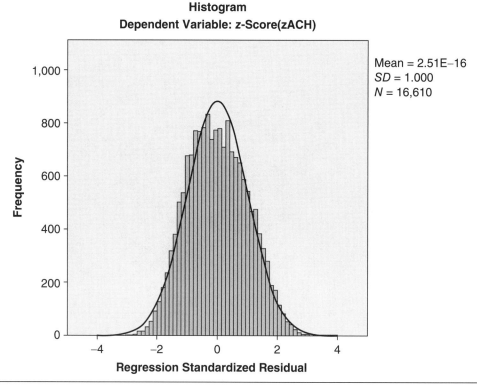

SOURCE: National Education Longitudinal Study of 1988 (NELS88) from the National Center for Education Statistics (http://nces.ed.gov/surveys/nels88/).

Finally, an important assumption is independence of observations, or independence of the originals. We usually assume that this is met unless there are clustered or nested data, or where we have repeated measures of the same individuals over time (or more esoteric research, such as on identical twins). However, most statistical packages include the Durbin-Watson test, which tests for a particular type of nonindependence (lag 1 autocorrelation). Where the test is near 2.0, the assumption is met. If it strongly diverges from 2.0, this assumption may not be met and other procedures (e.g., hierarchical linear modeling) should be employed. In this analysis, the Durbin-Watson estimate was 1.84, confirming the original assumption of independence of observations.

Summary of Results

As you can see in Table 3.3, the results of the regression analysis match exactly with the results of the correlation analysis, but with more detail extracted from the data. First, we will usually examine the overall model fit, which is represented by R, which is 0.505, with $R^2 = 0.255$. The overall model is significant, which is summarized in the

Table 3.3   Regression Analysis of Socioeconomic Status and Achievement Data

Model Summary

Model	R	R^2	Adjusted R^2	SE of the Estimate	Durbin-Watson
1	0.505[b]	0.255	0.255	0.86336531	1.843

ANOVA

Model		Sum of Squares	df	Mean Square	F	p
1	Regression	4,229.402	1	4,229.402	5,674.006	.000[b]
	Residual	12,379.598	16,608	0.745		
	Total	16,609.000	16,609			

Coefficients

Model		Unstandardized Coefficients		Standardized Coefficients	t	p	95% CI for B	
		B	SE	Beta			Lower Bound	Upper Bound
1	(Constant)	−1.004E-013	0.007		0.000	1.000	−0.013	0.013
	zSES	0.505	0.007	0.505	75.326	.000	0.491	0.518

SOURCE: National Education Longitudinal Study of 1988 (NELS88) from the National Center for Education Statistics (http://nces.ed.gov/surveys/nels88/).

ANOVA table. An F is a statistic that is similar to t in that it is a ratio of signal to noise (and, in fact, $t^2 = F$). In Table 3.3, you can see that the regression sum of squares (the variance accounted for by the IV) is 4,229.402, the residual (unexplained variance) sum of squares is 12,379.598, and the total is 16,609.000. In a literal sense, the variance accounted for is the ratio of variance explained by the regression equation to the total variance. If you do the calculations, that is 0.255, which is also equal to the R^2 value in Table 3.3 and is also equal to the r^2 from earlier in the chapter when we were exploring correlations (which is also identical to the R). Moving to the third part of the table, you can see the unstandardized regression coefficients in the first column, with the intercept very close to zero, and the unstandardized regression coefficient for zSES equal to 0.505 (as is the standardized regression coefficient, because the original variables were converted to z-scored variables; in general, unless you convert variables to z-scores, these statistics are different). You can further see that the intercept is not significantly different from zero (which is not surprising in light of the centering of both variables). However, the statistical test for the slope does lead us to reject the null hypothesis. Finally, we have 95% confidence intervals (95% CIs), which are calculated as follows:

$$95\% \text{ CI} = statistic \pm 1.96 \ (standard \ error \ of \ the \ statistic) \tag{3.6}$$

Thus, we are 95% confident that the population intercept in this case is between −0.013 and +0.013, and we are 95% confident that the unstandardized regression coefficient for this variable in the population is between 0.491 and 0.518. This helps us understand how precise our estimates are.

An Example Summary of Analysis in APA Format

A simple linear regression analysis examined the relationship between family SES and student achievement test scores in the eighth grade. Both variables were converted to z-scores prior to analysis. Assumptions (homoscedasticity, normal distribution of residuals, etc.) were met. Results (summarized in Table 3.3) indicate that achievement was positively related to SES ($b = 0.51$, $SE_b = 0.007$, $\beta = 0.51$, $t = 75.33$, $p < .001$). The 95% CIs around the regression coefficient were narrow [0.49, 0.52], indicating good precision. Finally, this effect was relatively strong, with 25.5% of the variance in achievement accounted for by SES.

Does Centering or z-Scoring Make a Difference?

Converting variables to z-scores does not influence the basic distribution of the variable, only the scaling (SD is changed to 1.0 from whatever it was) and that the distribution is now centered on the mean at a value of zero. If one were to use variables that were not converted to z-scores, the analysis should be identical except that the unstandardized regression coefficient would now be different (because the unstandardized regression coefficient is a function of the correlation, r_{xy}, as well as the SDs of the two variables, as we know from Equation 3.4). Briefly, the results of the analyses with variables in the original scales are presented in Table 3.4.

Note that although the unstandardized regression coefficients are now in the scales of the original variables, the effect sizes (standardized regression coefficient, R, R^2, t, p) have not changed. Thus, the only difference between the two is the ability to constitute the regression line equation in the original metric, as in Equation 3.7a, or in the z-scored metric, as in Equation 3.7b:

$$\hat{Y} = 0.103 + 0.631(\text{SES}) \tag{3.7a}$$

$$\hat{Y} = 0.00 + 0.505(z\text{SES}) \tag{3.7b}$$

Table 3.4   Regression Analysis of Socioeconomic Status and Achievement in Original Variable Scales

Model Summary

Model	R	R^2	Adjusted R^2	SE of the Estimate	Durbin-Watson
1	0.505[b]	0.255	0.255	0.86646	1.843

Coefficients

Model	Unstandardized Coefficients B	Unstandardized Coefficients SE	Standardized Coefficients Beta	t	p	95% CI for B Lower Bound	95% CI for B Upper Bound
1 (Constant)	0.103	0.007		15.344	.000	0.090	0.116
SES	0.631	0.008	0.505	75.326	.000	0.614	0.647

SOURCE: National Education Longitudinal Study of 1988 (NELS88) from the National Center for Education Statistics (http://nces.ed.gov/surveys/nels88/).

At this point, I hope you are seeing that correlation is redundant with what you can get from a regression analysis and is far more limited. Thus, my recommendation is that we not spend much time on it as a separate analysis. In the 21st century, I see little need for that anachronistic analysis.

Some Simple Multivariate Data Cleaning

We covered simple univariate data cleaning in Chapter 2, where we discussed the simple concept of the outlier, or more generally put, inappropriately influential data points. We discussed these as potentially concerning if they are the result of data entry errors, or if they come from individuals not likely to be part of the population of interest. As we established in Chapter 2, as data points fall farther outside ± 3 SD from the mean, they are increasingly unlikely to have legitimately been sampled from the population of interest (assuming of course a normal distribution for the variable).

Now we can look at bivariate outliers, and in later chapters, multivariate outliers. In general, we will use the same conservative philosophy regarding these cases: if we can develop a reasonable case that they are unlikely to be from the population of interest *or* they are inappropriately influencing the analysis, it is legitimate to remove them from the analysis. The goal, of course, is to come to estimates of population parameters that are optimally generalizable and replicable.

What Is a Bivariate Outlier?

Consider a simple example of height and sex.[7] You can have a nice normal distribution of height in, say, adults in the United States and have no data points in your sample that are more than 3 SD from the mean, but once we consider another variable, such as sex, we could have bivariate outliers. Specifically, a data point might not stand out as relatively extreme when everyone is lumped together. For example, the height of a woman who is 6 feet, 2 inches tall is probably extreme when considered solely in the context of women, and a man who is 4 feet, 11 inches tall might be considered extreme relative to other men, but neither value is extreme when both male and female heights are considered collectively. When we have two continuous variables, we can look at how well or how poorly individual data points fit the model across the range of observed values and we can potentially identify cases that do not fit. There are several general classes of indicators of influence, which will be discussed below.

Back to our example of SES and achievement, I have inserted two points in Figure 3.5 that could be excellent examples of the type of data point we are talking

---

7   At this point, I will pause and inform you that I disagree strongly with the APA and other guidelines that encourage authors to use the term *gender* to indicate biological male or female status. This is biological sex. Gender is a psychological variable that includes masculinity and femininity (and other nonbinary dimensions present in the diverse lesbian, gay, bisexual, transgender, and queer [LGBTQ] spectrum) that can have little to do with biological assignment of sex. Therefore, I will use the correct term *sex* to indicate a variable that identifies individuals as male or female, and anyone uncomfortable with seeing the word "SEX" in print can come back when they have grown up. Sex Sex Sex Sex Sex. This ends my nonstatistics-related rant for this chapter. Thank you for your patience. Sex.

about: they are not extreme in any univariate distribution, but when the combination of scores on two variables is examined, they are clearly not part of the overall distribution.

Standardized Residuals

Just as we can have variables that are "standardized" or converted to z-scores, we can also do this for residuals. Residuals, you recall, are merely the observed value minus the predicted value. Depending on the type of variable we are looking at, these can range from minuscule to massive, and there is no easy way to gauge what is a normal range and what is outside the normal range. However, converting residuals to a standard z metric, where 0 is the mean (and also is exactly on the regression line), ±1 is 1 SD from the regression line, and ±2 is 2 SD from the regression line, we can look at data points with standardized residuals outside the ±3 range. Using the same logic as in univariate data cleaning, those cases with standardized residuals more than 3 SD from the regression line are unlikely to be members of the population we desire to generalize to. More important, they are disproportionately adding error to the analysis and might be disproportionately influential (meaning that those data points might have more influence on the regression equation than others, which is not generally desirable). You will remember that when we were examining our regression

Figure 3.5   Example of Bivariate Outliers

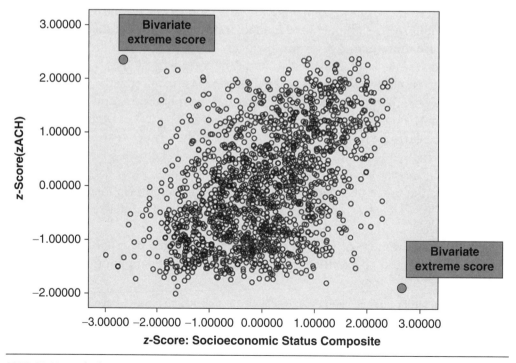

SOURCE: National Education Longitudinal Study of 1988 (NELS88) from the National Center for Education Statistics (http://nces.ed.gov/surveys/nels88/).

analysis above, there were a few residuals farther than 3 *SD* from the regression line. Specifically, they ranged from −3.40 to +3.66, and these are evident in Figure 3.3. By asking SPSS (or your software package of choice) to save standardized residuals, you can then select only those cases that have residuals within a reasonable range (e.g., −3 to +3) and rerun the analysis to see whether that small number of cases had an appreciable effect on the analysis.

The principle I use in data cleaning is that I want to improve the generalizability, replicability, and accuracy of the population estimates as much as possible by doing as little data cleaning as possible—in other words, I want to ideally remove no case from the data, but if I do, it should be a small number of extremely influential cases. In this example, by removing cases with residuals of more than ±3, we removed 14 cases, dropping our sample from 16,610 to 16,596 (0.08% of the sample). This data cleaning had a small effect on the parameter estimates (e.g., $\beta = 0.509$, rather than 0.505). A little bit more aggressive data cleaning (removing cases beyond 2.5 *SD*) brings our *N* to 16,516, and $\beta = 0.522$, with variance accounted for at 27.3% (up from 25.5%). I would not recommend doing more extreme data cleaning, as the goal is never to inappropriately inflate the population estimates. These data were relatively clean already and benefit little from the data cleaning. We will see other examples in which there are larger effects. Equation 3.8 shows the general formula for the standardized residual, where SD_{res} is the *SD* of the residuals.

$$ZRE = \frac{(Y - \hat{Y})}{SD_{res}} \tag{3.8}$$

Studentized Residuals

Studentized residuals are a refinement of the general notion of standardized residuals. The problem with standardized residuals, briefly, is that if you have extreme scores in the data, the error variances (*SD*) are larger than they should be, and thus the standardized residuals are smaller than they should be. Studentized residuals are calculated as the residual divided by an adjusted *SD* for the residuals, where SD_{res} is the *SD* of the residuals *without that particular case in the analysis*. Thus, they are more appropriately scaled to detect those cases that might be otherwise overlooked. However, I tend not to use these often, as standardized residuals are artificially conservative. I tend to recommend erring on the side of being more conservative than more aggressive when cleaning data.

Global Measures of Influence: DfFit or Cook's Distance (Cook's D)

While standardized or studentized residuals are a good start in helping to identify inappropriate cases, they actually do not directly assess what I think is most important: the extent to which a particular case might be inappropriately influential over the analysis. In general, those types of relatively extreme bivariate cases in Figure 3.5 will be easily identified through examination of standardized or studentized residuals.

However, another example of an influential case will not (as shown in Figure 3.6). In this fictional example, you can see three cases arrayed in such a way that they might substantially influence where the regression line is placed. Absent these three lines, those three cases in the top left and bottom right quadrants might pull the regression line into a very different (negative) slope. However, with the regression line arrayed in this way, none of these three circled cases would have large standardized residuals. What we often use is an indicator of *influence* or *leverage*—indicators of how much particular cases are influencing the parameter estimates, which is what I think is most central to the question of data cleaning.

SPSS and many other statistical packages report DfFit and Cook's Distance (Cook's D) as global indicators of influence.[8] If you observe the DfFit values (standardized to z-scores) in Figure 3.7, you can see there are some cases that seem inappropriately influential; at this point, we do not know if they are inflating the regression estimates (as the fictitious points in Figure 3.6 would) or suppressing them (as the cases in Figure 3.6 would). However, removing them is defensible if you use conservative, consistent guidelines. In this case, it seems to me that cases more 5 *SD* from the mean are both very conservative and defensible. As you can see in Table 3.5, removal of 46 cases (out of 16,610) led to a slight adjustment in the regression coefficients

Figure 3.6   Influential Cases That Would Not Be Detected by Standardized Residuals

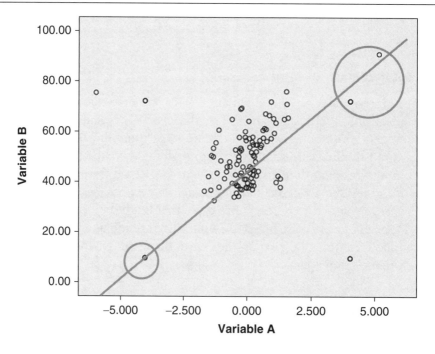

8   There are also other indicators, such as Mahalanobis's D, or DIFCHISQ, or other indicators depending on the statistical software used.

Figure 3.7   zDfFit Scores Converted to *z*-Scores: zSES Predicting zACH

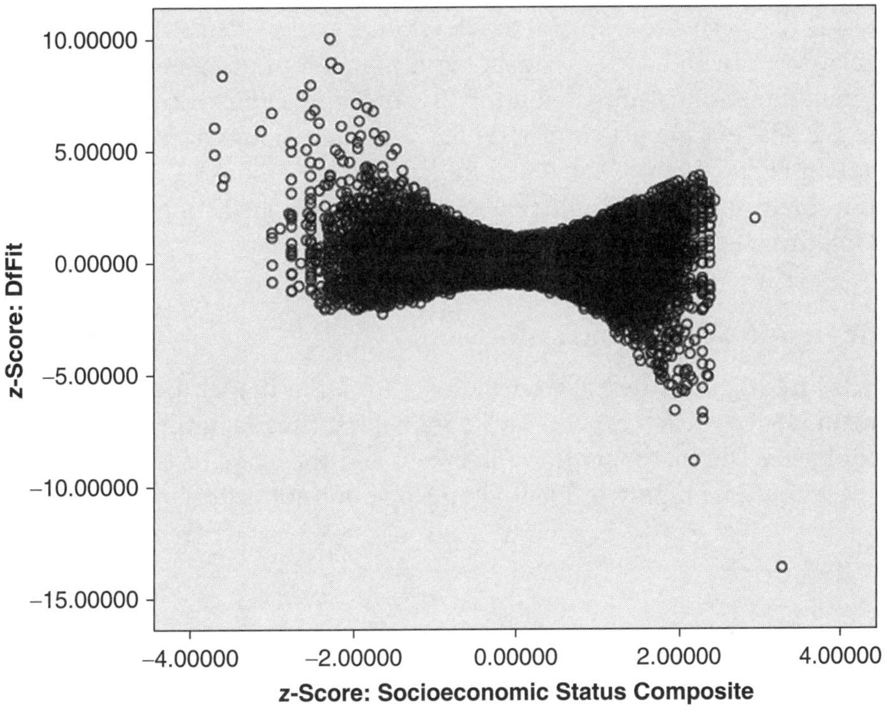

SOURCE: National Education Longitudinal Study of 1988 (NELS88) from the National Center for Education Statistics (http://nces.ed.gov/surveys/nels88/).

Table 3.5   Regression Results After Removing zDfFit > |5|

Coefficients[a]

Model	Unstandardized Coefficients		Standardized Coefficients	t	p	95% CI for B	
	B	SE	Beta			Lower Bound	Upper Bound
1 (Constant)	−0.001	0.007		−0.122	.903	−0.014	0.012
zSES	0.517	0.007	0.515	77.254	.000	0.504	0.530

[a]Dependent variable : zACH

(from 0.505 to 0.515; squaring each regression coefficient shows us the percent variance accounted for increased slightly from 25.5% to 26.5%). In other words, removing 0.277% of the sample resulted in improvement of 4% of the effect size.[9] In

---

9   Let me digress for a moment. Percent variance accounted for increased from 25.5% to 26.5%, which is an increase of 1%. So why did I say it was an "improvement of 4%"? Because this increase represents 4% of the observed effect. If variance accounted for was 1% and increased to 2%, that would be a 100% increase in effect.

my calculation, this means that group of cases were more than 14 times more influential than they ought to be.

Cook's D (converted to z-scores) has the similar goal of identifying inappropriately influential cases (with a single statistic) as a function of distance from the centroid of the multidimensional distribution. It is similar to DfFit in concept, and, in the case of this analysis, we get similar results (although Cook's D, like all distances, is positive). In Figure 3.8, you can see a similar pattern of Cook's D to those we saw in the DfFit results in Figure 3.7. Removing cases with Cook's D of more than $z = 5$, we see an improvement in the statistics, presented in Table 3.6.

Specific Measures of Influence: DfBetas

DfBetas are slightly different from Cook's D or DfFit in that these indicators measure the influence of each case on each parameter estimate. In this example analysis, we would get a DfBeta for both the intercept and the slope. It is certainly possible that a case could be highly influential on slope but not intercept, or vice versa. As

Figure 3.8   Cook's Distance, Converted to z-Scores

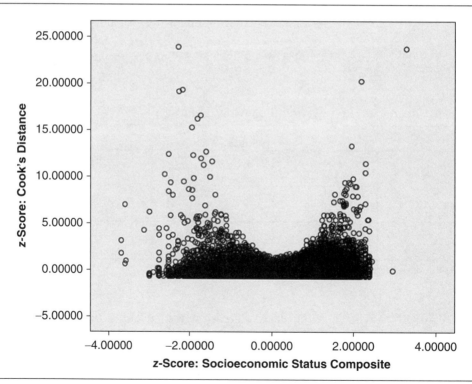

SOURCE: National Education Longitudinal Study of 1988 (NELS88) from the National Center for Education Statistics (http://nces.ed.gov/surveys/nels88/).

Table 3.6   Regression Results After Removing Cook's Distance (converted to z-Scores) > 5

| | | | Coefficients[a] | | | | | |
| Model | Unstandardized Coefficients | | Standardized Coefficients | | | 95% CI for B | |
	B	SE	Beta	t	p	Lower Bound	Upper Bound
1 (Constant)	0.000	0.007		0.030	.976	−0.013	0.013
zSES	0.528	0.007	0.525	79.241	.000	0.515	0.541

[a]Dependent variable: zACH.

you can see in Figure 3.9a, there are not any highly influential cases related to b_0 (all cases are within 4 SD of the mean for DFB0). However, in Figure 3.9b, you can see that there were some substantially influential cases related to b_1 (there are cases well beyond 5 SD). Removing cases with DfBetas for slope below −5 SD produced a small improvement in effect size (Table 3.7) similar to some of the other data cleaning examples above. The problem with DfBetas is that the number of DfBetas to evaluate gets larger as models get more complex.

Figure 3.9a   zDfBeta for the Intercept

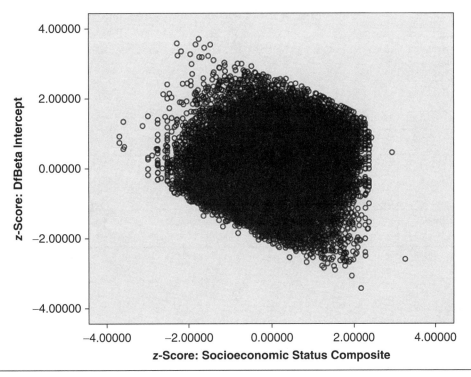

SOURCE: National Education Longitudinal Study of 1988 (NELS88) from the National Center for Education Statistics (http://nces.ed.gov/surveys/nels88/).

Figure 3.9b   zDfBetas for the Slope

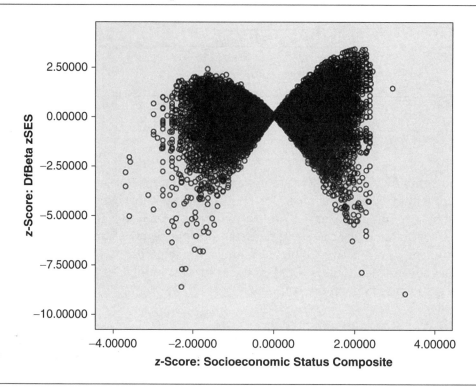

SOURCE: National Education Longitudinal Study of 1988 (NELS88) from the National Center for Education Statistics (http://nces.ed.gov/surveys/nels88/).

Table 3.7   Regression Results After Removing zDfBeta for Slope < −5

		Coefficients[a]						
		Unstandardized Coefficients		Standardized Coefficients			95% CI for *B*	
Model		*B*	*SE*	Beta	*t*	*p*	Lower Bound	Upper Bound
1	(Constant)	−0.002	0.007		−0.234	.815	−0.015	0.011
	zSES	0.515	0.007	0.513	77.008	.000	0.502	0.528

[a]Dependent variable: zACH.

SUMMARY

In this chapter, we explored a basic building block of GLM: the simple linear regression analysis. We explored how correlation and regression produce identical results, although correlation is much more limited in flexibility and scope. Thus, I hope I have persuaded you that it is not necessary to spend time learning simple correlation; rather, we can skip over that completely and move to simple regression.

We talked about important issues such as effect sizes and confidence intervals, correct interpretation of what we are seeing in regression analyses, how to test the assumptions of OLS regression, and finally, some simple multivariate data cleaning for regression models. In Chapter 4, we will explore how we can incorporate simple categorical variables (polytomous or unordered categorical variables) into simple regression models; along the way, we will prove that ANOVA is merely a special case of regression. We will also explore why the venerable *t*-test is not useful (again, except as an impressive historical feat).

At the end of this chapter, you will find some more enrichment exercises to help ensure that you have mastered these important concepts. As always, you will find answers to these enrichment exercises, as well as syntax, pointers on how to perform these analyses in SPSS, and data sets used in the chapter (and in the enrichment exercises) on the book website (study.sagepub.com/osbornerlm).

ENRICHMENT

1. Download the NELS88 data set from the book website and replicate the results found in the chapter.

2. Download the AAUP data set we first encountered in Chapter 2 (available from the book website) and analyze SAL_ALL (average salary of all faculty members at a university in hundreds of dollars) as a function of NUM_TOT (total number of faculty at an institution). Do faculty members at larger institutions earn more on average? When performing the analysis, be sure to do the following:

   a. Test assumptions and perform appropriate data cleaning/transformations.
   b. Summarize results in APA format, paying particular attention to effect size and confidence intervals.
   c. Write out a regression line equation for the model.
   d. Compare your results with what I have online in my Chapter 3 answer key.

3. Download the NHANES2001 data from the book website and investigate whether family income (FAM_INC; range is 1–13, with higher numbers meaning higher family income range) is significantly related to body mass index (BMI; higher numbers relate to higher weight to height ratio).

   a. Test assumptions and perform appropriate data cleaning.
   b. Summarize results in APA format, paying particular attention to effect size and confidence intervals.
   c. Write out a regression line equation for the model.
   d. Compare your results with what I have online in my Chapter 3 answer key.

REFERENCES

Anderson, N. H., & Shanteau, J. (1977). Weak inference with linear models. *Psychological Bulletin, 84*(6), 1155–1170. doi: 10.1037/0033-2909.84.6.1155

Cohen, J. (1988). *Statistical power analysis for the behavioral sciences* (2nd ed.). Hillsdale, NJ: Lawrence Erlbaum.

Cohen, J., Cohen, P., West, S., & Aiken, L. S. (2003). *Applied multiple regression/correlation analysis for the behavioral sciences*. Mahwah, NJ: Lawrence Erlbaum.

DeGroot, A. D., & Spiekerman, J. A. (1969). *Methodology: Foundations of inference and research in the behavioral sciences*. The Hague, The Netherlands: Mouton.

Kaplan, A. (1964). *The conduct of inquiry: Methodology for behavioral science*. San Francisco, CA: Chandler.

Osborne, J. W. (2011). Best practices in using large, complex samples: The importance of using appropriate weights and design effect compensation. *Practical Assessment Research & Evaluation, 16*(12), 1–7.

Pedhazur, E. J. (1997). *Multiple regression in behavioral research: Explanation and prediction*. Fort Worth, TX: Harcourt Brace College Publishers.

Scriven, M. (1959). Explanation and prediction in evolutionary theory. *Science, 130*(3374), 477–482. doi: 10.1126/science.130.3374.477

4
SIMPLE LINEAR MODELS WITH CONTINUOUS DEPENDENT VARIABLES

Simple ANOVA Analyses

Advance Organizer

In this chapter, we will cover basic aspects of linear modeling when you have continuous dependent variables (DVs) and an unordered categorical independent variable (IV). This is typically viewed as a simple analysis of variance (ANOVA) chapter in any statistics textbook. In our world of the generalized linear model (GLM), we will view this type of analysis not as a completely separate and alien creature compared with regression/correlation; rather, we will examine it as a simple variant on what we were doing in Chapter 3. By the time you have mastered Chapters 3, 4, and 5, you will be well prepared to take on a broad array of more complex analyses in the social and behavioral sciences (and many others!). The concepts following in all of the other chapters are more or less simple generalizations or evolutions arising from the basic concepts learned here.

In this chapter, we will cover

- Basic concepts in handling unordered polytomous IVs
- Dummy (and effects) coding of unordered polytomous IVs
- Why *t*-test analyses are unnecessary in the 21st century
- Confirmation that ANOVA and GLM/regression analyses of the same data produce identical outcomes, in estimating means for each group as well as effect sizes and so forth
- Assumptions and data cleaning when you have unordered polytomous IVs
- Confidence intervals and why they are valuable for these types of analyses
- Examples of American Psychological Association (APA)–compliant summaries of analyses

Guidance on how to perform these analyses in various statistical packages will be available online at study.sagepub.com/osberlmn.

It's All About Relationships! (Part 2)

In Chapter 3, we covered the most basic type of simple regression, in which both the IV and DV are continuous variables. I assume that you have been exposed to the basics

Table 4.1 Examples of the Generalized Linear Model as a Function of Independent Variable and Dependant Variable Type

	Continuous DV	Binary DV	Unordered Multicategory DV	Ordered Categorical DV	Count DV
Continuous IV	OLS regression	Binary logistic regression	Multinomial logistic regression	Ordinal logistic regression	OLS, Poisson regression
Mixed continuous and categorical IV					
Binary/ categorical IV only	ANOVA and t-test	Log-linear models	Log-linear models		Log-linear models

ANOVA, analysis of variance; DV, dependent variable; IV, independent variable; OLS, ordinary least squares.

of the *t*-test or ANOVA in an introductory class and thus will cover some of the details of these analyses in more of a cursory fashion. In the modern era of the 21st century, we (as a field) can talk about *t*-tests as a special case of ANOVA, and ANOVA as a special case of the GLM in which the IV is categorical (unordered polytomous) and the DV is continuous. In Chapter 3, we talked about the Pearson product-moment correlation (*r*) as a special, and more restricted, case of linear (ordinary least squares [OLS]) regression, and we now take this model one step further to incorporate what would otherwise be considered ANOVA-type analyses. Once you understand that the goal of the GLM is to examine and quantify relationships, regardless of the type of variable, you will be more flexible and effective in your statistical analyses. In reality, we want to know whether one variable relates to, or affects, another (regardless of the type of variable being examined).

In this chapter, we are dealing with the bottom left quadrant of Table 4.1: an unordered polytomous (categorical) IV and a continuous DV. During this chapter, I will make the argument that we should remove the *t*-test and include regression in that lower left box.

In this chapter, we will begin with another example from the National Education Longitudinal Study of 1988 (NELS88) data set[1] we were using in previous chapters. Let us examine the hypothesis of whether students who are retained and graduate from high

---

1 These data were from the NELS88 from the National Center for Education Statistics (NCES; http://nces .ed.gov/surveys/nels88/), a survey of students in eighth grade in the United States in 1988. Participants in this study were followed for many years on thousands of variables, similar to other studies from the NCES. This and all data sets used in this book will be available on the book's website so that you can replicate the results if desired. However, there is one note of caution. Because this is a book on statistics and not a book on student achievement, I purposefully refuse to appropriately weight the data sets we will use in this book in order to prevent people from drawing substantive conclusions based on these analyses. All analyses within this book are intended to illustrate particular techniques or topics within quantitative methods, and readers should not draw any conclusions based on their results.

school (GRADUATE; 0 = no, 1 = yes) have different overall achievement test scores in eighth grade (zACH) than those who do not graduate from high school.

Analyzing These Data via *t*-Test

As I have stated previously, the GLM is a rich statistical environment that includes many types of analyses from disparate traditions. Similar to the duality of correlation and regression, *t*-tests and ANOVA represent another unnecessary redundancy in the pantheon of analytic techniques. ANOVA is a generalization and extension of the original ideas and mathematics underlying the various types of *t*-tests; as such, *t*-tests are anachronistic (although quite common even in modern research and textbooks). Although this book is not focused on these techniques, I will take a moment to demonstrate that *t*-tests are more limited special cases of ANOVA and that analyses that would otherwise be labeled as "ANOVA"-type analyses fit easily into the GLM.

With ANOVA and *t*-tests, we are testing hypotheses that group means are generally equal or not equal:

$$H_0: \bar{X}_1 - \bar{X}_2 = 0; \ \bar{X}_1 = \bar{X}_2$$

$$H_a: \bar{X}_1 - \bar{X}_2 \neq 0; \ \bar{X}_1 \neq \bar{X}_2$$

When calculating a *t*-test, we essentially compare the difference between the two means by the "noise" or variance within each group, as you can see in Equation 4.1:

$$t = \frac{\bar{X}_1 - \bar{X}_2}{\sqrt{\dfrac{s_1^2}{N_1} + \dfrac{s_2^2}{N_2}}} \tag{4.1}$$

Intuitively, you should see a few things from the *t*-test formula. First, the larger the difference between the two means, the larger *t* will be. Second, the larger the variance of each variable is, the smaller *t* will be. Finally, larger samples will reduce the effect of the variance. These basic principles are consistent throughout a broad range of statistical techniques and conform to the general principles that our statistical tests (of which *t* is one of the most venerable) tend to be quantifications of signal (effect) to noise (variance).

This test is simple to calculate without advanced statistical software, which is what might make it so popular in textbooks. However, I will continue to insist that no practicing statisticians or researchers do calculations without software of some sort; therefore, ease of use is not a suitable rationale for including the *t*-test in a book.

Because we will not retain this test in our statistical toolbox, I will not delve deeply into assumptions. In general, they match the GLM assumptions discussed in Chapter 3, with the exception of the linear relationship. When discussing *t*-tests or ANOVA, we modify the assumption of homoscedasticity to be the assumption of equal variances (i.e., the variance for group if both groups are roughly equal). Most other assumptions are similar to those discussed already.

The results from the *t*-test, shown in Table 4.2, do not support the assumption of equal variances, by virtue of a significant Levene's test, which leads us to reject the null hypothesis that the variances are equal. By examining the standard deviations (*SD*s) for the two groups, we can see that one group has a *SD* substantially larger than the other. However, this is not an issue because SPSS provides output that compensates for this shortcoming, allowing us to avoid errors of inference that would otherwise occur as a result of violation of this assumption. Moving on in the second part of the table, we will examine the bottom row, labeled "Equal variances not assumed," and can see there are significant differences in eighth-grade achievement test scores between those who will graduate from high school or not four years later. In eighth grade, those who will fail to graduate score almost a full *SD* below those who will graduate ($t_{(2,170.50)} = -50.44$, $p < .0001$).

Mathematically, *t* is equal to the square root of *F* (or $t^2 = F$); therefore, we should not be surprised when we see ANOVA produce identical results with an *F* that is the square of the *t* that does assume equal variances (−36.627 squared is 1,341.537; see Table 4.3, below).

Analyzing These Data via ANOVA

I hope you are not disappointed that I have foreshadowed the outcome of this analysis throughout the earlier parts of this chapter. ANOVA is a more flexible and general analysis framework, as it can handle more than two groups and more than one IV. The hypothesis tests for ANOVA generalize to

Table 4.2   Differences in Eighth-Grade Achievement for Students Completing or Not Completing High School, *t*-Test Results

Group Statistics					
	GRADUATE	N	Mean	SD	SE Mean
zACH	.00 (no)	1477	−0.8750457	0.66385830	0.01727369
	1.00 (yes)	15133	0.0854056	0.98619828	0.00801681

Independent-Samples Test									
		Levene's Test for Equality of Variances		t-Test for Equality of Means					
		F	p	t	df	p (2-tailed)	Mean Difference	SE Difference	95% CI of the Difference
									Lower	Upper
zACH	Equal variances assumed	519.803	.000	−36.627	16,608	.000	−0.96045	0.026222	−1.011850	−0.909052
	Equal variances not assumed			−50.435	2,170.498	.000	−0.96045	0.019043	−0.997796	−0.923106

DATA SOURCE: National Education Longitudinal Study of 1988 (NELS88) from the National Center for Education Statistics (http://nces.ed.gov/surveys/nels88/).

Table 4.3   Differences in Eighth-Grade Achievement for Students Completing or Not Completing High School, ANOVA Results

GRADUATE[a]

GRADUATE	Mean	SE	95% CI	
			Lower Bound	Upper Bound
0 (no)	−0.875	0.025	−0.924	−0.826
1 (yes)	0.085	0.008	0.070	0.101

[a]Dependent variable: zACH.

Tests of Between-Subjects Effects[a]

Source	Type III Sum of Squares	df	Mean Square	F	p	Partial Eta Squared
Corrected model	1,241.328[b]	1	1,241.328	1,341.516	.000	0.075
Intercept	839.063	1	839.063	906.783	.000	0.052
GRADUATE	1,241.328	1	1,241.328	1,341.516	.000	0.075
Error	15,367.672	16,608	.925			
Total	16,609.000	16,610				

[a]Dependent variable: zACH.

[b]$R^2 = 0.075$ (adjusted $R^2 = 0.075$).

$$H_0: \bar{X}_1 = \bar{X}_2 = ... = \bar{X}_k \,; \text{ or}$$

$$H_a: \bar{X}_1 \neq \bar{X}_2 \neq ... \neq \bar{X}_k \,.$$

I will spare you the details of the calculations of the F test, but they follow the same principle, dividing the "signal" (mean square for the effect of GRADUATE, $MS_{GRADUATE}$) by the "noise" (mean square for error, MS_{error}). As you can see in Table 4.3, we get identical results to the prior t-test when subjecting the same data to ANOVA.

Note that in this analysis, we see identical group means, significance testing, p values, and so forth (within rounding error) compared with the previous t-test. Furthermore, we see that the F is indeed within a small rounding error of the square of t. We also get eta squared (η^2), which is an effect size for ANOVA indicating the percentage of the variance in the DV explained by group membership, the IV. In this example, with no thought to data cleaning, the IV GRADUATE accounts for 7.5% of the variance in student achievement test scores (zACH). This is the same concept as R^2 or r^2 from the previous chapter on correlation and regression. Thus, since ANOVA is more versatile (as regression is more versatile than correlation), we should, as a field, dispense with the t-test altogether except as an important and invaluable historical development (Student, 1908).

ANOVA Within an OLS Regression Framework

Now that we have established that correlation and regression analyses produce identical results, and t-tests and ANOVA analyses produce identical results, let us return to the overarching thesis of this chapter: that all of these analyses are special cases of the GLM, and they should not be treated as completely separate analytic techniques, but rather as appropriate GLM analyses for the types of variables present to be analyzed. My original argument with my adviser, that regression is a fine framework for analyzing data with dichotomous or categorical IVs, can now be examined. If both ANOVA and regression are part of the same analytic framework, then they should produce identical results when the same data are subjected to appropriate analysis in both frameworks. Indeed, Table 4.4 shows the results of OLS regression when zACH is entered as the DV and GRADUATE is entered as the IV, mirroring what we performed in the previous ANOVA analysis.[2] In Table 4.4, I present a slightly expanded overview of the regression output to make the point that both ANOVA and regression analyses produced identical results in all relevant statistics.

First, let us start with overall effect size, percent variance accounted for (η^2 in ANOVA, R^2 in regression). Both analyses produce identical estimates of variance accounted for: 0.075 or 7.5% of the variance in student achievement is estimated to be explained by (or overlaps with, if you like less "causal" language) group membership—whether the student will ultimately be retained or not through graduation from high school. The sum of squares (1,241.33 for effect, 15,367.67 for error), mean square (1,241.33), and F statistics (1,341.52) are identical in both analyses. Even the t-value for the GRADUATE coefficient is identical to that of the t-test presented in Table 4.2.

Other statistics are also identical, including the group means we can reproduce from the regression line equation derived from the unstandardized coefficients in Table 4.3. This might seem weird, as we rarely think about ANOVA-type analyses as

Table 4.4   Regression Analysis Predicting Eighth-Grade Achievement Test Scores From GRADUATE

Model Summary

Model	R	R^2	Adjusted R^2 Square	SE of the Estimate	Durbin-Watson
1	0.273	0.075	0.075	0.96193421	1.601

ANOVA

Model		Sum of Squares	df	Mean Square	F	p
1	Regression	1,241.328	1	1241.328	1,341.516	.000[b]
	Residual	15,367.672	16,608	.925		
	Total	16,609.000	16,609			

Coefficients

Model		Unstandardized Coefficients B	Unstandardized Coefficients SE	Standardized Coefficients Beta	t	p	95% CI for B Lower Bound	95% CI for B Upper Bound
1	(Constant)	−0.875	0.025		−34.960	.000	−0.924	−0.826
	GRADUATE	0.960	0.026	0.273	36.627	.000	0.909	1.012

2   There is a long tradition and history concerning using categorical variables in regression analyses. We will explore this later in this chapter. At this point, accept my assertion that in this case, it is acceptable to enter a binary variable as an IV in this analysis.

having a slope or intercept; however, think about the essence of a regression line equation, such as Equation 4.2:

$$Y = b_0 + b_1X_1 + e \tag{4.2}$$

In Chapter 1, I discussed that this equation simply means that we can decompose any person's score on a variable (Y) into several ingredients, as follows:

1. A starting point (b_0)—usually the intercept, or the place where a line intercepts an axis.

2. Slope—the expected change in the variable (b_1) as the result of where the person is on another variable (X_1).

3. The leftover stuff that is not explained by either of the other two ingredients (e)—often referred to as the residual, the unexplained variance, or error. These are all synonymous terms.

Furthermore, that we can conceptualize ANOVA through the lens of this equation as an individual's score on one variable (i.e., the DV) is the function of a starting point, and then the effect of another variable (the IV), and then often containing an unexplained portion. If we call b_0 (the starting point) the "grand mean" (or mean of all scores in the sample) and b_1 the "effect of group membership" (or the difference between a group mean and the overall "grand" mean), then we have a simple (one-way) ANOVA. In addition, we will always have individual residuals (e) where this equation does not exactly reproduce observed scores on Y (in ANOVA, these residuals are the difference between individual scores and the group mean).

Note that the intercept in the model above, −0.875, is the value of zACH when the IV = 0 (in this case, GRADUATE = 0 is the group of students who were not retained through graduation) and is identical to the group mean for GRADUATE = 0 in the ANOVA (−0.875). Furthermore, the slope (0.960) is identical (within rounding error) to the mean difference between the two groups (also present in the t-test output). Let's put this into regression equations and predict the group means that way so we are perfectly clear on the fact that these analyses produced absolutely identical results. Taking the unstandardized coefficients (intercept, slope) from the regression output in Table 4.3, we create the following regression line equation (Equation 4.3):

$$\hat{Y} = -0.875 + 0.960(\text{GRADUATE}) \tag{4.3}$$

Inserting the group values for GRADUATE into the regression equation, we get predicted values of −0.875 for those not retained (GRADUATE = 0) and 0.085 for those retained to graduation (GRADUATE = 1), matching our observed group means from the ANOVA and t-test analyses. Also note that the 95% confidence intervals (95% CIs) around the estimates are identical across groups. For example, for the mean of the GRADUATE = 0 group (−0.875), the 95% CI ranges from −0.924 to −0.826 in both analyses. This means we are 95% confident that the population mean for this group should fall between −0.924 and −0.826, a reasonably narrow confidence interval. For the other group, we would have to predict the mean and then calculate the 95% CI around that predicted value,[3] which produces the same results as the ANOVA (mean = 0.085, 95% CI = 0.07–0.10).

Thus, in the GLM, we can use regression line equations to predict group means that are identical to those derived from t-test and ANOVA analyses. For those of you having difficulty thinking of ANOVA in a regression framework, you can think of the slope as a mean difference between two groups. Again, the point for now is that these analyses are all part of the same GLM, and all produce identical results.

What if you have more than two groups in your IV? This is something we have known how to deal with for decades: dummy or effects coding.

When Your IV Has More Than Two Groups: Dummy Coding Your Unordered Polytomous Variable

"Dummy" variables are binary variables (usually coded either "0" or "1') that computer programmers have used for decades to represent the presence or absence of a state or condition. We in statistics have adopted the language and technique to code a set of dummy variables that represent the variable of interest. There are usually options to have your statistical software package perform the coding automatically, but I think there are good reasons to take a few moments to create your own dummy variables. Foremost is the desire to know exactly how the variables are being coded so that you know how to interpret the results. As we will see, the way the dummy-coded variables are created can have a large influence on your results. From a conceptual point of view, when we have a variable with k levels or groups (like in our GRADUATE variable, where we had two possible groups), we will be able to completely account for the effect of the variable with $k − 1$ dummy variables. Thus, my earlier assertion that one binary variable can represent GRADUATE is put into context.

In addition to the typical steps of performing a regression analysis, performing this type of analysis adds two additional steps. First, you must determine what group is going with serve as the reference group: the group all other groups are compared with. Second, you must create the dummy variables. Third, you perform the analysis by entering the dummy variables that represent the effects of the IV, and then proceed with data cleaning, testing of assumptions, interpretation of results, and so forth.

Our example will come from the National Health Interview Survey of 2010 (NHIS2010).[4] In this example, we will look at the age of diagnosis of diabetes (DIBAGE) for those aged >20 years as a function of smoking status (SMOKE; 0 = never smoked, 1 = former smoker, 2 = occasional smoker, and 3 = daily smoker).

---

3 Group mean ± (1.96 multiplied by the *SE*).

4 Public data are available from http://www.cdc.gov/nchs/nhis/nhis_2010_data_release.htm. Data and more information are available on the book's website.

Table 4.5   Distribution of the Smoking Status Variable

SMOKE				
	Frequency	*Percent*	*Valid Percent*	*Cumulative Percent*
0 (Nonsmoker)	1,206	49.2	49.2	49.2
1 (Former smoker)	818	33.3	33.3	82.5
2 (Occasional smoker)	92	3.8	3.8	86.3
3 (Daily smoker)	337	13.7	13.7	100.0
Total	2,453	100.0	100.0	

There is no real rationale behind this analysis other than I thought it was interesting.[5] The distribution for SMOKE is provided in Table 4.5.

Define the Reference Group

The first step when you have multiple groups like our smoking variable is to decide what group is the reference group—the group with which others will be compared. This is essentially a planned comparison (like you would do in ANOVA), meaning there are $k - 1$ degrees of freedom (*df*) for comparisons. In our case, with four groups, we can make three independent comparisons and no more.[6] What makes the most sense? That depends on your research question. When studying race, for example, I often use Caucasian individuals as the comparison group because they are often the majority and traditionally represent the societal *status quo*. If I am comparing the effects of several interventions, I might choose the traditional or more standard condition as the comparison group. Hardy (1993) suggested that the reference group

a. Should make sense, being a control group, for example, or a group that represents some sort of standard reference, routine, or dominant group;

b. Should be coherent and well defined, not an "other group" or heterogeneous catch-all group (e.g., "multiracial" or "other race" in the context of a race group, or "not married" in the context of an analysis of marital status); and

c. Should not have a small sample size compared with the other groups.

---

5   As with all analyses from public use data we present in this book, these results are purposefully not adjusted for sampling and so forth because the goal of the book is not to study a particular phenomenon, but rather to present real data in real analyses.

6   When talking about this issue, people always ask me if it is legitimate to rerun the analysis with a different coding scheme to fully explore the data. My response is that it is *not* legitimate to do that: if you take our example and perform four comparisons on this variable and then recode and perform four more, you are essentially performing eight comparisons with only 4 *df*. Furthermore, I don't think it is necessary if you are thoughtful in matching the coding scheme to the questions you want to ask. Other coding schemes (discussed below) will allow different questions to be asked.

Table 4.6 Setting Up Dummy-Coded Variables

Smoking Status	DUM1	DUM2	DUM3
0 (Nonsmoker)	0	0	0
1 (Former smoker)	1	0	0
2 (Occasional smoker)	0	1	0
3 (Daily smoker)	0	0	1

Using these guidelines, let us propose the first group, nonsmokers, as the comparison group, because this meets the criteria of being a common group (a large portion of the sample, well defined, and so forth). The other three groups will then be compared with this first group.

Set Up the Dummy-Coded Variables

Knowing that we have $k - 1$ possible comparisons, represented by $k - 1$ dummy variables, we need to create them and set them up appropriately. I often encourage students learning this technique to use a table such as Table 4.6.

Once we have identified the comparison group, it is arbitrary what group is assigned a 1 on what dummy variable: as long as each of the other groups is assigned a 1 on one and only one dummy variable, things will work out fine. We will set up three dummy variables to account for the three contrasts that make the most sense. In this case, we will use nonsmokers as the comparison group and compare each other group with that one.

When all three dummy variables are entered into the regression analysis at the same time, they completely capture the effect of SMOKE and also allow for simple contrasts between the designated comparison group and each of the other groups. In fact, in OLS regression, the unstandardized regression coefficient(s) for each dummy variable represents the actual difference between the two group means. Creating these variables will be different depending on the software package you use. I will have a guide online to show you how to do this in SPSS via point-and-click if you are interested. The most efficient way I know of to do this is to create all dummy variables via syntax, assigning each individual who is a member of one of these groups a 0 on all variables,[7] as most of the matrix above is populated by zeros:

```
compute DUM1=0.

compute DUM2=0.
```

---

7 Make sure to exclude any cases with missing data from this process; otherwise they will be lumped in together.

```
compute DUM3=0.
execute.
```

With four more commands, you can then assign the value of 1 to members of the four nonreference groups as follows:

```
if(SMOKE=1)DUM1=1.
if(SMOKE=2)DUM2=1.
if(SMOKE=3)DUM3=1.
execute.
```

Your three dummy variables are now ready to be used!

Evaluating the Effects of the Categorical Variable in the Regression Model

What we want to see is that the overall model is significant (R for the regression model is significantly different from zero at $p < .05$), and if that is the case, we can examine the individual effects of the dummy variables. A significant effect for any of the dummy variables indicates a significant mean difference between the reference group and the group that is assigned a 1 on that variable. If all of the dummy variables are entered into the equation and there is not a significant change in R, then we have to assume there is no overall effect of the variable and it is not useful (or legitimate) to examine the individual effects. This is similar to the norm of not performing pairwise comparisons in ANOVA if the overall effect of the variable is not significant.

Smoking and Diabetes Analyzed via ANOVA

For pedagogical purposes, we will first analyze these data in the classic manner through a simple univariate ANOVA. As you can see in Figure 4.1, the DV (DIBAGE) is relatively normal in terms of distribution (skew = −0.03, kurtosis = −0.55), but we still have to check for assumptions (e.g., homogeneity of variance).

Results of the ANOVA indicate that the assumption of homogeneity of variance was not supported (with a Levene's test that was significant, leading us to reject the null hypothesis that the variances of the different groups are equal; $F_{(3, 2,449)} = 5.291$, $p < .001$). As you can see in Table 4.7, this is likely due to the substantial differences in group sizes in combination with relatively modest differences in variance across groups. If this were a research project we wanted to publish, we would attempt to normalize the variance in some way, often through transformation of the DV or examination of residuals (looking for outliers or influential cases). In this example, there are no cases with standardized residuals outside ±3, so that option is not available. Let us continue to examine the results, as this is merely a methodological example rather than a study of the substantive topic.

You can see that SMOKE is significant, with an η^2 of 0.047, meaning that approximately 4.7% of the variance in age of diabetes diagnosis is attributable to smoking status for some reason. Finally, at the bottom of the table, you see the results of a pairwise comparison between the first group (nonsmokers) and each other group, comparable to what we are going to do in the next section with the dummy-coded variables. Note that according to this last part of the ANOVA results, groups 1 and 2 (nonsmokers versus former smokers) are significantly different (mean difference = 4.926, standard error [SE] = 0.614, $p < .0001$), groups 1 and 3 are not significantly different (mean difference = −2.678, $SE = 1.466$, $p < .068$), and groups 1 and 4 are significantly different (mean difference = −3.567, $SE = 0.835$, $p < .0001$). The results of this analysis would lead you to conclude that smoking is a significant factor in age of onset of diabetes, former smokers have a significantly later age of onset of diabetes than nonsmokers, there is no difference between nonsmokers and occasional smokers, and daily smokers have a significantly earlier age of onset than nonsmokers. Of course, this is a relatively modest effect, accounting for less than 5% of the variance in onset of diabetes. However, that is not important, as the goal of this part of the chapter is to demonstrate that you will reach exactly the same conclusions using regression analysis.

Figure 4.1   Distribution of DIBAGE

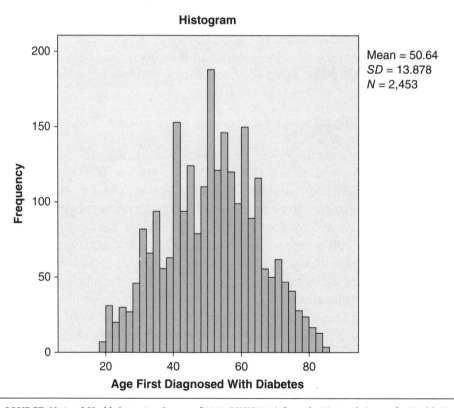

DATA SOURCE: National Health Interview Survey of 2010 (NHIS2010) from the National Center for Health Statistics (http://www.cdc.gov/nchs/nhis/nhis_2010_data_release.htm).

Table 4.7   Analysis of Smoking Status and Age of Diabetes Onset via ANOVA

Dependent variable: DIBAGE

SMOKE	Mean	SD	N
.00	49.59	14.140	1206
1.00	54.52	13.188	818
2.00	46.91	12.385	92
3.00	46.02	12.581	337
Total	50.64	13.878	2453

Tests of Between-Subjects Effects

Dependent variable: DIBAGE

Source	Type III Sum of Squares	df	Mean Square	F	p	Partial Eta Squared
Corrected model	22082.041[a]	3	7,360.680	40.044	.000	.047
Intercept	2443686.730	1	2,443,686.730	13,294.157	.000	.844
SMOKE	22082.041	3	7,360.680	40.044	.000	.047
Error	450166.842	2,449	183.817			
Total	6763564.000	2,453				
Corrected total	472248.882	2,452				

[a]$R^2 = 0.047$ (adjusted $R^2 = 0.046$).

Contrast Results (K Matrix)

SMOKE Simple Contrast[a]		Dependent Variable
		DIBAGE
Level 2 versus Level 1	Contrast estimate	4.926
	Hypothesized value	0
	Difference (estimate − hypothesized)	4.926
	SE	0.614
	p	.000
	95% CI for difference Lower bound	3.722
	Upper bound	6.130
Level 3 versus Level 1	Contrast estimate	−2.678
	Hypothesized value	0
	Difference (estimate − hypothesized)	−2.678
	SE	1.466
	p	.068
	95% CI for difference Lower bound	−5.554
	Upper bound	.197
Level 4 versus Level 1	Contrast estimate	−3.567
	Hypothesized value	0
	Difference (estimate − hypothesized)	−3.567
	SE	0.835
	p	.000
	95% CI for difference Lower bound	−5.206
	Upper bound	−1.929

[a]Reference category = 1.

Smoking and Diabetes Analyzed via Regression

Now let us demonstrate that all of the other groups are significantly different. Using the exact same data, with the dummy variables constituted as described above, we will identify EXERCISE as the DV and the four dummy variables as the IVs. First, we must establish that we met the assumptions of OLS regression. The residuals had reasonable homoscedasticity and were reasonably normally distributed, as shown in Figure 4.2. The skewness of the residuals was −0.036 and the kurtosis was −0.471, which is not perfect but reasonable.

Figure 4.2   Distribution of Residuals From Regression Analysis

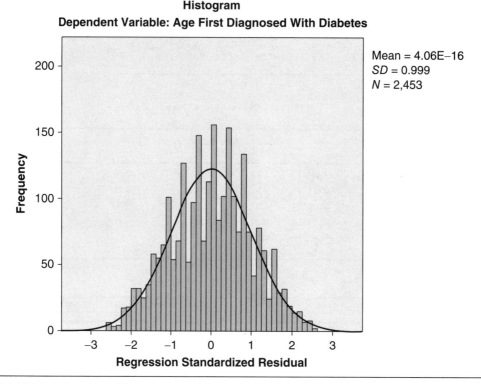

DATA SOURCE: National Health Interview Survey of 2010 (NHIS2010) from the National Center for Health Statistics (http://www.cdc.gov/nchs/nhis/nhis_2010_data_release.htm).

The results of the regression are presented in Table 4.8. Quickly, we will note that the regression sum of squares matches the ANOVA sum of squares, R^2 matches η^2, and almost every other statistic is identical within rounding error. Thus, we can establish that this is the same analysis in terms of overall model significance and variance accounted for.

Not only are the overall statistics identical, but the statistics for the dummy variables, including the 95% CIs, and the significance levels are as well. Furthermore, the unstandardized regression coefficients match the mean differences from the ANOVA (as do the *SE*s of those estimates). Finally, we can exactly reproduce the group means from the ANOVA by creating a regression line equation and entering the appropriate values from the dummy variables.

As you can see in the bottom of Table 4.8, the intercept is 49.591, which represents (and matches) the mean for the nonsmoker group (SMOKE = 0) from the previous ANOVA. From the last part of the table, we can create a regression line equation for this analysis (Equation 4.4a) as follows:

$$\hat{Y} = 49.591 + 4.926(DUM1) - 2.678(DUM2) - 3.567(DUM3) \qquad (4.4a)$$

Thus, we can predict each group mean as a function of the intercept and slope of a particular dummy variable. If we want the mean for the first group (former smokers),

Table 4.8   Regression Analysis of Smoking Status and Age of Diabetes Onset

Model Summary[a]

Model	R	R^2	Adjusted R^2	SE of the Estimate
1	0.216[b]	0.047	0.046	13.558

[a]Dependent variable: DIBAGE.
[b]Predictors: (constant), DUM3, DUM2, DUM1.

ANOVA[a]

Model		Sum of Squares	df	Mean Square	F	p
1	Regression	22,082.041	3	7,360.680	40.044	.000[b]
	Residual	450,166.842	2,449	183.817		
	Total	472,248.882	2,452			

[a]Dependent variable: DIBAGE.
[b]Predictors: (constant), DUM3, DUM2, DUM1.

Coefficients[a]

Model		Unstandardized Coefficients		Standardized Coefficients	t	p	95% CI for B	
		B	SE	Beta			Lower Bound	Upper Bound
1	(Constant)	49.591	0.390		127.024	.000	48.826	50.357
	DUM1	4.926	0.614	0.167	8.021	.000	3.722	6.130
	DUM2	−2.678	1.466	−0.037	−1.826	.068	−5.554	0.197
	DUM3	−3.567	0.835	−0.089	−4.270	.000	−5.206	−1.929

[a]Dependent variable: DIBAGE.

DATA SOURCE: National Health Interview Survey of 2010 (NHIS2010) from the National Center for Health Statistics (http://www.cdc.gov/nchs/nhis/nhis_2010_data_release.htm).

we can substitute in the values for each dummy variable for that group (a "1" on DUM1 and "0" on each other dummy variable), giving us Equation 4.4b:

$$\hat{Y} = 49.591 + 4.926(1) - 2.678(0) - 3.567(0)$$
$$\hat{Y} = 49.591 + 4.926$$
$$\hat{Y} = 54.517 \tag{4.4b}$$

Note that this matches the group mean from the ANOVA analysis. You can perform the other calculations to verify that the other means turn out to be estimated very closely to that of the ANOVA analysis as well.

What If the Dummy Variables Are Coded Differently?

What would happen if we had created the dummy variables differently? Try it. The variance accounted for by the overall model will be identical (if you follow the rules), but the comparisons will yield different results individually because you will be making different comparisons. However, at the end of it, if you take the new regression line equation and appropriately estimate the means using this new equation, you will still get the same means as in the ANOVA analysis. Thus, no matter how you create the four independent contrasts, you will end up with the same results for the overall model. Table 4.9, for example, presents the same analysis with the last group (daily smokers) set as the comparison group. Note that the model statistics (e.g., R, R^2, $SS_{regression}$, etc.) are the same as before. Smoking as a variable will always account for the same amount of variance no matter how you divide it up.

Because a different group is now the reference group, the unstandardized regression coefficients and the intercept will be different. Note that the intercept is identical to the mean of the new reference group from the ANOVA analysis; if you use the regression line equation to predict group means, they will also be identical to the previous analyses.

Table 4.9 Analysis of Smoking and Age of Onset of Diabetes Coded With Daily Smokers as Reference Group

Model Summary[a]

Model	R	R^2	Adjusted R^2	SE of the Estimate
1	0.216[b]	0.047	0.046	13.558

[a]Dependent variable: DIBAGE.
[b]Predictors: (constant), DUM13, DUM12, DUM11.

ANOVA[a]

Model		Sum of Squares	df	Mean Square	F	p
1	Regression	22,082.041	3	7,360.680	40.044	.000[b]
	Residual	450,166.842	2,449	183.817		
	Total	472,248.882	2,452			

[a]Dependent variable: DIBAGE.
[b]Predictors: (constant), DUM13, DUM12, DUM11.

Coefficients[a]

Model		Unstandardized Coefficients		Standardized Coefficients	t	p	95% CI for B	
		B	SE	Beta			Lower Bound	Upper Bound
1	(Constant)	46.024	0.739		62.317	.000	44.576	47.472
	DUM11	3.567	0.835	0.129	4.270	.000	1.929	5.206
	DUM12	8.493	0.878	0.289	9.678	.000	6.772	10.214
	DUM13	0.889	1.595	0.012	0.558	.577	−2.238	4.017

[a]Dependent variable: DIBAGE.

DATA SOURCE: National Health Interview Survey of 2010 (NHIS2010) from the National Center for Health Statistics (http://www.cdc.gov/nchs/nhis/nhis_2010_data_release.htm).

Unweighted Effects Coding

Another common coding strategy is called "effects coding," which uses −1, 0, and 1 to represent the groups. The difference between effects and dummy-coded analyses is one of interpretation. When using dummy-coded variables, we interpret the intercept as the mean for the comparison group, and the unstandardized regression coefficients are the differences between the comparison group mean and the mean for the group that has a "1" on that dummy variable. When using effects coding, the intercept is the unweighted mean of the sample (the mean of all groups, not the average of all values for the sample—that is, the *weighted mean*) rather than being the mean of a particular group, and each effects variable then compares each group with that unweighted sample mean rather than a specific other group. This of course has implications for how we interpret the results and may not even answer the question we wanted to ask. Rarely in my research, for example, did I want to know whether a particular group was significantly different from the unweighted average of each group, which is only a good estimate of the population mean if the groups are relatively equal in number within the sample and within the population. In this case, there are very few occasional smokers compared with the other groups, but an unweighted mean treats that group mean as equal in value to all other group means, which may or may not be appropriate.

Using unweighted effects coding also does not give you a direct comparison between the comparison group and the sample mean. Be careful when selecting a coding scheme to ensure that it matches the question you wish to have answered.

Procedurally, in effects coding, the comparison group gets a −1, the group of interest gets a 1, and all other groups get a zero. Thus, the coding for SMOKE would be as shown in Table 4.10.[8]

Table 4.10 Setting Up Effects Coded Variables

Smoking Status	EFF1	EFF2	EFF3
0 (Nonsmoker)	−1	−1	−1
1 (Former smoker)	1	0	0
2 (Occasional smoker)	0	1	0
3 (Daily smoker)	0	0	1

The procedural aspects of creating the effects coded variables are only slightly more complex with the addition of one group getting −1 on each variable. My syntax for SPSS would be as follows:

```
do if (SMOKE ge 0).
compute eff1=0.
compute eff2=0.
compute eff3=0.
end if.
do if (SMOKE =0).
```

---

8 I also tend to name the variables so they are immediately obvious whether I used dummy or effects coding: DUM for dummy variables and EFF for effects coded variables. You should find a system that works for you.

```
compute eff1=-1.

compute eff2=-1.

compute eff3=-1.

end if.

if (SMOKE =1) eff1=1.

if (SMOKE =2) eff2=1.

if (SMOKE =3) eff3=1.

execute.
```

The results of this analysis are presented in Table 4.11. As you can see, we still account for the same variance in age of onset, and we have the same sum of squares. The change here is in the unstandardized regression coefficients and significance tests. In this case, the intercept, 49.261, is the average of the four group means, not the average of all scores on the variable. Each regression coefficient is then the mean difference between that unweighted average and the group mean, with the significance test telling us whether that difference is significant. Note particularly that EFF2 is significant, although it was not significant in the dummy-coded analysis because the comparison was different.

Table 4.11   Analysis of Smoking and Age of Onset of Diabetes via Unweighted Effects Coding

Model Summary

Model	R	R^2	Adjusted R^2	SE of the Estimate
1	0.216[a]	0.047	0.046	13.558

[a]Predictors: (Constant), EFF3, EFF1, EFF2.

ANOVA[a]

Model		Sum of Squares	df	Mean Square	F	p
1	Regression	22,082.041	3	7,360.680	40.044	.000[b]
	Residual	450,166.842	2,449	183.817		
	Total	472,248.882	2,452			

[a]Dependent variable : DIBAGE
[b]Predictors: (constant), EFF3, EFF1, EFF2.

Coefficients[a]

Model		Unstandardized Coefficients		Standardized Coefficients	t	p	95% CI for B	
		B	SE	Beta			Lower Bound	Upper Bound
1	(Constant)	49.261	0.427		115.300	.000	48.423	50.099
	EFF1	5.256	0.543	0.339	9.679	.000	4.191	6.321
	EFF2	−2.348	1.087	−0.096	−2.160	.031	−4.480	−0.217
	EFF3	−3.238	0.675	−0.166	−4.798	.000	−4.561	−1.914

[a]Dependent variable : DIBAGE

DATA SOURCE: National Health Interview Survey of 2010 (NHIS2010) from the National Center for Health Statistics (http://www .cdc.gov/nchs/nhis/nhis_2010_data_release.htm).

The regression equation (Equation 4.5a) can be used to reproduce the group means, just as in the dummy-coded analyses:

$$\hat{Y} = 49.261 + 5.256(EFF1) - 2.348(EFF2) - 3.238(EFF3) \qquad (4.5a)$$

Using this equation to predict the mean for the second (former smoker) group, we get the identical group mean as in prior analyses (in Equation 4.5b):

$$\hat{Y} = 49.261 + 5.256(1) - 2.348(0) - 3.238(0)$$

$$\hat{Y} = 49.261 + 5.256$$

$$\hat{Y} = 54.517 \qquad (4.5b)$$

What if we want to calculate the mean for the first (nonsmoker) group, because that is no longer the intercept? If we use the same regression line equation and enter −1 for each variable, we get the following (Equation 4.5c):

$$\hat{Y} = 49.261 + 5.256(-1) - 2.348(-1) - 3.238(-1)$$

$$\hat{Y} = 49.591 \qquad (4.5c)$$

which is identical to the mean we calculated for this group in prior analyses. Remember that any conclusions from this analysis only make sense if use of an unweighted mean is appropriate.

Weighted Effects Coding

In unweighted effects coding, the reference for all the comparisons is the *unweighted mean*, or the mean of all the group means. This may not answer the question you wish to answer. What if your question is more about comparing the means of each group to the mean you expect in the population? From the group sizes in Table 4.5, you can see there are vast differences in the size of each group. The unweighted mean in this case is 49.26, but the weighted mean (the mean of all individuals, or each mean weighted by the size of each group) is 50.64. If you think this represents the relative ratios within the population and want to compare each group with the estimated population average, then weighted effects coding might be a better choice.

To produce weighted effects coding, we will replace the −1 for the comparison group with $-(n_k/n_1)$, where n_k is the number in that group and n_1 is the number in the comparison group (Cohen, Cohen, West, & Aiken, 2002, pp. 328–329). Again, only use this if your goal is to model a population mean with uneven proportions of the various groups.

Let us assume you want to keep the first group (nonsmokers) as the comparison group but you want to calculate weighted effects. Referring to Table 4.5, we can divide

Table 4.12 Setting Up Weighted Effects Coded Variables

Smoking Status	WEFF1	WEFF2	WEFF3
0 (Nonsmoker)	−0.678	−0.0763	−0.279
1 (Former smoker)	1	0	0
2 (Occasional smoker)	0	1	0
3 (Daily smoker)	0	0	1

each group size by 1,206, the size of the nonsmoker group. Thus, each weight for the nonsmoker group will be different, as you see in Table 4.12.

To easily calculate these variables, we simply modify the previous syntax to calculate the weighted effect coded variables:

```
do if (SMOKE ge 0).
compute WEFF1=0.
compute WEFF2=0.
compute WEFF3=0.
end if.
do if (SMOKE =0).
compute WEFF1=-0.678.
compute WEFF2=-0.0763.
compute WEFF3=-0.279.
end if.
if (SMOKE =1) WEFF1=1.
if (SMOKE =2) WEFF2=1.
if (SMOKE =3) WEFF3=1.
execute.
```

What should we expect if this works as I have indicated? The intercept should equal the weighted mean of the sample, and the variance accounted for should be the same. In addition, the contrasts should now be different than those in the unweighted effects coded analysis, because the comparison is different; however, the regression equation should be able to reconstruct the group means accurately.

Let us first confirm that we are still meeting assumptions of OLS regression, as we did in the first dummy-coded analysis. Residuals are relatively even across the four groups, and the residuals are relatively normally distributed (skew = −0.036,

Figure 4.3   Distribution of Residuals From Weighted Effects Coded Analysis

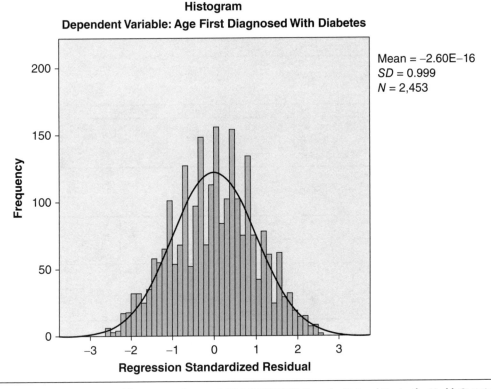

DATA SOURCE: National Health Interview Survey of 2010 (NHIS2010) from the National Center for Health Statistics (http://www.cdc.gov/nchs/nhis/nhis_2010_data_release.htm).

kurtosis −0.47) as you can see in Figure 4.3. You might note that the distribution of these residuals is similar to the other analyses.

As with previous analyses, the results for the overall model (variance accounted for, etc.) are identical to prior analyses (Table 4.13), and the rest of the details match what we would expect (e.g., the intercept matches the overall mean for the sample).

The regression line equation is similar to others described above and is presented in Equation 4.6a:

$$\hat{Y} = 50.64 + 3.873(\text{WEFF1}) - 3.731\ (\text{WEFF2}) - 4.620(\text{WEFF3}) \tag{4.6a}$$

Once again, if we compute any group mean, it will match what we observed in previous analyses. For example, the mean for the first group (nonsmokers) would be as presented in Equation 4.6b:

$$\hat{Y} = 50.64 + 3.873(-0.678) - 3.731\ (-0.0763) - 4.620(-0.279)$$

$$= 49.588\ (\text{which rounds to } 49.59, \text{matching our expected mean}) \tag{4.6b}$$

Table 4.13 Weighted Effect Coded Analysis

Model Summary[a]

Model	R	R^2	Adjusted R^2	SE of the Estimate
1	0.216[b]	0.047	0.046	13.558

[a]Dependent variable: DIBAGE.
[b]Predictors: (constant), WEFF3, WEFF2, WEFF1.

ANOVA[a]

Model		Sum of Squares	df	Mean Square	F	p
1	Regression	22,082.041	3	7,360.680	40.044	.000[b]
	Residual	450,166.842	2,449	183.817		
	Total	472,248.882	2,452			

[a]Dependent variable: DIBAGE.
[b]Predictors: (constant), WEFF3, WEFF2, WEFF1.

Coefficients[a]

Model	Unstandardized Coefficients		Standardized Coefficients	t	p	95% CI for B	
	B	SE	Beta			Lower Bound	Upper Bound
(Constant)	50.644	0.274		185.004	.000	50.107	51.181
WEFF1	3.873	0.387	0.209	10.008	.000	3.114	4.632
WEFF2	-3.731	1.387	-0.054	-2.690	.007	-6.450	-1.011
WEFF3	-4.620	0.686	-0.140	-6.734	.000	-5.965	-3.275

[a]Dependent variable: DIBAGE.

DATA SOURCE: National Health Interview Survey of 2010 (NHIS2010) from the National Center for Health Statistics (http://www.cdc.gov/nchs/nhis/nhis_2010_data_release.htm).

Example Summary of This Final Analysis

To examine the relationship between smoking status (0 = nonsmoker, 1 = former smoker, 2 = occasional smoker, 3 = daily smoker) and the age of onset of diabetes (DIBAGE) for participants aged >20 years, smoking status was converted to three weighted effects coded variables (with nonsmokers as the comparison group). Each effects coded variable (WEFF1–WEFF3) was weighted for the comparison group by the ratio of the size of each other group (N = 818, 92, and 337, respectively) to that of the comparison group (N = 1,206). This produces comparisons between each of the other groups (former smokers, occasional smokers, and daily smokers) and the estimated population mean.

Assumptions of OLS regression were checked and were reasonably met. Specifically, the variances of the different groups were reasonably close, and the residuals were normally distributed (skew = -0.036, kurtosis = -0.47).

Overall, the model was significant, and smoking status accounted for about 4.7% of the variance in age of onset (R = 0.216, R^2 = 0.047, $F_{(3, 2,449)}$ = 40.04, p < .0001). Each comparison, as shown in Table 4.13, was significant, indicating that each of the latter three groups (former smokers mean = 54.52, occasional smokers mean = 46.91, daily smokers mean = 46.02) were significantly different from the estimated population mean of 50.64. Interestingly, former smokers had a substantially later age of onset than any other group, whereas occasional and daily smokers had a substantially earlier age of onset of diabetes. If I had any clinical training, I might be able to make sense of this interesting finding.

Common Alternatives to Dummy or Effects Coding

It is perfectly fine to create dummy-coded variables and enter them into the regression equation as I did in the example above. The nice aspect of this is that through taking control of the situation, you know exactly what the comparisons are. However, some statistical software packages have some features in some GLM routines that attempt to make these sorts of coding efforts easier. For example, in SPSS logistic regression and GLM (what used to be the ANOVA menu), the contrast options include the following: simple, difference, Helmert, repeated, polynomial, and deviation. Indeed, these options can be used to perform regression analyses and these automatic contrasts can be utilized for polytomous variables. Each software package will have different options, so if you choose to use these options, be sure to understand what the software is doing so that you can appropriately interpret the results. In all cases, the overall model statistics should be identical to those we already presented.

Simple Contrasts

In SPSS GLM, *simple* contrasts essentially perform dummy coding. However, your only option is to have the first or last category be the reference group, whereas you can use any group as the reference group if you calculate them yourselves.

Difference (Reverse Helmert) Contrasts

In this contrast, the effects of each category are compared with the average of all of the previous categories (with the exception, of course, of the first group, which is the reference group in this scheme). Thus, in our example, nonsmokers (which are coded as 0) are necessarily the reference group. Former smokers (coded 1) are compared with nonsmokers, and then occasional smokers (coded 2) are compared with the *average of nonsmokers and former smokers*. Finally, the daily smokers will be compared with the average of the other three groups, each of which is very different. I honestly do not see the value of this particular type of contrast in most contexts, but I am sure it was invented to solve a particular problem. The issue, as I see it, is that of making sure the contrasts again make sense.

Helmert Contrasts

These are contrasts in which the effects of each category are compared with the mean of subsequent categories. In other words, it is the opposite of the previous (difference or reverse Helmert) contrasts, which is why they are named as they are. In this case, the first group (nonsmokers) is compared with the average of all other groups (former, occasional, and daily smokers), the second group (former smokers) is compared with the average of the following groups (occasional and daily smokers), and so forth. The last group is, by default, the reference. Again, I worry that combining groups in this way without careful thought could lead to overlooking real effects. Averaging combinations of very different groups might lead to a nonsignificant contrast, but that does not mean there is no difference, just that the coding scheme is masking the true effects.

Repeated Contrasts

Repeated contrasts look at adjacent categories, comparing group 1 versus 2, group 2 versus 3, group 3 versus 4, and so forth. Whether this produces the desired comparisons and masks or shows the real differences between groups depends on how the groups are ordered. In some cases, this pattern of comparisons might be sensible (e.g., if you have ordered groups with increasing exposure to some intervention). In unordered groupings, it is not necessarily the case that adjacent groups are the contrast that makes sense. Again, you, the researcher, must be thoughtful about your choice of contrast.

SUMMARY

Thus far, we have explored the idea that in generalized linear modeling, the OLS regression framework is a good model that can deal with either categorical or continuous variables as IVs, as we initially proposed in Table 4.1 at the beginning of this chapter. ANOVA is a special case of the GLM, and whether you analyze the data via typical ANOVA or regression software routines should be moot.[9] However, let me be clear that when the data are *only* composed of unordered categorical IVs, the ANOVA routines make life easier. I am not advocating for their disuse, but rather for a broader understanding that they are essentially the same analysis, just made simpler for a particular class of IVs. As we will see in future chapters, there are some analyses (like interactions between continuous and categorical variables) that are probably more easily modeled in a regression framework.

All of our examples thus far have dealt with continuous DVs and were all *parametric* analyses, indicating that we hold certain assumptions about distributions and variances. In Chapter 5, we will explore another aspect of the GLM: logistic (and briefly, probit) regression, both of which deal with binary (yes/no or 0/1) DVs. These techniques are only recently becoming well known in the social and behavioral

---

9   Also, I really don't think we should waste time with *t*-tests or simple correlation, and we *really* shouldn't waste weeks of students' lives covering these limited and unnecessary techniques except as historical context.

sciences, although they have been common in the health sciences for many years. They are important, because many of our most critical research can often come down to a binary outcome, such as whether someone graduates or whether someone gets sick. Logistic (and probit) regression appropriately models these types of data using concepts that will by now be familiar.

After Chapter 6, we will look at the opposite of this chapter: What happens if you have a polytomous (multinomial) dependent variable, either ordered or unordered? What if your outcome variable is not yes/no, but perhaps something like one of these:

1. Has not made a plan to quit smoking

2. Has made a plan to quit smoking but has not quit

3. Quit smoking but restarted later

4. Quit smoking permanently

We will deal with these and other interesting variables in Chapter 6.

ENRICHMENT

1. Download the NHIS2010 data used in the second half of the chapter and replicate the analyses and results. Also,

   a. Try different dummy coding schemes with different comparison groups to confirm my assertion that the overall model statistics will not change and that the group means will all be available through the regression line equation regardless of how the model is constituted.

   b. Try another coding scheme with the smoking and diabetes data to see whether another scheme makes sense. Also confirm that model fit and effect sizes do not change.

2. Download the second NHIS2010 data set from the book website and

   a. Perform an ANOVA testing whether individuals with diabetes (DIABETES; 0 = no, 1 = yes) have different body mass index (BMI; a measure of how heavy one is relative to height, where lower numbers are better than higher numbers). Be sure to test assumptions.

   b. Perform a regression analysis (testing assumptions) predicting BMI from DIABETES, and confirm that you get the identical results, such as

      i. Percent variance accounted for

      ii. Significance test

    c. Write out the regression line equation from 3b, and predict mean BMI scores for those with and without diabetes. Do they match (within rounding error) the means for the groups produced from the ANOVA analysis?

    d. Summarize the analysis in APA format, particularly paying attention to whether the effect was a large or small effect.

3. Download the AAUP data for this chapter and explore whether average assistant professor salary (SAL_ASST; coded in hundreds of dollars) varies significantly as a function of the older Carnegie Classification (CARNEGIE; 1 = I, 2 = IIA, 3 = IIB).

    a. Test assumptions (transform if necessary).

    b. Write up results in APA format.

REFERENCES

Cohen, J., Cohen, P., West, S., & Aiken, L. S. (2002). *Applied multiple regression/correlation analysis for the behavioral sciences.* Mahwah, NJ: Lawrence Erlbaum.

Hardy, M. A. (1993). *Regression with dummy variables* (Quantitative Applications in the Social Sciences, Vol. 93). Newbury Park, CA: SAGE.

Student. (1908). The probable error of a mean. *Biometrika, 6*(1), 1–25. doi: 10.2307/2331554

5

SIMPLE LINEAR MODELS WITH CATEGORICAL DEPENDENT VARIABLES

Binary Logistic Regression

Advance Organizer

What if your research question doesn't fall neatly into one of the models we have discussed already? What if you are interested in understanding what predicts whether a student will graduate, whether someone will get sick or be healthy, whether someone saves for retirement, or whether an individual voted in the last election?

In this chapter, we will cover the basic aspects of linear modeling when you have a categorical (binary yes/no or 1/0) dependent variable (DV). We will efficiently move through basic logistic regression modeling when there is either a binary or continuous independent variable (IV; it is also possible to perform a logistic regression with a categorical IV as in Chapter 4). In Chapter 6, we will explore modeling with more complex DVs such as polytomous (multinomial) unordered DVs or count variables. Our logistic regression models will be similar procedurally to models we have explored in Chapters 3 and 4.

The concepts following in all of the other chapters are more or less simple generalizations or evolutions arising from the basic concepts learned in the first five chapters of this book. One thing I am reluctant to include here is probit regression, which is very similar to logistic regression (except for using a different link function). Both types of regression answer the same questions (dealing with the same types of variables), and their use seems more to be linked to tradition in various fields. This chapter includes a brief discussion of probit regression in Appendix 5A; however, probit regression really seems to be unnecessary in the modern era if you have mastered logistic regression.

In this chapter, we will cover

- Basic concepts in handling binary IVs
- Why we cannot use linear probability models (using ordinary least squares [OLS] regression with a binary DV)
- Assumptions and data cleaning when you are using logistic regression
- Confidence intervals (CIs) and effect sizes in logistic regression
- Examples of American Psychological Association (APA)–compliant summaries of analyses

Guidance on how to perform these analyses in various statistical packages will be available online at study.sagepub.com/osbornerlm.

It's All About Relationships! (Part 3)

In Chapters 3 and 4, we demonstrated that analysis of variance (ANOVA) and OLS regression (and the much less useful *t*-tests and simple correlation) are part of the same generalized linear model (GLM). Although they have traditionally been (and often, currently are) covered as separate families of procedures (often in completely different courses), we have demonstrated that the GLM using OLS estimation can adequately deal with data in which DVs are continuous and IVs are either continuous or categorical. Now that we have dispensed with that false dichotomy, we can modify Table 5.1 to be simpler. If your DV is continuous, your choice will be OLS regression or ANOVA, depending on your preference and whether most of the variables are categorical or continuous.

In this chapter, we will focus on the second column, logistic regression, which can easily handle both categorical and continuous variables the same as OLS regression. Binary logistic regression can easily handle models with only binary IVs and DVs, as you will shortly see. However, log-linear models will be covered in Chapter 12 and can also appropriately handle these types of models.

Why Is Logistic Regression Necessary?

Over the years, I have seen a variety of "kludgy" attempts at analyzing data with binary (or categorical) outcomes. Many of these involve flipping the IV and DV so that *t*-tests (or ANOVA) can be utilized to explore where groups differ on multiple variables

Table 5.1 Examples of the Generalized Linear Model as a Function of Independent Variable and Dependant Variable Type

	Continuous DV	Binary DV	Unordered Multicategory DV	Ordered Categorical DV	Count DV
Continuous IV	OLS regression	Binary logistic regression	Multinomial logistic regression	Ordinal logistic regression	OLS, Poisson regression
Mixed continuous and categorical IV	OLS regression	Binary logistic regression	Multinomial logistic regression	Ordinal logistic regression	OLS, Poisson regression
Binary/ categorical IV only	ANOVA	Binary logistic or log-linear models	Log-linear models	Ordinal logistic regression	Log-linear models

ANOVA, analysis of variance; DV, dependent variable; IV, independent variable; OLS, ordinary least squares.

in an attempt to build theory or understanding. For example, one could look at differences between people who contract a disease and those who do not across variables such as age, race/ethnicity, education, body mass index (BMI), smoking and drinking habits, participation in various activities, and so on. Perhaps we would see a significant difference between the two groups in BMI and number of drinks per week on average. Does that mean we can assume that those variables might be causally related to having this illness? Definitely not. Furthermore, it might also be the case that neither of these variables is really predictive of the illness at all. Being overweight and drinking a certain number of drinks might be related to being a member of a certain sociodemographic group; this may, in turn, be related to health habits (e.g., eating fresh fruits and vegetables [or not] and exercising), stress levels, commute times, or exposure to workplace toxins, which might in fact be related to the actual causes of the illness.

I mean no disrespect to all those going before me who have done this exact type of analysis—historically, there were few other viable options. In the modern world of statistics, let us think for a minute about some drawbacks to the approach I just mentioned. One issue is that researchers can have issues with power if they adjust for Type I error rates that multiple univariate analyses require (or worse, they might fail to do so). In addition, using this group-differences approach, researchers cannot take into account how variables of interest covary. This issue is similar to performing an array of simple correlations rather than a multiple regression. To be sure, you can glean some insight into the various relationships among variables, but at the end of the day, it is difficult to figure out which variables are the *strongest* or *most important* predictors of a phenomenon unless you model them in a multiple regression (or path analysis or structural equation modeling) type of environment, which we will explore in Chapter 8.

Perhaps more troubling (to my mind) is the fact that this analytic strategy prevents the examination of interactions, which are often the most interesting findings we can come across. Let us imagine that we find sex differences,[1] as well as differences in household income, between those who graduate and those who do not. That might be interesting, but what if, in reality, there is an interaction between the sex of the student and family income in predicting graduation or dropout rates? What if boys are much more likely than girls to drop out in more affluent families, and girls are more likely to drop out in more impoverished families? That finding might have important policy and practice implications, but we are unable to test for that sort of interaction using the method of analysis described above. Logistic regression (like OLS regression) models variables in such a way that we get the unique effect of the variables, controlling for all other variables in the equation. Thus, we get a more sophisticated and nuanced look at what variables are uniquely predictive (or related to) the outcome of interest.

I have also seen aggregation used as a strategy, particularly with binary outcomes (e.g., graduation from high school). Instead of looking at individual characteristics and individual outcomes, researchers might aggregate to a classroom or school level.

---

1   This is another gentle reminder that I will use the term *sex* in this book to refer to physical or biological sex—maleness or femaleness. *Gender*, conversely, refers to masculinity or femininity of behavior or psychology. The two terms are not synonymous, and it diminishes these concepts to conflate them (Mead, 1935; Oakley, 1972). Please write your political leaders and urge them to take action to stop this injustice!

Researchers might thus think they have a continuous variable (0%–100% graduation rate for a school) as a function of the percentage of boys or girls in a school and the average family income. In my opinion, this does a tremendous disservice to the data, losing information and leading to potentially misleading results. In fact, it changes the question substantially from "What variables contribute to student completion?" to "What school environment variables contribute to school completion rates?" Furthermore, the predictor variables change from, say, sex of the student to the percentage of students who are male or female, from race of the student to the percentage of students who identify as a particular race, and from family socioeconomic status (SES) to average SES within the school. These are fundamentally different variables; thus, analyses using these strategies answer a fundamentally different question. Furthermore, in my own explorations, I have seen aggregation lead to wildly misestimated effect sizes—double that of the appropriate analysis and more (for an example of how aggregation can substantially distort relationships, see Osborne, 2000). Thus, aggregation changes the nature of the question and the nature of the variables, and it can lead to inappropriate overestimation of effect sizes and variance accounted for. Therefore, we cannot consider it a best practice.

The Linear Probability Model

I am sure some of you have also wondered why we cannot just model an OLS regression equation with a binary outcome as the DV. This is a real procedure often discussed in older regression texts, and it is referred to as the linear probability model (LPM). The LPM is not the same as a probit model (which we will cover later in Appendix 5A to this chapter). The LPM would be a step forward, in that the LPM has advantages over the previous kludgy strategies and we could simultaneously estimate the unique effects of several IVs and examine relative importance in predicting the outcome, unlike the above-described approach. In fact, the statistical software you use will perform this analysis if you tell it to. However, there are issues with this approach. First, predicted scores (which are supposed to be predicted probabilities) can range outside the theoretically bounded 0.00–1.00 range. Second, residuals are neither normally distributed nor homoscedastic. For these and other reasons, this simply is not an appropriate analysis.

To illustrate this issue, I used the National Education Longitudinal Study of 1988[2] (NELS88; Ingels, 1994) to predict student graduation (GRADUATE, coded 0 = dropped out before graduation, 1 = graduated from high school) from standardized family socioeconomic status (zSES; z-scored version of eighth-grade composite family SES variable BYSES). As you can see in Table 5.2, statistical software will produce output indicating that increasing SES predicts better graduation.

You can see in Figure 5.1 that performing this analysis in an OLS framework produced the expected violation of assumptions. For example, the residuals are not close to being normally distributed (skewness = −4.90, kurtosis = 33.32).

---

2 *Data Source:* National Education Longitudinal Study of 1988 (NELS88), National Center for Educational Statistics (http://nces.ed.gov/surveys/nels88/).

Table 5.2   Linear Probability Model Analysis of Binary GRADUATE Variable Predicted From zSES

Model Summary

Model	R	R^2	Adjusted R^2	SE of the Estimate
1	0.265[a]	0.070	0.070	.27446

ANOVA

Model		Sum of Squares	df	Mean Square	F	p
1	Regression	94.586	1	94.586	1,255.631	.000[b]
	Residual	1,251.075	16,608	0.075		
	Total	1,345.662	16,609			

Coefficients[a]

Model		Unstandardized Coefficients		Standardized Coefficients	t	p
		B	SE	Beta		
1	(Constant)	0.911	0.002		427.816	.000
	zSES	0.075	0.002	0.265	35.435	.000

[a]Dependent variable: GRADUATE.

Figure 5.1   Residuals From an Ordinary Least Squares Analysis With Binary Outcome

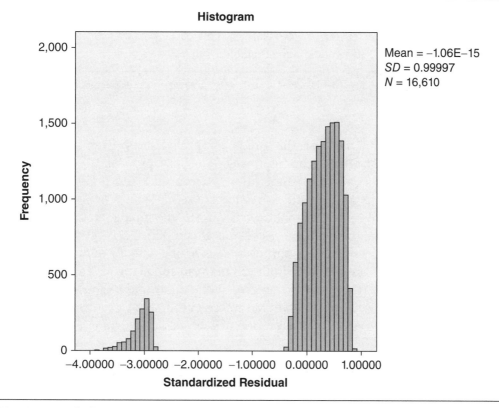

SOURCE: National Education Longitudinal Study of 1988 (NELS88) from the National Center for Education Statistics (http://nces.ed.gov/surveys/nels88/).

Figure 5.2   Predicted Values From an Ordinary Least Squares Analysis With Binary
Outcome

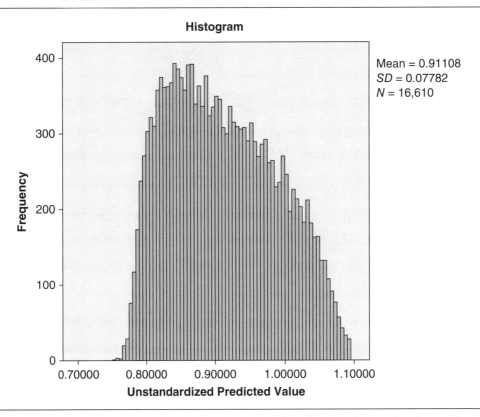

SOURCE: National Education Longitudinal Study of 1988 (NELS88) from the National Center for Education Statistics
(http://nces.ed.gov/surveys/nels88/).

Furthermore, about 17% of the predicted probabilities from this example
(Figure 5.2) exceed 1.0, which is an impossible value for a probability.

Thus, we come to conceptual similarities between OLS and logistic regression.
Procedurally, both OLS and logistic regression are set up with a single DV and one
(or more) IV. Both allow us to simultaneously assess the unique effects of multiple
predictor IVs (and their interactions or curvilinear components, if desired), and both
allow examination of residuals for purposes of screening data for outliers, follow-up
analyses, or testing of assumptions. Both can perform simultaneous entry, hierarchical or blockwise entry (groups of IVs entered at one time), and various stepwise procedures.[3] With both, we have the ability to assess a group of IVs to determine which
predictor is the strongest unique predictor of a particular outcome and to answer

---

3   Some of you reading this will have been trained to have visceral negative reactions to stepwise procedures,
which is ironic because they were heralded as important tools just a generation earlier. I am of the mind
that stepwise procedures have their place in the pantheon of statistical tools and that we should be knowledgeable of them and use them *when appropriate*. For most of you reading this, the answer to when these
procedures are appropriate is "almost never." We will cover methods of entry in Chapter 8 when we discuss
multiple IVs.

many of the types of questions that have made regression a valuable tool in quantitative methods. We will cover more on this issue in Chapter 8.

How Logistic Regression Solves This Issue: The Logit Link Function

Statisticians reasoned that we can get around the issue of impossible predicted values in the LPM if we look at odds rather than probabilities. There are drawbacks to odds—such as being difficult to interpret and inflation of effect sizes compared with probabilities—but they are not constrained by the 0–1 range as probabilities are. Theoretically, odds can range from zero to infinity, and they thus solve half the problem faced by the LPM. Odds are similar to probabilities with an important distinction: the odds of something happening (e.g., being graduated) is the probability of that event (π) divided by the probability of the event not happening ($1 - \pi$), as in Equation 5.1:

$$\text{Odds}_{(GRADUATE)} = \text{probability of being graduated/probability of not being graduated.}$$

$$\text{Odds}_{(GRADUATE)} = \pi / (1 - \pi) \tag{5.1}$$

However, this does not entirely solve the problem. The one drawback of OLS regression with this type of data remains that we could see predicted conditional odds below 0.00 (which are impossible values). Therefore, this approach eliminates one issue—predicted odds over 1.0 are legitimate, but the problem of negative odds still remains. Therefore, the solution is to take the natural logarithm of the odds, which has the benefit of having no restriction on minimum or maximum values. Thus, the "logit" (or "log unit") is the natural logarithm of the conditional odds of an outcome, as shown in Equation 5.2:

$$\text{Logit} = natural \ log\left(\frac{\pi}{1-\pi}\right) \tag{5.2}$$

The regression equation we calculate is, as presented in Equation 5.3, a typical simple regression equation, but it predicts the logit of the DV rather than simply the DV:

$$\text{Logit}(\hat{Y}) = b_0 + b_1 X_1 \tag{5.3}$$

A "probit" (or probability unit) is a different mathematical transformation of the same data, with the same goal: to allow us to predict binary or categorical outcomes without violating assumptions. Probit regression uses the inverse standard normal cumulative distribution function (the area under the curve less than z, which is the boundary). If you are familiar with a z-score and the z-table associated with it, you are familiar with the basics of probit regression. Essentially, probit regression uses a transform to a z-value, which solves the same problem as the logit transform in a slightly different way. Although probit is much less common in modern behavioral sciences, we can explore it in Appendix 5A in this chapter.

In generalized linear modeling, we can use almost any *link function*. A link function is any mathematical transformation that modifies the regression equation. Technically, in OLS regression, we had the multiplicative identity (1) as the link function; however, because 1 multiplied by Y is Y, we do not explicitly model it. To effectively and appropriately deal with prediction of binary outcomes in logistic regression, we use the *log unit* or *logit* link function, which involves computing the natural log of the odds of Y. In probit regression, we use the probability unit (or probit) link function rather than the logit link function, which is a transform to the inverse standard normal cumulative distribution.

Both logistic and probit regression simply use different *link functions* in the same GLM as OLS regression. Therefore, almost anything one can do with OLS linear regression can be done with logistic or probit regression. This chapter is but a brief introduction to logistic (and probit) regression, and these topics will reappear throughout this book.

Many other types of regression link functions are possible beyond the log odds or standard normal cumulative distribution function, but we will confine ourselves to the logit link function for the purposes of this book.

Let us continue examining the same data we started this chapter with: attempting to understand graduation and graduation from high school as a function of student achievement. This time we can create a model that (at least to my mind) makes more sense: we can predict GRADUATE as the DV and use zACH (standardized achievement test scores converted to *z*-scores) as the IV. In essence, we are flipping the question being tested. In the previous analyses, we asked whether there are differences in eighth-grade achievement between students who ultimately graduate high school or not. Because there is a temporal component to the data (achievement was gathered in eighth grade, and graduation was assessed 4 years later), it makes more sense to ask the question of whether eighth-grade achievement predicts (or explains) whether a student will ultimately be retained through graduation or not. With simple analyses such as these, it really leads to the same conclusions (as you will soon see), but when we get into multiple IVs (and interactions and curvilinearity), the advantage of having this tool in your toolbox will become more apparent.

Let's start with the hypotheses being tested. When we have two groups, we are testing whether the conditional probabilities of an outcome are the same or different across the two groups. The null hypothesis is that the probabilities are not different, which means that the odds for each group is the same. This means that the ratio of the two odds (*the odds ratio* [OR]) should be 1.0 (i.e., the same number divided by itself should be 1):

$$H_0: \text{Odds ratio} = 1.0$$

$$H_a: \text{Odds ratio} \neq 1.0$$

Because the natural log of 1.0 is 0.00, we can also write these hypotheses in terms of logits:

$$H_0: \log (\pi_{(Y=1)} / (1 - \pi_{(Y=1)})) = 0$$

$$H_a: \log (\pi_{(Y=1)} / (1 - \pi_{(Y=1)})) \neq 0$$

Thus, depending on whether you are talking about the hypothesis testing of the regression coefficient or the OR (which is a slope in odds rather than logits), you will be testing whether it is significantly different from 0 or 1. But don't worry: the decision is the same because both of these numbers are dependent on the same data and always lead to the same conclusion regarding significance tests.

A Brief Digression Into Probabilities, Conditional Probabilities, and Odds

With a little hand calculation, we can easily see how simple counts can turn into powerful statistics that we will discuss throughout the rest of the book. I find this instructive and important to understanding what is happening in logistic regression. Let us start with the simplest logistic regression analysis: two binary variables. To do this, we will split zACH into "high-achieving" and "low-achieving" students. Note that I do this only for the sake of providing an example. We should never split continuous variables into categorical or binary variables except when writing textbooks (and perhaps not even then!).

Table 5.3 contains the simple cross-tabulation of the variable GRADUATE and the binary variable based on our variable zACH, split into two groups at the mean for this artificial example (the new variable is ACH_LOHI). The probability of an event is calculated as the frequency of the event (graduating) divided by the total observations (in this case, 15,133 graduates out of 16,610 total students).

$$\text{Probability of graduation } (P_{graduate}) = \text{number graduation / total students}$$

$$P_{graduate} = 15,133 / 16,610$$

$$P_{graduate} = 0.911$$

When there are two categories (as with this graduated/not graduated variable), the probability of a student falling into the "not graduated" category is $(1 - P_{graduate}) =$

$$\text{Probability of graduated } (1 - P_{graduate}) = \text{number not graduated / total students } or$$
$$1 - P_{graduate}$$

$$1 - P_{graduate} = 1,477 / 16,610 \ or \ 1 - 0.911$$

$$1 - P_{graduate} = 0.0889$$

According to these data, the overall probability that any student in the eighth grade will drop out is 0.0889, and the probability that any random student will complete a high school education is 0.911. Knowing nothing else about any student, the best we could do is guess that each student has about a 91% chance of completing a high school education if he or she is enrolled in eighth grade. Of course, we know that the probabilities are not the same for all students. For example, looking at Table 5.3, it is relatively obvious that eighth-grade achievement level has an effect on the probability of graduating. Thus, we can calculate *conditional probabilities,* a fancy phrase that simply means

Table 5.3   Figuring Odds and Odds Ratios by Hand

| | | GRADUATE | | | Conditional | | OR |
		0.00	1.00	Total	Probabilities	Odds	(Change in Odds)
ACH	High (1)	169	7,621	7,790	0.978306	45.09467	7.851948
	Low (0)	1,308	7,512	8,820	0.851701	5.74312	
Total		1,477	15,133	16,610	0.911078		

DATA SOURCE: National Education Longitudinal Study of 1988 (NELS88) from the National Center for Education Statistics (http://nces.ed.gov/surveys/nels88/).

the probability of something happening within a condition. Thus, we can see that for a student with above-average eighth-grade achievement, the probability of graduating is 7,621/7,790, or about 97.8%. Almost all students with above-average achievement in eighth grade stick it out to graduate in twelfth grade in the United States, according to these data. For those with below-average achievement in eighth grade, the story is different. Only 7,512/8,820 graduated from this group, or about 85.1%. Although this is still high, the conditional probabilities are different (significantly so). As we discussed previously, in logistic regression, we convert these statistics to odds:

$$\text{Odds}_{(graduate)} = \text{probability of graduate / probability of not graduate.}$$

$$\text{Odds}_{(graduate)} = \pi / (1 - \pi)$$

In the case of these data, the conditional odds of graduating are 45.09 for those with high achievement and about 5.74 for those with low achievement. You can see that these odds look very different depending on what group you are in (odds calculations tend to magnify effects like this). Finally, the ratio of the two odds, or the change in odds from one group to the other (the slope of the effect, literally), is

$$\text{Odds ratio} = \text{odds}_{(graduate)} / \text{odds}_{(not\ graduated)}$$

$$= 45.09 / 5.74$$

$$= 7.85$$

Thus, we can say that the odds of those with "high" achievement graduating are 7.85 that of those with "low" achievement.

Simple Logistic Regression Using Statistical Software

Now let us confirm that our hand calculations match with SPSS calculations. Using the same data (available on the book website), I predicted GRADUATE from ACH_ LOHI (the terrible binary variable we calculated from a beautiful continuous variable

Table 5.4   Logistic Regression Predicting GRADUATE From Binary Achievement
Variable

Omnibus Tests of Model Coefficients

		Chi-Square	df	p
Step 1	Step	933.810	1	.000
	Block	933.810	1	.000
	Model	933.810	1	.000

Model Summary

Step	–2LL	Cox and Snell R^2	Nagelkerke R^2
1	9,033.432[a]	0.055	0.121

[a]Estimation terminated at iteration number 6 because
parameter estimates changed by less than .001.

Variables in the Equation

		B	SE	Wald	df	p	Exp(B)	95% CI for Exp(B) Lower	Upper
Step 1[a]	ACH_LOHI	2.061	0.083	611.391	1	.000	7.852	6.669	9.245
	Constant	1.748	0.030	3403.916	1	.000	5.743		

[a]Variable(s) entered on step 1: ACH_LOHI.

only for exploration). Logistic regression provides some different statistics and output
because it uses maximum likelihood (ML) estimation rather than OLS. The results of
this analysis are presented in Table 5.4.

Indicators of Overall Model Fit

As with OLS regression, overall model fit is an important first step in assessing your
model. In OLS regression, we have measures of overall model fit such as R, R^2, and
significance tests for them in the form of an F test. In logistic regression, there is no
simple, substantively interpretable measure of overall model fit, but there is a statistic
that yields a significance test for the overall model.

In general, overall model fit looks at several pieces of information, such as omnibus
chi-square test, classification tables, and pseudo-R^2 indices. None of these are important and sufficient on their own, but in aggregate they can begin the initial exploration
into whether a logistic regression model is useful.

As you can see in Table 5.4, we get an overall chi-square test for the model (and for
the current step and block, all of which are identical in the case of one IV) testing the
null hypothesis that there is no association between the IV and DV (i.e., that the probability of experiencing the outcome is the same regardless of status on the IV). In this
case, a $\chi^2_{(1)} = 933.81$ is significant ($p < .0001$).[4]

---

4   SPSS tends to truncate p values, but you cannot have a probability of exactly 0, so I always add a 1 to the end if I do
not calculate an exact probability. Calculating an exact probability for a chi-square test is simple in Excel with
the following command: =CHIDIST(x, degrees of freedom). Using this, our p value would be
$p < 4.3861 \times 10{-205}$. For those of you not familiar with scientific notation, that is a 0. [205 zeroes] and
then 43,861 (in other words, a really small probability).

Based on this outcome, we reject the null hypothesis that there is no difference in probability of the outcome variable as a function of the IV, allowing us to conclude there *is* a difference in the probability of experiencing the outcome depending on where you score on the IV. Keep in mind that this example has a very large sample. Chi-squared statistics (and all inferential statistics) are strongly influenced by sample size, making it possible to have very small *p* values even when there is a very small overall effect.

What Is a –2 Log Likelihood?

In Table 5.4, you will see a statistic called –2 log likelihood (which we will abbreviate as –2LL), which is a generally a measure of *lack of fit*, or *error variation*, and is conceptually the opposite of R^2 in regression. Thus, the smaller the –2LL gets, the better the model fits. Unlike R^2, –2LL is not appropriate for comparing non-nested models (models that are not based on the same data set, same DV, and subsets of IVs). It is not meaningful in its absolute value; rather, –2LL is used for comparison only within nested models. The –2LL is the product of –2 and the individual log likelihood for each individual in the sample for a particular model. In practical situations, we generally care about the comparison between the initial (null/empty) model with only the intercept in the model and the final model with all predictors in the model.

Most of us don't care about exactly how this is calculated, but it is instructive for some of you with a more nerdly[5] mindset. Conceptually, it is –2 times the natural log of the conditional probability of each group multiplied by the number of individuals in each group, as you can see in Equation 5.4:

$$-2LL = -2\,[N_{y=1}\,{}^*\ln[(p_{(y=1)})] + N_{y=0}\,{}^*\ln[(p_{(y=0)})]] \tag{5.4}$$

Thus, –2LL is affected by two different things: conditional probabilities for each group and the number of individuals in each group. This is why very large samples tend to have very large –2LLs and small samples tend to have small –2LLs.

The likelihood of the model is used to test whether all of the predictors' regression coefficients in the model are simultaneously zero. The larger the initial –2LL, the less tenable that hypothesis is, meaning the more likely it is that your model is explaining some of the DV. This –2LL is also used in tests of nested models to see whether additional variables significantly improve model fit. The chi-square of 933.81 is the difference between the –2LL with no variables in the model and with the one IV in the model, and the –2LL listed in Table 5.4 (9,033.43). When –2LL is compared across nested models, the number of parameters estimated is subtracted as well to obtain the degrees of freedom (*df*) for the chi-square. In this case, the model went from zero parameters to one parameter estimated; thus, the chi-square has *df* = 1. If we had added three variables to the equation (as we will do in later chapters), the chi-square would have 3 *df*. The –2LL is not generally interpreted in any conceptual or practical sense and is highly influenced by sample size in addition to the goodness of the model fit, unlike *R* and R^2, which are on standardized, substantively interpretable metrics.

---

5   The highest compliment one can get in the field of quantitative methods in my world.

The Logistic Regression Equation

As you can see in Table 5.4, the binary achievement variable ACH_LOHI is significant, with a regression coefficient of 2.061 (the slope of the line, in log units), with a standard error (SE) of 0.083. The SE of the estimate (the b) can also be instructive, because it is an indicator of how precise that point estimate of b is (and it is used in constructing CIs[6]). Higher SEs indicate less precision and will lead to larger CIs.

Another interesting bit of trivia is that the Wald statistic (which is interpreted as a chi-square) is calculated as shown in Equation 5.5:

$$\text{Wald} = \left(\frac{b}{SE_b} \right)^2$$

$$= (2.060761 / 0.083343)^2 = 611.39[7] \tag{5.5}$$

As you can calculate, a Wald statistic of that magnitude and 1 df will yield a p value that is very small. Thus, in this example, we would reject H_0 and conclude the regression coefficient is significantly different than 0 (meaning that the OR is also significantly different than 1.0).

The OR in SPSS is labeled "Exp(B)" because the OR is a direct mathematical transform of the logit. If you exponentiated the regression coefficient [exponentiation is the reverse, mathematically, of a logarithm; in Excel, you can type "=EXP(2.060761)"], you would get the OR, 7.852, and the natural log of 7.852 is then 2.060761. If you go back to Table 5.3, where we did hand calculations, you would find that this value is also identical to the hand-calculated OR.

CIs are important to report because they give the reader an estimate of precision of the point estimate (in this case, the OR or slope). SPSS easily provides these if requested, and in this example, the OR varies across a rather large range, although the effect is strong throughout the range of the 95% CI. This indicates that although we have a strong effect, our precision in estimating the effect is somewhat less than we would like. However, if we look at the logits for the slope, the 95% CI would be [1.898, 2.224]. This is not a terribly broad range in terms of precision, particularly when there is only one increment possible because the IV is binary.

Interpreting the Constant

The constant (or intercept) in the regression equation is the value of the DV (Y) when the IV (X) is 0. In the case of this binary IV, where low achievement is coded 0 (low

---

6   95% CIs are constructed as $b \pm 1.96{*}SE_b$. Thus, in this example, the CI in logits is 2.061 ± 1.96*.083. If you do the math, the 95% CI for this b is [1.898081, 2.223441]. If you exponentiate each of those CI boundaries, you get 6.67 and 9.24, within rounding error of the 95% CIs in Table 3.10 for the OR.

7   SPSS truncates decimals in the output, as you can see in the tables. If you double-click on the number, you can see it with much greater precision, or you can adjust the output settings in the software to give more decimals.

achievement), the predicted logit is 1.748, which is about 5.743 after exponentiation (converting back to conditional odds). In Table 5.3, we calculated the conditional odds of graduation for students with low achievement to be about 5.743, which turns out to be the same number as in Table 5.4 (the exponentiated intercept in the SPSS output). To reconstruct the conditional odds of graduation for students in the high achievement group, we can use the following regression equation (Equation 5.6):

$$\text{Logit}(\hat{Y}) = 1.748 + 2.061(\text{ACH\_LOHI}) \tag{5.6}$$

When the binary achievement variable is "high" the variable is coded as 1. We get a predicted logit of 3.809. When exponentiated, this reveals a conditional odds of 45.10, which is within rounding error of the conditional odds we calculated for this group in Table 5.4.

What If You Want CIs for the Constant?

If you know the formula for CIs (B ± 1.96*SE), this is simple, even though SPSS does not provide it (and other software packages may vary). Our intercept is 1.748, and the SE is 0.030, which leads to a 95% CI of [1.689, 1.807] around our point estimate. Thus, we are 95% certain that the log odds of graduation for "low-achieving" students is between 1.689 and 1.807, which converts to probabilities of 0.844 and 0.859. This is not a bad level of precision, and note that the probabilities do not overlap at all with the CI for the other groups.

Summary So Far

In this section, we showed that we can use a log link function to model binary DVs within the GLM. Although this alters the math a bit, the general model is the same as OLS regression (and ANOVA). Indeed, although the information is slightly different than the same data analyzed through OLS regression, the conclusion is the same: students with higher achievement scores in eighth grade are *much more likely* to graduate from high school than those with low achievement in eighth grade. This section is presented mostly so you can understand the relatively simple math that creates our ability to do logistic regression, since it is not ideal to convert continuous variables to binary or categorical variables for the convenience of the researcher. However, it is not usually a good idea to convert continuous variables to categorical variables, so we will now turn to logistic regression with a continuous IV.

Logistic Regression With a Continuous IV

Our example includes ORs and logits that reflect the change in probability (or odds, or logits) that a change of 1.0 in the IV makes. Thus, in our first analysis, student achievement scores were recoded into a binary variable, with a 0 meaning low

Table 5.5   Logistic Regression With a Continuous Independent Variable

Omnibus Tests of Model Coefficients

		Chi-Square	df	p
Step 1	Step	1,491.382	1	.000
	Block	1,491.382	1	.000
	Model	1,491.382	1	.000

Model Summary

Step	−2LL	Cox and Snell R^2	Nagelkerke R^2
1	8,475.860[a]	0.086	0.190

Variables in the Equation

		B	SE	Wald	df	p	Exp(B)	95% CI for Exp(B) Lower	Upper
Step 1[a]	zACH	1.359	0.043	1,003.838	1	.000	3.891	3.577	4.232
	Constant	2.955	0.045	4,326.004	1	.000	19.194		

[a]Variable(s) entered on step 1: zACH.

achievement and 1 meaning high achievement. Thus, the change from 0 (low achievement) to 1 (high achievement) was a large step, and a dramatic comparison.[8] We saw that the odds of graduating for those with high achievement in eighth grade was about 7.8 times that of the odds of graduating for those with low achievement. Using a continuous variable is a simple extension of this special case. When an IV has more than two values, the statistics reflect the change in odds corresponding to a change of 1.0 in the IV. The only difference is that in this case, more than one of these increments is present in the IV. Thus, the effect will take into account how many increments there are. In our example data, we have z-scored the achievement variable (zACH), which ranges from −2.01 to 2.36, giving us over four increments of 1.0. The abbreviated results from these analyses are presented in Table 5.5.

In my recent book on logistic regression (Osborne, 2015), I devoted an entire chapter to how much information we lose when we convert continuous variables to categorical or binary variables. If you think about this variable and what we did when we converted it to binary, we essentially introduced a massive amount of error by claiming that students scoring 2 standard deviations (SD) below the mean are the same as those scoring 0.02 SD below the mean. It is tough to argue that those students are similar. Likewise, we considered students scoring 0.02 SD above the mean the same as students scoring more than 2 SD above the mean. Again, these are very different. Finally, we created a stark and false distinction: those who scored slightly below the mean (−0.02 SD) are treated as very different from those scoring slightly above the mean (0.02 SD) when, in fact, they are probably separated only by random error. Furthermore, because the majority of the sample is clustered around the mean, the majority of these distinctions are false, creating error.

---

8   And also a dramatic disservice to the data.

In Table 5.5, we instantly see that when we use the continuous variable zACH instead of the binary achievement variable (ACH_LOHI), we have a stronger change in −2LL (1,491.38 versus 933.81 in the previous example). This in itself is a testament to the value of keeping continuous variables continuous. With a significant improvement in model fit (we can reject the null hypothesis that the change in model fit is zero), we can now look at the effect of the variable in the equation. Although the change in model fit for the continuous variable was substantially stronger than for the binary version of that variable, you can also see that the effect appears smaller—which is why many researchers lean toward binary IVs. They inflate the apparent effect size, but in reality are creating a substantially inferior model (as measured by improvement in model fit when the variable is entered into the equation). The slope for achievement is 1.359 (much less than the binary variable); remember, *this is the effect for an incremental change of 1.0*, and there are about four increments in this IV, whereas there was only one in the binary example. The OR looks relatively small as well (3.89 versus 7.85), but consider the following: would you rather have one increment of 7.85 or four increments of 3.89? As you can see in Figure 5.3, it is not a trivial difference.

Some Best Practices When Using a Continuous
Variable in Logistic Regression

There are some good reasons to convert continuous variables to the standard normal distribution (*z*-scores) as a standard practice. First, you will see in later chapters that this will be a necessary step when we explore curvilinear or interaction effects. But more relevant at this point is the fact that effect sizes in logistic regression (like an OR or unstandardized regression coefficient) are based on an increment of 1.0 in the IV.

Figure 5.3   Comparison of the Effect Size of Binary Versus Continuous Independent Variables

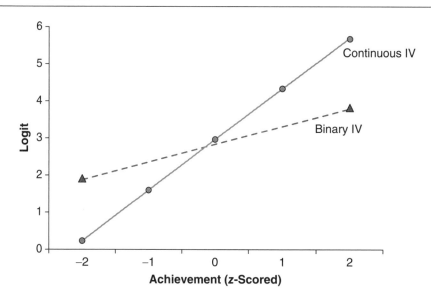

Thus, binary variables look like they have a much larger effect because the entire effect is subsumed in the 1.0 increment between the two groups, whereas a continuous variable can have dozens or hundreds of increments in its natural metric. In addition, different variables can have different numbers of increments of 1.0. A 5-point Likert-type scale might have four increments, and family income could have many thousands if reported in dollars. However, if all continuous variables are converted to *z*-scores, all variables are compared on the same metric and readers have an easier time of interpreting the effects. Because the metric is a *SD*, only a few will be present, and the interpretation for all variables will be roughly identical. Although different variables might have a different number of *SD* in any given sample, at least the interpretation (a change of 1 *SD*) will remain constant.

This rationale is similar to how users of OLS regression standardize regression coefficients: the beta is the same as the analysis with variables converted to *z*-scores. An additional advantage to this strategy is that we place the Y intercept at 0, which is now the mean of the IV, rather than often an arbitrary or irrational point on the distribution.

Another best practice that I encourage is to routinely pick two points along the distribution (e.g., 1 *SD* below the mean and 1 *SD* above the mean) and calculate predicted probabilities for each so the reader can get a more concrete idea of the effect. You will see an example of how that works in the example write-up below.

Testing Assumptions and Data Cleaning in Logistic Regression

Logistic regression uses ML estimation and thus (as discussed in Chapter 2) there are different assumptions than in models using OLS estimation. Specifically, we are not now assuming normally distributed residuals, nor are we required to assume equal variance across the range of observed values.

We still assume that (a) the model is correctly specified, (b) variables are measured without error, (c) the residuals (observations) are independent, (d) the additive model is accurate (that effects are additive), and (e) effects are *linear on the logit*, meaning that the relationship between the IV and the logit-transformed DV is linear. We will discuss this more in Chapter 8 when we discuss general curvilinear modeling, but note that the assumption is *not* that the relationships are linear, but rather only that they are linear after the log transform.

We previously discussed simple data cleaning with respect to OLS regression, and the same principles apply to logistic regression, although there is one small caveat. I tend to examine standardized residuals, DfBetas, or other indicators of influence after standardizing them, and I examine them separately for each DV group. I have found that they can have very different ranges, and if you don't examine them separately, you can accidentally delete an entire group.

For example, if you examine the normalized or standardized residual plot below in Figure 5.4, you can see that one group has a relatively narrow range of residuals, whereas the other group has a broader range; if you selected all cases with residuals greater than −3, you might eliminate most of the nongraduated group. You can see that the nongraduated group has a broad range of residuals, some relatively large. What these residuals indicate is that as student achievement increases, those who

Figure 5.4   Standardized Residuals From zACH Analysis

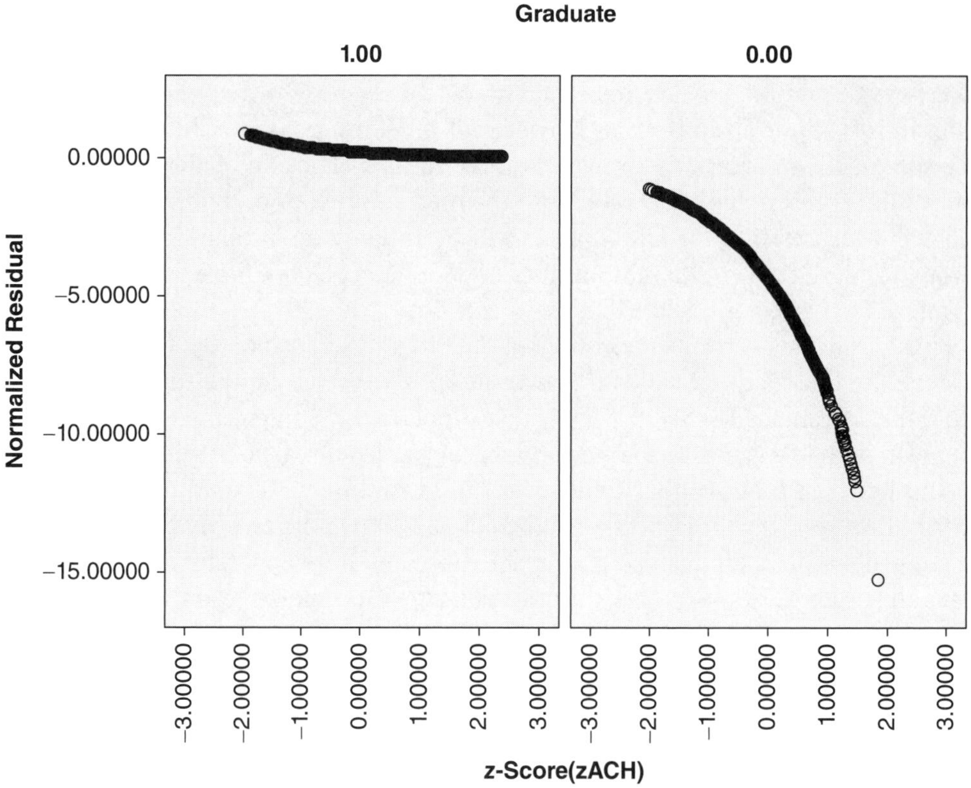

Graduate

1.00   **0.00**

Normalized Residual (y-axis): 0.00000, −5.00000, −10.00000, −15.00000

z-Score(zACH) (x-axis): −3.00000, −2.00000, −1.00000, 0.00000, 1.00000, 2.00000, 3.00000

DATA SOURCE: National Education Longitudinal Study of 1988 (NELS88) from the National Center for Education Statistics (http://nces.ed.gov/surveys/nels88/).

dropped out have increasingly large residuals, because they do not fit the model of increasing achievement being associated with higher probability of graduation. The question becomes, from a data cleaning perspective, at what point (if any) do these cases fail to represent the group of students we wish to generalize to? As cases have residuals passing 10 *SD* from the mean or larger, it is increasingly challenging to argue that they represent the population of interest. They also add disproportionate error to the model, as well as disproportionate levels of influence. To illustrate how influential these cases can be, when we picked a highly conservative cutoff point, cases with standardized residuals below −10 *SD* were removed from the analysis. The sample size for this analysis dropped from the original of 16,610 to 16,597 ($N = 13$ removed), and the model chi-square improved from 1,491.38 to 1,556.17 (Table 5.6a). You can also compare the parameter estimates and see that the effect size also increased.

Let me point out an important point. Many researchers are uncomfortable removing cases, but these 13 cases contributed a −2LL of 64.79, or an average level of mis-fit of 4.98 each. Evaluated as a chi-square with 1 *df*, the removal of each one of these cases leads to a significant improvement in model fit.

More realistically, one might wish to remove the 1% most serious offenders in this analysis. The worst 1% of the standardized residuals fall at −4.50 or lower. Thus, an analysis removing these 159 cases (Table 5.6b) resulted in a much better fitting model, with a change in −2LL of 2,028.49 and a much stronger slope (the original OR was 3.89 and is now estimated at 6.77).

Deviance Residuals

As we discussed before, there are several different types of statistics you can examine concerning influential or mis-fitting cases. Standardized residuals capture one type of information. We just identified another casewise contribution to overall model fit. Cook's D and other similar indicators (already discussed in Chapter 3) attempt to capture overall influence on the model for each case. Another interesting statistic available only in logistic regression (and other procedures that use ML estimation) is the change in model fit (e.g., change in −2LL) that removal of each

Table 5.6a   Analysis of GRADUATE and zACH With Standardized Residuals Below −10 Removed

Omnibus Tests of Model Coefficients

		Chi-Square	df	p
Step 1	Step	1,556.166	1	.000
	Block	1,556.166	1	.000
	Model	1,556.166	1	.000

Variables in the Equation

		B	SE	Wald	df	p	Exp(B)	95% CI for Exp(B) Lower	Upper
Step 1[a]	zACH	1.412	0.044	1,027.142	1	.000	4.106	3.766	4.476
	Constant	3.009	0.046	4,196.098	1	.000	20.271		

[a]Variable(s) entered on step 1: zACH.

Table 5.6b   Analysis of GRADUATE and zACH With Standardized Residuals Below −4.5 removed

Omnibus Tests of Model Coefficients

		Chi-Square	df	p
	Step	2,028.489	1	.000
Step 1	Block	2,028.489	1	.000
	Model	2,028.489	1	.000

Variables in the Equation

		B	SE	Wald	df	p	Exp(B)	95% CI for Exp(B) Lower	Upper
Step 1[a]	zACH	1.913	0.057	1123.906	1	.000	6.772	6.056	7.574
	Constant	3.591	0.064	3168.419	1	.000	36.268		

[a]Variable(s) entered on step 1: zACH.

individual case would cause. This is captured in DIFCHISQ in SAS and in deviance residuals in SPSS. These are more holistic indicators of the influence of each case and can help detect cases that are particularly poor fits to the model. In SPSS, each deviance residual squared gives us an indication of how much the model would change if that case were eliminated. In this example, the minimum value is −3.304 (rounded) and thus would result in roughly a change (improvement) of more than 10 in −2LL. The distribution of deviance residuals (Figure 5.5) ranges from −3.304 to 1.073. Given that change in −2LL is evaluated as a chi-square distribution, we could propose that any case with a deviance residual over |2| might be fair game for examination and deletion, because a chi-square of 3.85 is significant at $p < .05$. For our data, this would result in about 3.2% of our sample being deleted, primarily from the smaller group; thus, I might select a more conservative target, such as −2.5, which removes just 1% of the sample. With 149 cases removed, our −2LL is 7,227.288, compared with 8,475.86, a reduction in −2LL of 1,248.57, or an average reduction of −2LL of 8.38 *per case removed*. I think this is defensible to even the most strident data cleaning critic. After this data cleaning, the change in −2LL is 2,004.57, a markedly stronger effect than in previous analyses, as you can see in Table 5.7.

Figure 5.5   Deviance Residuals From GRADUATE and zACH Analysis

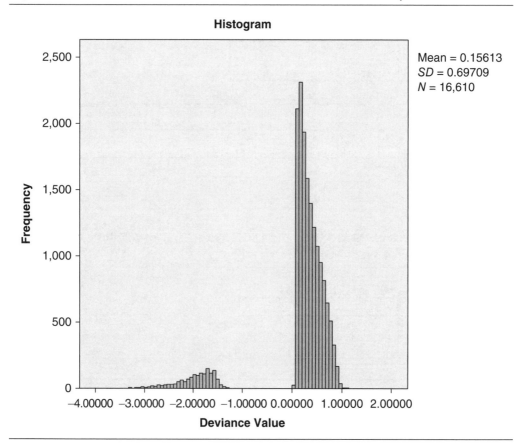

DATA SOURCE: National Education Longitudinal Study of 1988 (NELS88) from the National Center for Education Statistics (http://nces.ed.gov/surveys/nels88/).

Table 5.7   Effect of zACH on GRADUATE After Removing Deviance Residuals < −2.5

		B	SE	Wald	df	p	Exp(B)	95% CI for Exp(B)	
								Lower	Upper
Step 1[a]	zACH	1.880	0.056	1,122.579	1	.000	6.556	5.873	7.319
	Constant	3.550	0.063	3,224.701	1	.000	34.816		

[a]Variable(s) entered on step 1: zACH.

DfBetas

A third type of information is the individual contribution to parameter estimates. The entire concept of influence and leverage, which we introduced in previous chapters, allows us to identify cases that can have undue (or disproportionate) influence on the parameter estimates. In this example, we will explore DfBetas, which measures the effect of each case on each parameter. In this simple example, we have cases that can potentially influence the intercept or slope of zACH, both, or neither. It happens to be the case that both graphs look similar (which is not always the case), so we will examine the DfBetas for b_1. These are presented in Figure 5.6.

Figure 5.6   Unstandardized DfBetas for b_1 From zACH Analysis

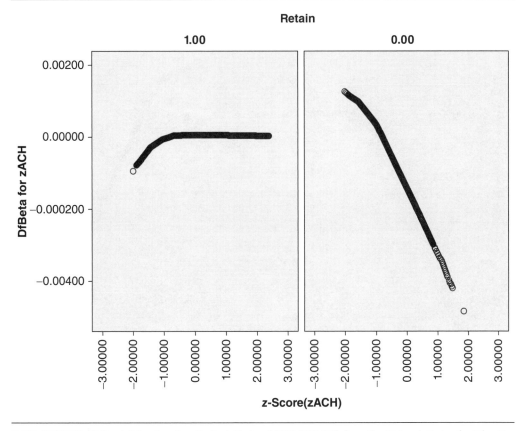

DATA SOURCE: National Education Longitudinal Study of 1988 (NELS88) from the National Center for Education Statistics (http://nces.ed.gov/surveys/nels88/).

As we discussed previously, DfBetas are the change in a parameter attributable to each individual case. Because we are discussing the slope (b_1), each value is the change in the slope if that case is removed. With more than 16,000 cases in the sample, these are understandably small. For example, some of the more extreme cases might change the slope −0.004 if removed. Is that a large or small effect? One challenge with interpreting raw DfBetas, like raw residuals, is that it is difficult to know what might represent disproportionate or inappropriate influence. Unfortunately, in logistic regression, evaluating DfBetas is more complex. As you can see in Figure 5.7, the standardized DfBetas (raw DfBetas converted to z-scores) have a relatively narrow range for the graduated group and a large range for the nongraduated group.

There is no objective way to evaluate these, in my experience. My guidance for you is to examine the graphical representations and the quantitative data, and if you feel that it is appropriate, remove the fewest number of cases that results in the maximum improvement to the model fit. If we try to remove less than 1% of the most inappropriately influential cases, about 1% of the DfBetas for slope, once converted to z-scores, is below approximately −4.50; therefore, let us also use that as a cutoff and recalculate the analysis.

Figure 5.7   DfBetas for b_1 Converted to z-Scores for Easier Interpretation

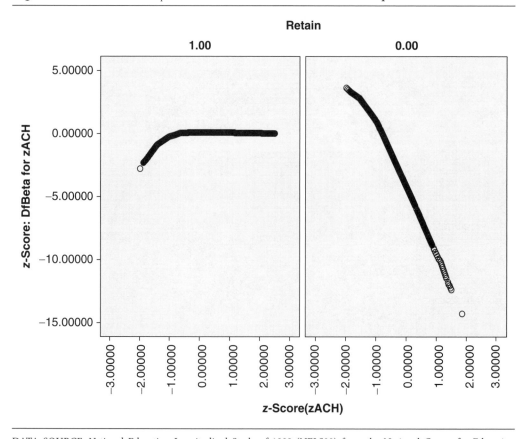

DATA SOURCE: National Education Longitudinal Study of 1988 (NELS88) from the National Center for Education Statistics (http://nces.ed.gov/surveys/nels88/).

As you can see in Table 5.8, removing the 153 cases with the most substantial DfBetas for slope results in a similarly improved model as removing a similar number of cases with extreme standardized residuals or deviance residuals. This is because, in this simple model, these are mostly the same cases being removed, regardless of how we identify them. Indeed, it is sometimes desirable to identify cases that are influential according to multiple criteria. However, once we get into models with multiple IVs, we will begin to see differences in the cases identified through these different mechanisms.

Hosmer and Lemeshow Test for Model Fit

Once we get a variable with more than two values, we can use the Hosmer and Lemeshow test, which looks at lack of fit in the same way that the Pearson chi-square test does, except this test looks at deciles rather than all individual cells in an attempt to more accurately model lack of fit where continuous variables are concerned. The Hosmer and Lemeshow goodness-of-fit statistic is more robust than the traditional goodness-of-fit statistic used in logistic regression, particularly for models with continuous variables and studies with small sample sizes. It is based on grouping cases into deciles of risk and comparing the observed probability with the expected probability within each decile.

This test examines the hypothesis that predicted probabilities are different from the observed probability (again, a lack-of-fit index):

$$H_0: \text{predicted probabilities} = \text{observed probabilities}$$

$$H_a: \text{predicted probabilities} \neq \text{observed probabilities}$$

Therefore, significant results mean that the predicted probability is significantly different from the observed probability. In Table 5.9, I present the high school test results from the original analysis predicting graduation from zACH (with no data cleaning). As you can see, the model actually does a relatively decent job of tracking the general pattern of where cells should be small and where they should be large.

Table 5.8   Results of Analysis After Removal of DfBetas for Slope < −4.50

Omnibus Tests of Model Coefficients

		Chi-Square	df	p
	Step	2,014.266	1	.000
Step 1	Block	2,014.266	1	.000
	Model	2,014.266	1	.000

Variables in the Equation

		B	SE	Wald	df	p	Exp(B)	95% CI for Exp(B)	
								Lower	Upper
Step 1[a]	zACH	1.893	0.056	1,123.176	1	.000	6.642	5.946	7.420
	Constant	3.566	0.063	3,201.992	1	.000	35.389		

[a]Variable(s) entered on step 1: zACH.

DATA SOURCE: National Education Longitudinal Study of 1988 (NELS88) from the National Center for Education Statistics (http://nces.ed.gov/surveys/nels88/).

Table 5.9a Hosmer and Lemeshow Test for Goodness of Fit From GRADUATE and zACH Analysis

Hosmer and Lemeshow Test

Step	Chi-Square	df	p
1	17.487	8	.025

Contingency Table for Hosmer and Lemeshow Test

		GRADUATE = .00		GRADUATE = 1.00		Total
		Observed	Expected	Observed	Expected	
Step 1	1	493	464.035	1,171	1,199.965	1,664
	2	303	321.896	1,358	1,339.104	1,661
	3	236	232.423	1,425	1,428.577	1,661
	4	137	165.542	1,525	1,496.458	1,662
	5	117	114.031	1,544	1,546.969	1,661
	6	73	75.818	1,589	1,586.182	1,662
	7	57	48.471	1,606	1,614.529	1,663
	8	35	29.814	1,627	1,632.186	1,662
	9	23	16.905	1,637	1,643.095	1,660
	10	3	8.066	1,651	1,645.934	1,654

Table 5.9b Hosmer and Lemeshow Test for Goodness of Fit From GRADUATE and zACH Analysis After Data Cleaning

Hosmer and Lemeshow Test

Step	Chi-Square	df	p
1	79.152	8	.000

Contingency Table for Hosmer and Lemeshow Test

		GRADUATE = .00		GRADUATE = 1.00		Total
		Observed	Expected	Observed	Expected	
Step 1	1	488	520.711	1,158	1,125.289	1,646
	2	299	319.164	1,344	1,323.836	1,643
	3	238	203.681	1,408	1,442.319	1,646
	4	138	126.704	1,511	1,522.296	1,649
	5	118	75.124	1,530	1,572.876	1,648
	6	47	42.312	1,598	1,602.688	1,645
	7	0	22.294	1,646	1,623.706	1,646
	8	0	11.143	1,645	1,633.857	1,645
	9	0	5.041	1,647	1,641.959	1,647
	10	0	1.825	1,646	1,644.175	1,646

The overall significant effect from Hosmer and Lemeshow indicates that there are significant discrepancies between observed and expected groups (and hence, probabilities) across the 10 deciles of zACH, but the conditional probabilities match what we would expect: higher achievement produces higher probabilities of being retained through graduation. The reason we have a significant Hosmer and Lemeshow statistic is because we have a good deal of power to detect differences.

Interestingly, the Hosmer and Lemeshow test reveals a slightly worse fit after deviance residual cleanings (Table 5.9b). If you examine the contingency table below, you will see that the data cleaning removed cases from the highest deciles of zACH in the dropout group. This resulted in a much stronger model overall by all other measures, but decreased the goodness of fit according to this measure, because deciles 7–10 in the dropout group now have no cases where some are expected.

A Brief Example Summary of the Analysis in APA Format

The probability of completing high school (GRADUATE, coded 0 for nongraduates and 1 for graduates) was analyzed as a function of standardized achievement test scores (zACH; higher scores mean higher achievement) in a simple logistic regression equation. Assumptions were met, and various indicators of influence (standardized residuals, deviance statistics, and DfBetas) were examined. Of 16,610 cases, 149 were removed from the analysis due to inappropriate levels of influence (deviance residuals of more than $|2.5|$), resulting in a substantially improved model fit when the remaining 16,461 cases were analyzed.

Overall model fit was significantly improved when zACH was entered into the equation $(\chi^2_{(1)} = 2,004.57, p < .0001)$. The constant $(b_0 = 3.55$, Wald $= 3,224.70, p < .0001)$ indicates that when a student has average achievement (zACH = 0), the probability of graduation is 0.97. The slope $(b_1 = 1.880$, OR $= 6.56$, 95% CI for OR of [5.87, 7.32], Wald $= 1,122.58$, $p < .0001)$ was significant, indicating that as achievement increases, the probability of graduation increases. Specifically, those with achievement of 2 SD below the mean have only a 0.45 probability of graduating, those 1 SD below the mean have a 0.84 probability of graduating, and those with achievement at 1 SD above the mean have a probability of 0.99, according to these data.

How Should We Interpret Odds Ratios That Are Less Than 1.0?

One significant problem with ORs is that they are asymmetrical. They can theoretically range from 0.00 to ∞, but a value of 1.0 means there is no difference in risk or odds (i.e., there is no effect of the IV). Thus, the entire infinity of decreasing (inverse) relationships must fit between 0.000001 and 0.99999, whereas that same infinity of positive relationships fits in a much larger space between 1.00001 and ∞. As you saw in the DROPOUT and POOR example, that relationship was positive—odds of dropout increase as you move from nonpoor to poor households, and the OR was impressive at 5.711. What if we had reversed the variable so that we were studying *affluence* (0 = poor households and 1 = affluent households)? The OR would have been 0.175. Reversing the coding of a variable (swapping 0 and 1) merely inverts the OR (thus, 0.175 = 1/5.711). An OR of 20.00 is therefore equivalent to an OR of 0.05 in magnitude. However, from a psychological and interpretation point of view, ORs below 1.0 tend to seem less impressive and (in my experience) are more likely to be misinterpreted.

Two issues arise here. First is the use of directional language (e.g., "individuals in group 1 are X times *more likely* to experience a specific outcome than in group 2" or "individuals in group 2 are X times *less likely* ... "). Leaving for a moment the difficulty with cogently describing an OR, the difficulty here comes in the common mistake people make in describing decreasing ratios. If you have an OR of 5.71, as we did, it is straightforward to say something like "students from poor households are 5.71 times more likely to withdraw from school than students from more affluent households." If we had coded the variables differently, as suggested above, we would have gotten an OR of 0.175. This means the exact same thing—that students from affluent households are much less likely to drop out.

Yet these ORs less than 1.0 can be treacherous to interpret. Authors often want to say things like "students from affluent households are 0.175 times *less likely to drop out* than students from high-poverty households." In fact, that is not the case: 0.175 times less likely implies 0.815 times *as likely*—a much smaller effect than is actually the case. Psychologically, people do not generally view 0.175 and 5.71 as being equal in magnitude, just as people often do not view 0.01 and 100 as the same magnitude, although they are in terms of ORs.

What do we do about this? First, we need to be careful regarding how we talk about ORs. My advice has always been to use "as likely," rather than "less likely" or "more likely." Another way of staying clear of trouble is to use language like "the odds of GROUP X having an outcome is <odds ratio> that of GROUP Y." Using our previous example, saying "the odds that students from affluent households will drop out of school is 0.175 that of students from poor households" is preferable to using terminology like "less likely." However, this does not solve the problem of people perceiving that as a less impressive effect than 5.71. This brings us to the second issue: the psychological impact of ratios and accurately conveying effect sizes when the effect sizes themselves vary depending on whether they are increasing or decreasing odds.

Taking a more extreme example, imagine a drug that made the risk of experiencing a cancer relapse OR = 0.001 for patients taking the drug compared with people who do not take the drug. Mathematically, this is identical to saying that taking the drug makes you 1,000 times less likely to experience relapse, or not taking the drug makes you 1,000 times more likely to have a relapse. But are they perceptually identical? No. Further elaborating, let us say you have two drugs. One produces an OR = 0.001 and one an OR = 0.01. Even the most technically proficient of us will view these as reasonably close in magnitude (i.e., both are really small). However, if the direction of the IV were arbitrarily reversed, the ORs would be 1,000 and 100, respectively, and most would interpret those as " relatively large"; however, in the latter case, it is much more apparent that the magnitude of the effect is a magnitude different (Figure 5.8).

Thus, another practical recommendation is to refrain from reporting ORs less than 1.0. It would make sense to standardize the reporting of this effect size so that all ratios would be reported as >1.0. Analyses that result in ratios less than 1.0 would take the inverse of the OR and reverse the categories or the description of the results to keep the conclusion consistent.[9]

---

9 Pedhazur (1997) suggests an alternative solution to this issue: taking the natural log (ln) of all ORs. This has the effect of moving the "null effect" point from 1.0 to 0.0 and removing the lower bound so that effect size distribution is symmetrical. The drawback to this elegant solution is that it changes the effect from substantively interpretable (relative risk or OR) to something much more complex to interpret: the natural log of a relative risk or an OR. Thus, I would not recommend this solution.

Figure 5.8   The Nonlinear Relationship Between Odds Ratios Over 1.0 and Below 1.0

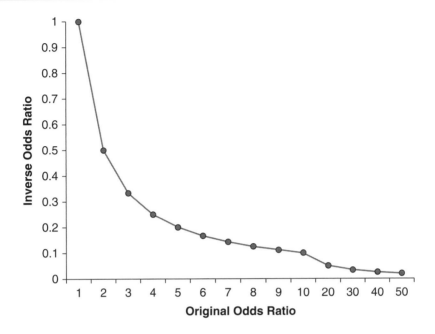

Another, perhaps preferable, method of dealing with this issue is to convert odds or logits to predicted probabilities for each group, as we did in Chapter 4 (and as we will do with more urgency in later chapters when looking at interactions and curvilinear effects). This has the laudable effect of reducing the exaggeration of effect size that ORs can have under certain circumstances and helping avoid issues like interpreting an OR of 0.175 as being "0.175 times *less likely*" for something to happen.

SUMMARY

We covered a lot of ground in this chapter—from assessing the overall goodness of a model to some of the benefits of utilizing continuous variables (in which your data has continuous variables) rather than engaging in the common practice of dichotomizing. Along the way, I argued that any continuous variable should be converted to the standard normal distribution before using it in logistic regression so as to standardize interpretation of those variables. Furthermore, I suggested that variables should be coded in such a way that ORs end up over 1.0 for ease of interpretation. I hope at this point you are feeling increasingly comfortable with generalized linear modeling approaches to data cleaning. At this point, we have covered simple univariate examples of some of the most common analyses you might encounter in the behavioral sciences. Almost everything from this point onward will be extensions and generalizations of generalized linear modeling covered in these first five chapters.

In Chapter 6, we will toy with the idea of a DV with multiple categories, which is usually referred to as multinomial logistic regression. It is the reverse of Chapter 4, in which we had a polytomous (unordered) categorical IV. However, we can also have polytomous (unordered or ordered) DVs such as the following:

Have you ever tried illegal drugs?

1. No, I have never tried any drugs.

2. I have tried marijuana.

3. I have tried cocaine.

4. I have tried heroin.

5. I have tried more than one of the above.

These types of DVs present interesting opportunities and challenges. Now that you have mastered the basics of linear modeling with binary DVs, this is merely an extension and elaboration of what we have already done.

ENRICHMENT

1. Convert the following probabilities to logits:

    a. 0.25
    b. 0.50
    c. 0.75
    d. 0.99
    e. 0.001

2. Convert the following examples from logits to predicted probabilities:

    a. −4
    b. −1.5
    c. 0
    d. 2
    e. 5

3. Download the NELS88 data from the book website and reproduce analyses included in the chapter.

4. Download the National Health Interview Survey of 2010 data from the book website. Analyze the risk of diabetes as a function of body mass index (BMI) using best practices. Then do a median split of BMI into "overweight" and "not overweight" and predict diabetes from this dichotomous variable. Discuss the differences in the two analyses.

5. Download the NELS88 data set that includes self-reported marijuana use (EVERMJ is coded 0 = never tried marijuana, 1 = tried marijuana at least once). For each of the following, summarize briefly in APA format.

   a. Does SEX (1 = male, 2 = female) predict reported marijuana use? If so, are males or females more likely to report having tried it?

   b. Does student race (RACEBW; 0 = Caucasian/White, 1 = Black/African American) predict reported marijuana use? If so, are White (Caucasian) or Black/African American students more likely to report having tried it?

   c. A long time ago, I calculated a variable I called "BADCNT," which counted the number of bad things a student reported having happened to them in school that year. These included things like being assaulted, having things stolen, being offered drugs, and experiencing other stressful events. This variable ranges from 0 to 6, with most students having experienced none of these or one of these types of events. According to common theories of marijuana (and other drug use), stressful events might make students more likely to try marijuana. Your final challenge in this chapter is to see whether this "continuous" variable predicts reported marijuana use. Contrary to my recommendation above, we will *not* convert this variable to the standard normal distribution because it is already in a substantively interpretable metric: number of events experienced.

      i. If there is a significant effect, describe the nature of the effect, keeping in mind that this is a continuous variable.

      ii. Calculate a predicted logit for BADCNT = 0 and for BADCNT = 4. Convert the predicted logits back to conditional probabilities for BADCNT = 0 versus 4.

Appendix 5A: A Brief Primer on Probit Regression

You may have encountered this creature called "probit" regression, which also deals with binary DVs, but in a slightly different way. Probit applies a different *link function* to appropriately model binary DVs.

Probit was developed in a separate tradition from logistic regression but shares many surface, procedural, and mathematical similarities to logistic regression. In this appendix, we will briefly explore the mathematical underpinnings of probit regression, contrast it with what you now know of logistic regression, and discuss pros and cons of this interesting technique.

What Is a Probit?

My research indicates that probit[10] regression was originally presented in the form we know today by Bliss (1934a, 1934b) as a method of dealing with an important nonlinear relationship in economic entomology—the relationship between pesticide toxicity and kill percentage.[11] In particular, Bliss (and others who preceded him) was trying to find a way of transforming the very nonlinear relationship between dosage and the percentage of pests the pesticide would kill (which asymptotes at low and high dosages) to something that would be linear if plotted against each other. You can imagine that at very low dosages (or concentrations) of a pesticide, there is very little effect (or a very low percentage of pests that are killed). Then, at some point, with increasing concentrations, the kill percentage increases markedly, and continues increasing until some later point at which effectiveness will asymptote and increased concentrations or dosages beyond this point may be less effective. The ability to easily calculate the optimal dosage via hand tools like graph paper was important in many scientific fields before the widespread availability of computing power in the field. In that original application, scientists used cross section paper that made calculations simple if a line was straight.

Bliss's and others' great contribution to practical toxicology was to convert this nonlinear relationship to something easily understood—the z-score of the probability or percentage killed—which then has a linear relationship with dosage or concentration, and then could be used with the common tools of the time: linear graph paper of various types. This practical breakthrough was subsequently adapted to other fields such as mental measurement (Finney, 1944) and is still in wide use today in diverse fields such as finance (Malkiel & Saha, 2005), medicine (Krentz, Auld, & Gill, 2004), and genetics (Zhou, Wang, & Dougherty, 2004). In fact, probit and logit analyses were so valuable that the US Department of Agriculture supported the development of POLO, a program specifically designed to analyze data obtained from insecticide

---

10 The word *probit* is short for "probability unit."

11 Yes, this is certainly a morbid topic, but without this groundbreaking work, we would not have the current food security that many of us enjoy. Let's not forget that modern epidemiology got the OR and framework for logistic regression from studying poor sewer sanitation in London (Snow, 1855)—although Brands (2010) reports that Benjamin Franklin may have been one of the earliest epidemiologists, systematically exploring the link between professions that provide exposure to lead and other metal toxins and maladies now known to be related to exposure to these substances. Wow, that footnote got away from me a little!

bioassays that was made available to anyone who requested a magnetic tape and had a Univac 1100 series computer or equivalent "large" scientific computer (Robertson, Russel, & Savin, 1980).[12]

Because probit is simply another link function within the broad umbrella of the GLM, almost anything we have done in previous chapters (or will do in subsequent chapters) can be done with probit regression. This appendix is but a brief introduction to probit regression, designed to be a reference in case you are interested in pursuing this topic further. I will not discuss it explicitly past this chapter, but as we move into curvilinear, multiple IVs, interactions, and so forth, these aspects of modeling are also available in probit models.

As the words *logit* and *probit* imply, if a logit is an attempt at quantifying a relationship between a dichotomous variable and a predictor variable in terms of log odds (in other words, working in a metric of log odds), probit works in a metric of probability. Therefore, we would consider probit regression a version of generalized linear modeling with a "probit link" connecting the predicted variable to the linear part of the regression equation (just as logistic regression uses a "logit link" to connect the predicted variable to the linear part of the regression equation).

The Probit Link

The probit link assumes that the observed binary outcome variable results from an underlying latent normally distributed random variable with a threshold that triggers a 0 turning into a 1.0. This is not dissimilar to what we do with high-stakes testing in the US educational system, where we have high-stakes test scores that vary, often along a normal distribution, and we set a particular cut score to signify whether the student has "passed" the test or not. The difference is that in the case of probit, the underlying normally distributed variable is not available to us.[13]

Probit regression uses the *inverse* standard normal cumulative distribution function (area under the curve greater than z, which is the boundary). The standard normal cumulative distribution function uses the standard normal distribution we are familiar with and accumulates all of the probabilities for values below a given value. This function is shown in Figure 5.9a; as you can see, it is a sigmoid function similar to the logit curve. This curve represents the percentage of the population *below* each z-value, which makes sense for looking at things like graduation. If your achievement is below average (e.g., 1 *SD* below the mean), few at or below your level of achievement will graduate. Conversely, if your achievement is 1 *SD* above the mean, most individuals around that level of achievement will graduate.

---

12 Of course, the cell phone in your pocket today probably has exponentially more processing power than the room-sized computers scientists used to run these programs. Back in the old days of UNIVAC, you were required to walk uphill both ways in the snow in order to use the machine.

13 You can play with the cumulative normal distribution in Excel by entering any z-value in the following formula: =NORM.S.DIST(z,TRUE), which then returns the probability a value will be below a particular point. For example, entering 0 will return 0.50, or the 50th percentile. Entering 1.96 returns 0.975, or the 97.5th percentile, −1.96 returns 0.025, or the 2.5th percentile, and so on.

Figure 5.9a   Standard Normal Cumulative Distribution

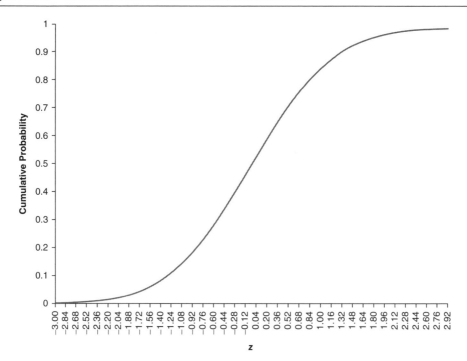

One interesting thing to note about the cumulative frequency distribution is that it is not a uniform, linear effect. The probabilities associated with a $z = 1.0$ change depend on the location of the initial threshold. If the first threshold is -3.0 and we increase by an increment of 1.0, the difference between $z = -3.0$ and -2.0 is the difference between 0.00135 probability and 0.0228, or a difference in probability of 0.0214. However, if the initial threshold is -0.50 and we increase by an increment of 1.0, the corresponding probabilities are 0.31 and 0.69, or a change in probability of about 0.39. Therefore, in probit regression (as in real estate), location matters.

The application this was originally developed for, however, was kill rate, or survival rate. Thus, it uses the inverse of this distribution, which you can see in Figure 5.9b.

Imagining a particular level of pesticide application, at a very low level, most of the population will survive; at a very high level, almost none of the population will survive. Thus, when we calculate the probit analysis in SPSS, you will see the inverse of what we expect. Using the probit link function, we can constitute a regression equation similar to that of logistic regression, as you can see in Equation 5.7:

$$\text{Probit } (\hat{Y}) = b_0 + b_1 X_1 \tag{5.7}$$

Similar to logistic regression, the hypotheses we are testing include the probit link:

$$H_0: b_1 = 0.0$$

$$H_a: b_1 \neq 0.0$$

Figure 5.9b   The Inverse Standard Normal Distribution

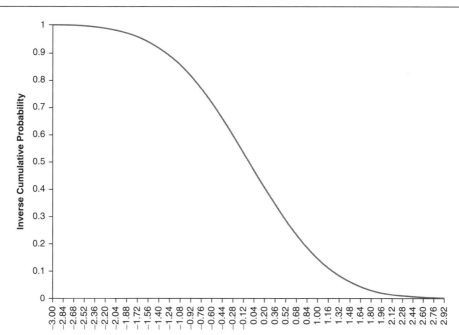

As discussed above, we also implicitly test hypotheses about the intercept, usually that it is different from zero:

$$H_0: b_0 = 0$$

$$H_a: b_0 \neq 0$$

A Real-Data Example of Probit Regression

We will use the same data as above, looking at our binary achievement (ACH_LOHI; 0 = low achievement, 1 = high achievement) variable predicting GRADUATE (0 = no, 1 = yes). You can see that all of the calculations are similar to that in Table 5.3.

The z-score of both of the conditional probabilities [available in any z-table or via the =NORMSINV(x) command in Excel] is presented in Table 5.10. The z-score that corresponds to a probability of 0.978 is 2.02, and the z-score for the intercept (low achievement) is 1.04, giving us a slope (change from 0 to 1) of 0.976.

Submitting the same data to SPSS and running probit analysis yields a significant overall improvement in the model, as you can see in Table 5.11. We also see that the effect of achievement is also significant, with the same values for intercept and slope (albeit negative, as it uses the inverse distribution). We also have to deal with the concept of "thresholds," which frankly adds a bit of confusion because they are another piece of information to process.

Table 5.10   Cross-Tabulation of Student Achievement and Graduation, Example Data From Earlier in This Chapter

| | | GRADUATE | | Total | Conditional Probabilities | z-Scores for Conditional Probabilities | Slope | Predicted Probit |
		0.00	1.00					
ACH	High (1)	169	7,621	7,790	0.978306	2.02	0.976	
	Low (0)	1,308	7,512	8,820	0.851701	1.04		
Total		1,477	15,133	16,610	0.911078			

SOURCE: National Education Longitudinal Study of 1988 (NELS88) from the National Center for Education Statistics (http://nces.ed.gov/surveys/nels88/).

Table 5.11   Probit Analysis of GRADUATE and ACH_LOHI

Model Fitting Information

Model	−2LL	Chi-Square	df	p
Intercept only	949.610			
Final	15.801	933.810	1	.000

Link function: Probit.

Parameter Estimates

| | Estimate | SE | Wald | df | p | 95% CI | |
						Lower Bound	Upper Bound
Threshold [GRADUATE = .00]	−2.020	0.032	4,029.142	1	.000	−2.082	−1.958
Location [ACH_LOHI = .00]	−0.976	0.036	744.413	1	.000	−1.046	−0.906

Link function: Probit.

The problem with interpreting all these numbers is that everything is an inverse from what we are used to looking at with logistic regression. We are using an inverse cumulative probability distribution, and looking at the inverse slope and inverse outcome (not graduating). This is analogous to understanding double negatives in the English language (I am not going to avoid punching you). But let us move forward regardless. If we look at the threshold for GRADUATE, the inverse cumulative probability for a z of −2.02 equates to 0.978, which is the same probability we calculated in Table 5.10. Furthermore, if we know that the location of low achievement is 0.976 away from this point estimate, we need to add 0.976 to −2.02 to get −1.04, which has a probability of 0.85, which also matches our hand calculations. These outcomes match our results obtained from logistic regression.

Why Are There Two Different Procedures If They Produce the Same Results?

As with many things in statistics, the different procedures evolved in parallel within different disciplines. Consider the simple case of ANOVA and OLS regression—for so many decades of our history, they were talked about as though they were radically different because they developed within different groups of researchers. Today, in the 21st century, we acknowledge the GLM, of which OLS regression and ANOVA are two examples (as are logistic and probit regression). Mathematically, one can get identical data to produce identical results regardless of which method is used to analyze the data. This is another case of the same situation.

Because both probit and logistic regression use essentially the same information to come to the same conclusions, in one sense it really does not matter which routine you use. Properly interpreted, you will come to largely the same conclusion. I demonstrated this more deeply (including using an example of an interaction) in my logistic regression book (Osborne, 2015).

Some Nice Features of Probit

As I have discussed in previous chapters, the logit (and logarithms in general) is difficult for people to interpret accurately, even if they are mathematically trained, because the log can exaggerate very small differences and can minimize very large differences once converted to logarithms. The metric for probit, z-scores, is more familiar and exaggerates effects less dramatically. Again, remember that this is all the same information as presented in logistic regression, just presented in different form. Despite this, some authors, such as Berkson (1951), express preferences for one over the other, although even Berkson admitted that there is little functional difference between the results the two provide.

Assumptions of Probit Regression

Interestingly (or perhaps not . . .), probit uses ML estimation and thus follows the same basic assumptions as logistic regression. As a result, this is one of the shortest discussions of assumptions I have ever written.

SUMMARY AND CONCLUSION

In summary, I think I have shown that probit regression is a simple link function that produces similar results to that of logistic regression. Because it uses the same ML estimation, it produces virtually identical results to logistic regression when best practices are followed and results are interpreted in light of predicted conditional probabilities. Thus, the decision to use logistic versus probit regression is one of style and preference. Probit regression is the preferred method in some fields, whereas logistic regression is the preferred method in others.

One factor might be the strength of the implementation of the two procedures in the software you use. In SPSS, which I use quite a bit, I find the logistic implementation to be more fun and versatile, but the probit is certainly adequate.

REFERENCES

Berkson, J. (1951). Why I prefer logits to probits. *Biometrics, 7*(4), 327–339.

Bliss, C. I. (1934a). The method of probits. *Science, 79*(2037), 38–39. doi: 10.1126/science.79.2037.38

Bliss, C. I. (1934b). The method of probits—a correction. *Science, 79*(2053), 409–410. doi: 10.1126/science.79.2053.409

Brands, H. W. (2010). *The first American: The life and times of Benjamin Franklin.* New York, NY: Anchor.

Finney, D. J. (1944). The application of probit analysis to the results of mental tests. *Psychometrika, 9*(1), 31–39. doi: 10.1007/BF02288711

Ingels, S. (1994). *National Education Longitudinal Study of 1988. Second follow-up: Student component data file user's manual.* Washington, DC: US Department of Education, Office of Educational Research and Improvement, National Center for Education Statistics.

Krentz, H. B., Auld, M. C., & Gill, M. J. (2004). The high cost of medical care for patients who present late (CD4< 200 cells/μL) with HIV infection. *HIV Medicine, 5*(2), 93–98. doi: 10.1111/j.1468-1293.2004.00193.x

Malkiel, B. G., & Saha, A. (2005). Hedge funds: Risk and return. *Financial Analysts Journal, 61*(6), 80–88. doi: 10.2469/faj.v61.n6.2775

Mead, M. (1935). *Sex and temperament in three primitive societies.* New York, NY: William Morrow and Company.

Oakley, A. (1972). *Sex, gender, and society.* London, England: Temple Smith.

Osborne, J. W. (2000). Advantages of hierarchical linear modeling. *Practical Assessment, Research & Evaluation, 7*(1), 1–3.

Osborne, J. W. (2015). *Best practices in logistic regression.* Thousand Oaks, CA: SAGE.

Pedhazur, E. J. (1997). *Multiple regression in behavioral research: Explanation and prediction.* Fort Worth, TX: Harcourt Brace College Publishers.

Robertson, J. L., Russel, R. M., & Savin, N. (1980). *A user's guide to Probit Or LOgit analysis.* Berkeley, CA: Pacific Southwest Forest and Range Experiment Station.

Snow, J. (1855). *On the mode of communication of cholera.* London, England: John Churchill.

Zhou, X., Wang, X., & Dougherty, E. R. (2004). Gene prediction using multinomial probit regression with Bayesian gene selection. *EURASIP Journal on Applied Signal Processing, 4*(1), 115–124. doi: 10.1155/S1110865704309157

SIMPLE LINEAR MODELS WITH POLYTOMOUS CATEGORICAL DEPENDENT VARIABLES

Multinomial and Ordinal Logistic Regression

6

Advance Organizer

Not all outcomes are continuous or simple dichotomous variables. In this chapter, we will explore outcomes that are a bit more complex. Think of career choice. There are many ways we can group careers, but whatever one you use, there is little rationale for ordering or prioritizing careers. Using the techniques covered in this chapter, you can model predictors of career choice.

Graduation from high school or college is also a complex phenomenon, although in the last chapter, we viewed it as a binary (yes/no) outcome. For example, some students graduate on time, whereas some graduate but not on time (some graduate early, some late). Some students drop out of school, and of those who drop out, some come back and some do not. This variable might be rationally ordered using value judgments from not graduating (as least desirable) to graduating on time (or early, which might be most desirable).

Tobacco smoking status is rarely as simple as whether you ever smoked cigarettes or not. Aside from the issue of whether smokeless alternatives (snuff, chewing tobacco, or even tobacco-less e-cigarettes) should be considered in the mix, there are different products (cigarettes, cigars, pipe tobacco, hookahs, etc.) and different patterns of use (never used, used only occasionally, used regularly, used regularly but stopped, etc.). In the previous chapter's exercises, we compared those students who admitted to using marijuana with those who indicated they had never used it. Later in this chapter, we will examine a more complex analysis regarding this same topic.

Multinomial logistic regression (also sometimes called polytomous logistic regression) allows us to leverage the power of binary logistic regression and generalize it to a categorical dependent variable (DV)—ordered or unordered—with more than two categories, as you can see in Table 6.1. It may be helpful to think of this type of analysis as similar to the analysis of variance (ANOVA) type of analyses we examined in Chapter 4, because this more complex procedure will require similar comparisons between groups and will introduce complexity to the analysis. However, the similarity ends there because the mathematics and estimation procedures will be closer to binary logistic regression.

Table 6.1 Examples of the Generalized Linear Model as a Function of Independent Variable and Dependent Variable Type

	Continuous DV	Binary DV	Unordered Multicategory DV	Ordered Categorical DV	Count DV
Continuous IV	OLS regression	Binary logistic regression	Multinomial logistic regression	Ordinal logistic regression	OLS, Poisson regression
Mixed continuous and categorical IV					
Binary/ categorical IV only	ANOVA	Binary logistic or log-linear models	Multinomial or log-linear models		Log-linear models

ANOVA, analysis of variance; DV, dependent variable; IV, independent variable; OLS, ordinary least squares.

We will examine not only the general case of multinomial (unordered) DVs but also the special case of ordered categorical DVs. In general, the multinomial logistic regression analyses you will use can handle either unordered or ordered DVs equally well. If you have an ordered variable, however, and it meets the assumptions of ordinal logistic regression, you can create some efficiencies and simplifications of the results.

Let us start with a quick note on terminology. Some books use the term *multinomial* and some use *polytomous* to refer to this group of variables. Linguistically, *polytomous* is a generalization of dichotomous, and *multinomial* is a generalization of binomial. There are probably subtleties that distinguish them as more or less appropriate for certain contexts, but let us agree that they are generally synonymous and move on. Unordered multinomial variables will have groupings that are not necessarily able to be placed into an objective order (e.g., people who drink wine, beer, or liquor are probably not defensibly ordered). Ordered multinomial variables are groupings that can be ordered in some way (e.g., nonsmoker, occasional smoker, daily smoker). The problem is that there are often complexities such that groupings are not always clearly one or the other. For example, in the smoking example, where do you place former smokers? Currently they are nonsmokers, but historically they might have been heavy smokers, which might alter long-term risk factors for disease. Later in this chapter, we will see how a simple test for ordinality will allow us to simplify some of these analyses if the groupings truly are ordered in some way.

In this chapter, we will cover

- Basic concepts in handling unordered multinomial DVs
- Multinomial logistic regression with categorical and continuous independent variables (IVs)
- Assumptions and data cleaning for multinomial logistic regression

- Confidence intervals and effect sizes in logistic regression
- Examples of American Psychological Association (APA)–compliant summaries of analyses
- Testing for ordinality
- Performing ordinal regression and interpreting the results

Guidance on how to perform these analyses in various statistical packages will be available online at study.sagepub.com/osbornerlm.

Understanding Marijuana Use

Of course, this book is *not* about understanding the complexities of substance use and abuse, but we ended Chapter 5 with an exercise analyzing the predictors of student self-reports of marijuana use. If you successfully completed that challenge from the last chapter, you might have wondered whether there are more interesting nuances in the data. There are often complexities to behavior that might not be satisfyingly captured in binary variables. Instead of the binary marijuana variable (yes/no), let us explore self-reported marijuana use as it was originally coded in the National Education Longitudinal Study of 1988 (NELS88) data, as a four-category variable (Table 6.2).

Table 6.2   In Lifetime, Number of Times R Used Marijuana

Original Category	Frequency	Label
0	13,487	Tried marijuana 0 occasions
1	1,578	Tried marijuana 1–2 occasions
2	1,094	Tried marijuana 3–19 occasions
3	751	Tried marijuana 20 or more occasions

SOURCE: National Education Longitudinal Study of 1988 (NELS88) from the National Center for Education Statistics (http://nces.ed.gov/surveys/nels88/).

I dislike having continuous variables broken into categories, but the data were provided this way and there is nothing we can do about it. To treat it as a continuous variable would be to violate basic tenets of measurement, but we can hope there was a good rationale for the survey designers to provide the data in this form and treat this variable as categorical.[1]

Because multinomial logistic regression can get complex, let's start with some basic concepts and the raw data, as we did in Chapter 5.

---

1   In other words, we will assume, for the moment, that each group is qualitatively different than the other.

Dummy-Coded DVs and Our Hypotheses to Be Tested

Conceptually, multinomial logistic regression is a series of binary logistic regressions, each comparing one particular group with each other group, similar to when we had dummy-coded IVs. Thankfully, we don't have to code the variables by hand, but we do have to be thoughtful about the comparison group, because it will affect the questions we can answer in this analysis, just as it did in the dummy coding analyses. You may recall from Chapter 4 that there are some criteria for a good comparison group, including that it should be relatively well populated, it should be conceptually coherent (not a catch-all "other" group), and it should represent some sensible reference for the other groups that serves to answer the key research questions. In this case, the "never tried marijuana" group will serve that purpose, creating an analysis that compares this group with each other group in the analysis.

As with binary logistic regression (and other analyses already covered in this book), our initial hypothesis is that the overall model is significantly improved by the entry of the IV (in this case, change in −2 log likelihood [−2LL], which is evaluated as a chi-square):

$$H_0: \chi^2\,(\Delta\text{−2LL}) = 0$$

$$H_a: \chi^2\,(\Delta\text{−2LL}) \neq 0$$

If this initial hypothesis is rejected, and we conclude that the IV has significantly improved the model fit, then we can examine our subsequent hypotheses for each pairwise comparison. Specifically, these hypotheses test whether the conditional probabilities differ between the comparison group and the other group as a function of some IVs. For example,

$$H_0: \text{Odds ratio} = 1.0$$

$$H_a: \text{Odds ratio} \neq 1.0$$

Alternatively, because the natural log of 1.0 is 0.00, we can also write these hypotheses in terms of logits and use standard slope notation as we would with any other linear model:

$$H_0: b_0 = 0$$

$$H_a: b_0 \neq 0,$$

and

$$H_0: b_1 = 0$$

$$H_a: b_1 \neq 0$$

If these look familiar, they should, because they are the same hypotheses from Chapter 5. The difference here is that there was only one comparison in binary logistic regression, whereas there will be $M - 1$ comparisons in multinomial logistic regression (where M is the number of groups in the DV). In the case of this analysis, we will actually have three sets of individual hypotheses:

Group 0 versus 1:

$$H_0\text{: Odds ratio} = 1.0 \ (\text{or } b_1 = 0)$$
$$H_a\text{: Odds ratio} \neq 1.0 \ (\text{or } b_1 \neq 0)$$

Group 0 versus 2:

$$H_0\text{: Odds ratio} = 1.0 \ (\text{or } b_1 = 0)$$
$$H_a\text{: Odds ratio} \neq 1.0 \ (\text{or } b_1 \neq 0)$$

Group 0 versus 3:

$$H_0\text{: Odds ratio} = 1.0 \ (\text{or } b_1 = 0)$$
$$H_a\text{: Odds ratio} \neq 1.0 \ (\text{or } b_1 \neq 0)$$

Note that if we fail to reject the null hypothesis for the overall model, then it is not legitimate to examine individual effects.

Basics and Calculations

We will begin this chapter the same way we did with Chapter 5—that is, by demonstrating that these calculations are relatively simple (at least in the case of binary IVs). Let us take a simple case of SEX as the IV and the expanded four-category MJ variable as the DV, presented in Table 6.3.

Table 6.3   SEX × MJ Cross-Tabulation

		MJ				1 Versus 0	2 Versus 0	3 Versus 0
		0	1	2	3			
SEX	1 (Male)	5,829	743	499	380	743 / (743 + 5,829) = 0.1131	499 / (499 + 5,829) = 0.0789	380 / (380 + 5,829) = 0.0612
	0 (Female)	6,465	679	503	266	679 / (679 + 6,465) = 0.0950	503 / (503 + 6,465) = 0.0722	266 / (266 + 6,465) = 0.0395
				Relative risk:		1.189	1.092	1.549

SOURCE: National Education Longitudinal Study of 1988 (NELS88) from the National Center for Education Statistics (http://nces.ed.gov/surveys/nels88/).

Before you try to interpret these results, think back to the case of binary logistic regression. We were examining the probability that an individual would be in the "1" category (compared with the "0" category) as a function of another variable (in this case, sex). This analysis is similar, except we are making three comparisons: the probability of being in each of the last three groups compared with the first group. Thus, we essentially (and literally) have the same results as if we had performed three binary logistic regressions with the same data, changing the "1" group each time to represent the new group. Remember also that at this point, we are not assuming any ordering of the groups. That will come later.

As you can see in Table 6.3, boys and girls do not have the same conditional probabilities of self-reported marijuana use across the categories. For example, girls are more likely to report having never tried marijuana, and boys are more likely to report having tried marijuana 20 or more times. In setting up this multinomial logistic regression analysis, we have four categories in our DV, and we thus will have three regression equations, each comparing the probabilities of being in one group versus being in a comparison group.

Referring back to Table 6.3, we can hand calculate the expected conditional probabilities of each of these groups and then calculate relative risk (RR) ratios. As you can see, we expect boys to be about 19% more likely to admit to trying marijuana 1–2 times, and we also expect little sex difference in the 3–19 times category and for boys to be about 55% more likely to admit to trying marijuana 20 or more times in their lives. As with the previous chapter, we expect to see identical results when these same data are analyzed with statistical software.[2] We also carry forward another theme from previous chapters: that more specificity in the variables is usually better. In Chapter 5, we showed how using a continuous variable produces a substantial improvement in model fit over that same variable reduced to a dichotomous variable. In the exercises for Chapter 5, you were asked to predict whether anyone had ever used marijuana, which was a simplified version of this four-category MJ variable. What we would expect to see in this chapter is a more nuanced and powerful model.

Multinomial Logistic Regression (Unordered) With Statistical Software

In performing this analysis using statistical software, we have $N = 15,364$ valid cases, the majority of which are in group 0. Sex produced a modest, statistically significant effect ($\chi^2_{(3)} = 42.14$, $p < .0001$).[3] This procedure gets more complex when we attempt to interpret the parameter estimates, presented in Table 6.4a.

As you can see, the output is presented in terms of three binary logistic regression equations, each with different intercepts and regression coefficients. The key

---

2   With some small tolerance for rounding errors, of course.

3   This is almost identical to the Pearson or likelihood ratio chi-square values that can be calculated from the contingency table. Using the same data with the binary marijuana variable as the DV, the change in model fit was $\chi^2_{(1)} = 28.89$, $p < .0001$, which is significant but smaller in magnitude. This indicates that there are interesting patterns in the data that are not captured by a simple yes/no variable.

Table 6.4a   Parameter Estimates From Multinomial Logistic Analysis Predicting MJ From SEX

MJ[a]		b	SE	Wald	df	p	Exp(B)	95% CI for Exp(B)	
								Lower Bound	Upper Bound
0 versus 1	Intercept	−2.254	0.040	3,120.515	1	.000			
	SEX	0.194	0.056	11.922	1	.001	1.214	1.087	1.355
0 versus 2	Intercept	−2.554	0.046	3,043.149	1	.000			
	SEX	0.096	0.066	2.115	1	.146	1.100	0.967	1.252
0 versus 3	Intercept	−3.191	0.063	2,600.951	1	.000			
	SEX	0.460	0.082	31.533	1	.000	1.584	1.349	1.861

[a]The reference category is: 0.

SOURCE: National Education Longitudinal Study of 1988 (NELS88) from the National Center for Education Statistics (http://nces.ed.gov/surveys/nels88/).

to deciphering this output is to remember that each one is a comparison, as you would see in a simple binary logistic regression (or an analysis with dummy-coded variables is a comparison between two groups). We essentially have three separate binary logistic regression analyses captured within a single analysis, summarized by a single set of overall model fit statistics. The first compares group 0 with group 1 (never tried marijuana versus tried once or twice). The second compares group 0 with group 2 (tried marijuana on 3–19 occasions). The third compares group 0 with 3 (tried marijuana 20 or more times).

To coherently explain the analysis, I find it simplest to walk through the analysis step by step (or equation by equation). The first comparison is significant, yielding a positive logit and odds ratio of 1.21 for SEX, which is in line with our RR calculation of 1.19. Indeed, if we construct a regression equation and create predicted probabilities, the three regression equations would be

Category 0 versus 1

$$\text{Logit}(\hat{Y}) = -2.254 + 0.194(\text{SEX}) \qquad (6.1a)$$

Category 0 versus 2

$$\text{Logit}(\hat{Y}) = -2.554 + 0.096(\text{SEX}) \qquad (6.1b)$$

Category 0 versus 3

$$\text{Logit}(\hat{Y}) = -3.191 + 0.460(\text{SEX}) \qquad (6.1c)$$

Looking at the first equation (Equation 6.1a), we calculate predicted logits of −2.254 and −2.06 for female and male, respectively. These convert to probabilities of 0.095 and 0.113. Girls have a 0.095 probability of trying marijuana 1–2 times in their life, whereas boys have a probability of 0.113, or about 19% higher probabilities (RR = 1.19), both of which match nicely with the observed probabilities. Note that each equation has a different intercept, which might be unexpected for some readers. In this analysis, the intercept is interpreted as the probability of the event happening when all predictors are 0. Thus, the –2.254 (which converts to a probability of 0.095) is the probability of trying marijuana 1–2 times in a student's life when all predictors (in this case, sex) are 0.

There is no significant effect of sex for the second equation ($p < .146$, and the 95% confidence interval [95% CI] includes 1.0), as you can see in Table 6.4a. This only means that there are not significant differences in the conditional probability of trying marijuana 3–19 times as a function of sex (i.e., boys and girls do not have different probabilities of trying marijuana this many times). This is one reason why the original binary logistic regression analysis from Chapter 5 was less powerful than this analysis: there are heterogeneous sex effects for different groups. Thus, we would interpret this result as no significant sex differences in the probability of trying marijuana 3–19 times. If we perform the same predictions for the second comparison (comparing students who claim to have tried marijuana 3–19 times versus 0 times in their lives), we get predicted probabilities of 0.072 and 0.079 for girls and boys, respectively, which gives us a RR = 1.09, identical to what we calculated in Table 6.4a from the observed probabilities.

Finally, in the third analysis, which compares the probability for girls and boys to have tried marijuana 20 or more times (versus 0 times in their lives), we get a significant and relatively strong effect. Predicted probabilities are 0.040 and 0.061 for girls and boys, respectively, giving us RR = 1.55, again in line with what we calculated in Table 6.4a.

These results indicate that there is a modest sex difference in the probability of trying marijuana, no significant sex difference in moderate users, and a strong sex difference in those students admitting to more extensive use of marijuana. I hope this reinforces the idea that when you have a variable with more specificity, unless you demonstrate that the effects are the same for all groups, it is beneficial to leave the complexity in the equation.

Multinomial Logistic Regression With a Continuous Predictor

As you can see in the first section of this chapter, multinomial logistic regression is a simple extension of binary logistic regression and, if each contrast is taken one at a time, is not much more complex to interpret, particularly when results are converted back to conditional probabilities. Let us look at the same analysis, but this time using z-scored student achievement scores (similar to the analysis we performed previously with the binary marijuana variable). Simply entering the continuous, z-scored variable as a covariate[4] produces a similar analysis. In this case, the change in model fit is significant

---

4   Recall in logistic regression that "covariates" are merely continuous variables and "factors" are categorical. Having too many categorical variables can lead to sparse data and should be avoided. Entering a continuous variable into the model as a factor can lead to substantial issues with sparseness and should be avoided.

Table 6.4b   Parameter Estimates for Multinomial Logistic Regression Predicting MJ From zACH

MJ[a]		b	SE	Wald	df	p	Exp(B)	95% CI for Exp(B)	
								Lower Bound	Upper Bound
1	Intercept	−2.150	0.029	5,625.854	1	.000			
	zACH	−0.241	0.030	65.496	1	.000	0.786	0.741	0.833
2	Intercept	−2.502	0.034	5,535.874	1	.000			
	zACH	−0.200	0.035	33.272	1	.000	0.818	0.764	0.876
3	Intercept	−2.949	0.042	5.040.454	1	.000			
	zACH	−0.292	0.044	44.839	1	.000	0.747	0.686	0.813

[a]The reference category is: 0.

SOURCE: National Education Longitudinal Study of 1988 (NELS88) from the National Center for Education Statistics (http://nces.ed.gov/surveys/nels88/).

$(\chi^2_{(3)} = 129.49$, $p < .0001)$, allowing us to reject the null hypothesis that the model does not improve with the addition of the IV and to interpret the individual effects.[5]

The parameter estimates are presented in Table 6.4b. As you can see, zACH is a significant predictor in each equation, although the magnitude of the effect varies a bit. In general, for all three comparisons, as student achievement increases, the probability of using marijuana decreases. To make this concept easier for your readers, let's calculate predicted probabilities for each comparison at ±1 standard deviation.

You can also see that the intercepts are again different for each analysis. In this case, because we followed best practices and converted the achievement variable to z-scores, the intercept is the probability of trying marijuana 1–2 times (or 3–19 or 20 or more times) for those with average achievement (zACH = 0).

Category 0 versus 1

$$\text{Logit}(\hat{Y}) = -2.150 - 0.241(\text{zACH}) \qquad (6.2a)$$

Category 0 versus 2

$$\text{Logit}(\hat{Y}) = -2.502 - 0.200(\text{zACH}) \qquad (6.2b)$$

Category 0 versus 3

$$\text{Logit}(\hat{Y}) = -3.191 - 0.292(\text{zACH}) \qquad (6.2c)$$

As you can see in Equations 6.2a–6.2c, higher student achievement is generally associated with lower probabilities of marijuana use, although the magnitude of the probabilities varies depending on the comparison. Students are more likely, overall, to try marijuana once or twice (first comparison) than more times (second and third comparison). However, the general pattern is similar for all three comparisons.

---

5   The same analysis in binary logistic regression with only the EVERMJ variable has a change in −2LL of 126.60, which is not substantially different than for this model. This indicates that the effect is relatively consistent across all groups, in contrast with the effect of SEX in the prior example. We will confirm this later in the chapter with an ordinal logistic regression analysis.

Summary of Previous Multinomial Logistic Regression Analysis

A multinomial logistic regression analysis predicted self-reported marijuana use (0 = never tried, 1 = tried 1–2 times, 2 = tried 3–19 times, 3 = tried 20 or more times) from student achievement test scores, which centered via conversion to z-scores prior to analysis (zACH). Entry of zACH significantly improved model fit ($\chi^2_{(3)}$ = 129.49, p < .0001), allowing us to examine the effects of the IV for each group. As you can see in Table 6.4, zACH was a significant predictor for each group, with unstandardized regression coefficients ranging from –0.20 to –0.29 (all p < .0001). These equate to odds ratios of 0.75–0.82. In general, what that means is that as student achievement increases 1 standard deviation, the odds of trying marijuana are 0.75–0.82 that of students who are 1 standard deviation lower. To illustrate these effects graphically, we predicted scores for students with achievement scores 1 standard deviation below the mean and 1 standard deviation above the mean, converting those predicted values to conditional probabilities, and we present them in Figure 6.1. As you can see in Figure 6.1, higher student achievement is associated with substantially lower probabilities of being in any of the three groups who have reported trying marijuana.

Figure 6.1   Predicted Probabilities of MJ Use Given Student Achievement

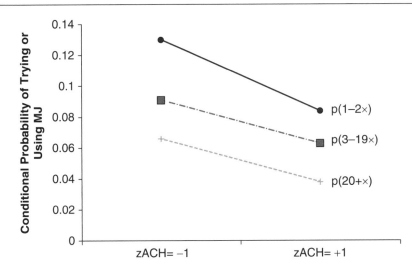

SOURCE: National Education Longitudinal Study of 1988 (NELS88) from the National Center for Education Statistics (http://nces.ed.gov/surveys/nels88/).

Multinomial Logistic Regression as a Series of Binary Logistic Regressions

I have made the assertion that multinomial dependent variables, like dummy-coded IVs, are essentially a series of binary comparisons. Like dummy-coded IVs, you also can split up the groups for separate analyses and replicate the results. For example, if

Table 6.5a   Simple Binary Logistic Regression With Only Groups 0 and 3 From MJ

		B	SE	Wald	df	p	Exp(B)	95% CI for Exp(B)	
								Lower	Upper
Step 1ª	zACH	−0.289	0.043	44.391	1	.000	0.749	0.688	0.815
	Constant	−2.949	0.042	5,041.433	1	.000	0.052		

ªVariable(s) entered on step 1: zACH.

Table 6.5b   Simple Binary Logistic Regression With Only Groups 0 and 1 From MJ

		B	SE	Wald	df	p	Exp(B)	95% CI for Exp(B)	
								Lower	Upper
Step 1ª	zACH	−0.238	0.030	64.707	1	.000	0.788	0.744	0.835
	Constant	−2.150	0.029	5,627.327	1	.000	0.117		

ªVariable(s) entered on step 1: zACH.

I select just the 0 and 3 group and perform a binary logistic regression analysis, I get a change in −2LL or 46.04 (which is significant at $p < .0001$) and the following results in Table 6.5.

Comparing the results in Table 6.5a with those in Table 6.4, you can see that the results are generally similar to within a couple of decimal points. I performed the same analysis with only groups 0 and 1 in the data set and also obtained similar results to the multinomial analysis (as you can see in Table 6.5b).

Data Cleaning and Multinomial Logistic Regression

Why is this important? First, for conceptual clarity we can understand that multinomial logistic regression is not completely different from binary logistic regression. It is merely an extension of the original. However, depending on what software you use for statistical analysis, the routines are often implemented differently. For example, when performing multinomial logistic regression in SPSS, there is no option to save residuals or indices of leverage or influence. Thus, if I am performing this type of analysis, I usually compute separate binary logistic regressions solely for the purpose of identifying troublesome cases for data cleaning. Once the data cleaning (if any) is performed, the final multinomial analysis should be performed. This final analysis gives us the likelihood ratio tests that are appropriate and the final parameter estimates that would be reported.

Testing Whether Groups Can Be Combined

Let us start with the concept of the likelihood ratio test (change in −2LL evaluated as a chi-square). In other types of analyses, like structural equation modeling, we use these types of likelihood ratio tests for evaluating hypotheses about different and competing models. Technically, we are also doing that with every analysis, but the competing models we are comparing are an empty model (a model with no IVs or with all coefficients constrained to zero) and the model with our variable(s) in it. What if we can use this procedure to test other interesting hypotheses regarding nested models? To note, this does not work for non-nested models or models using different samples.

Let us imagine that we want to test whether it is legitimate to keep our DV, marijuana use, as the original four categories, or whether it is acceptable to combine the three "used marijuana" groups into one. There are many reasons one might want to do this, including improving the simplicity of the model, combining two relatively small groups, or testing substantive hypotheses (e.g., whether keeping the groups separate improves the model fit).

Technically, to combine groups is to hypothesize that all regression coefficients for the two groups are equal. Thus, referring to the regression equations for the three groups in our analysis (Table 6.4, above), combining the last two groups would argue that the constants are equal ($b_{01} = b_{02} = b_{03}$) and the coefficients for zACH are equal ($b_{11} = b_{12} = b_{13}$). If any of these conditions are not tenable in the data, we will see a significant difference between overall model fit. In the language of structural equation modeling, we are constraining these parameters to be equal across groups. Although these coefficients *seem* to be different (e.g., the slopes in logits for the three groups range from −0.200 to −0.292), the question is whether they are *significantly* different. In our sample, we have tremendous power to detect small differences, so I suspect that the result of this analysis will be a significant difference in model fit. However, in smaller samples, small differences in coefficients will not be significantly different.

Setting up the analysis to test these hypotheses is relatively straightforward. These are nested models, with the constrained model nested within the nonconstrained model. First we make sure that neither of the two groups we are examining for combination is the reference group (in our example, 0 remains the reference group), and we perform an analysis with the groups not combined (not constrained to equality) and then with the groups combined, comparing the −2LL or deviance. If there is no significant change in goodness of fit between the nonconstrained and constrained models, we can plausibly conclude that the two groups are not significantly different (at least in terms of the variables modeled) and that it is defensible to combine them. On the other hand, if model fit is harmed by this constraint (if −2LL for the constrained model is significantly higher than −2LL for the nonconstrained model), then the hypothesis that the coefficients are equal across groups is not tenable and is rejected.

Let us continue with the example of MJ and zACH. For the unconstrained model, the −2LL was 13,120.935, which improved to 12,991.445 once zACH entered the equation ($\chi^2_{(3)} = 129.49$, $p < .0001$). If we combine the three groups and recompute this analysis, we have a $\chi^2_{(1)} = 126.603$ ($p < .0001$), which is 2 *df* smaller and 2.887 smaller. A chi-square of 2.887 with 2 *df* would not be significant ($p < .24$ using the Excel =CHIDIST function).

An alternative method for testing whether two groups could be legitimately combined would be to select the two categories in question and perform a binary logistic regression with only those two groups as the DV. If there are significant effects of any of the IVs in this context, this implies that the two groups differ significantly and thus are not eligible for combination. For example, if I select only groups 1 and 3 and perform a binary logistic regression analysis predicting these two outcomes, I see the results shown in Table 6.6.

These results indicate there is no significant difference in the effect of zACH on marijuana use across these two groups. However, there are different intercepts, meaning that the two groups might not have different relationships between achievement and marijuana use, but they do have significantly different base rates of usage. This leads me to conclude that it would not be defensible or advisable to combine them for your analysis. However, this does not mean that these groups are always the same. As you can see in Table 6.6b, the same analysis using SEX as the IV produces a significant effect for both slope and intercept, again leading us to conclude that it would not be defensible to combine these groups.

Thus, in the latter case, it would not be defensible to combine these groups. Note that one must have sufficient power to detect differences in order to perform any of these analyses, as well as large enough groups to reasonably detect differences. If one has only two or three cases in each group, there is obviously not enough power to detect differences, and groups that small should be removed from the analysis rather than risking the potential of inappropriately combining them.

Table 6.6a   Using Binary Logistic Regression to Test Whether Two Groups Can Be Combined Without Harming the Model Fit

Omnibus Tests of Model Coefficients

		Chi-Square	df	p
Step 1	Step	1.059	1	.304
	Block	1.059	1	.304
	Model	1.059	1	.304

Variables in the Equation

		B	SE	Wald	df	p	Exp(B)	95% CI for Exp(B) Lower	95% CI for Exp(B) Upper
Step 1[a]	zACH	−0.054	0.053	1.055	1	.304	0.947	0.854	1.050
	Constant	−0.800	0.049	269.501	1	.000	0.449		

[a]Variable(s) entered on step 1: zACH.

Table 6.6b   Using Binary Logistic Regression to Test Whether Two Groups Can Be Combined Without Harming the Model Fit

Omnibus Tests of Model Coefficients

		Chi-Square	df	p
Step 1	Step	7.765	1	.005
	Block	7.765	1	.005
	Model	7.765	1	.005

Variables in the Equation

		B	SE	Wald	df	p	Exp(B)	95% CI for Exp(B) Lower	95% CI for Exp(B) Upper
Step 1[a]	SEX	0.267	0.096	7.718	1	.005	1.306	1.082	1.576
	Constant	−.937	0.072	167.847	1	.000	0.392		

[a]Variable(s) entered on step 1: SEX.

Ordered Logit (Proportional Odds) Model

Another extension of the binary logistic regression model is the ordered logit, or proportional odds model. This model is designed to deal with ordinal (ordered, ranked) variables such as the marijuana variable we have been exploring through the first part of this chapter. Multinomial logistic regression can handle this variable just fine, as you have seen, but it is most specifically designed for handling *nonordered* variables (e.g., what type of diet one is going to attempt, which method of birth control an individual will use, what college major a student will matriculate into, etc.). Although I made a passing argument that the marijuana classification may represent qualitatively different groups (nonusers, people who tried it but never used it, occasional user, and frequent user), the multinomial analysis does not take advantage of the fact that in this variable, categories *might be* ordered. This variable meets the definition of an ordinal variable, in which there is rank ordering but there are no consistent intervals between groups.

In the social sciences, one of the most common applications of this model *could be* analysis of Likert-type scales, where responses are on a scale such as in Table 6.7.

Although these scales are often treated as interval or ratio variables by scholars, in reality we have no idea how much distance is between "strongly disagree" and "agree" and whether that is the same distance as between "neutral" and "agree." Another example of this type of scale is from the same National Health Interview Survey data we have been using for our diabetes analyses (Table 6.7b).

There are several types of ordinal logit models, each discussed thoroughly in Menard (2010). Although there are interesting choices in the literature, it seems the most commonly implemented choice in the statistical software is the *cumulative logit model*. This ordinal logistic regression model essentially performs a series of cumulative binary logistic regressions comparing all groups below a particular threshold with all groups above a threshold. For example, with the marijuana variable, the first binary comparison would be group 0 versus all other groups (1, 2, and 3). The next comparison would be groups 0 and 1 versus all other groups (2, 3). The third and final comparison would be groups 0, 1, and 2 versus 3. In this way, ordinal logistic regression provides three

Table 6.7a   Example Likert Scale

Value	Description
1	Strongly disagree
2	Disagree
3	Neutral
4	Agree
5	Strongly agree

Table 6.7b   In the Past 30 Days, How Often Have You Felt So Sad That Nothing Could Cheer You Up?

Value	Description
1	ALL the time
2	MOST of the time
3	SOME of the time
4	A LITTLE of the time
5	NONE of the time

estimates for the effect that each IV has on the response. Although this is interesting, because we are assuming that there is a continuous latent variable underlying the ordinal variable we are modeling, we would like to be able to summarize the model with a single set of parameter estimates that summarizes the effect of each predictor variable on the DV. To do this, we must make two assumptions: that the DV is *ordered*, and that the relationship of each predictor is constant across all possible comparisons for the DV. In essence, this is a systematized approach to some of the efforts to explore whether groups could be combined.

Other options (e.g., the continuation ratio logit model, or the adjacent category model) have slightly different assumptions about the nature of the data and answer slightly different questions. Because they are relatively rare compared with the cumulative logit model (which is also relatively rare in the behavioral sciences at this point, but I hope that will change as it gets more exposure in books such as this), I will focus on it in this chapter but encourage readers to explore other possibilities if their software supports those options.

The general form of this analysis is slightly different than prior logistic models and can be implemented differently across different software packages.[6] In general, the linear model for the cumulative logit model is presented in Equation 6.3a:

$$\ln(p_{(Y \le i)} / p_{(1-Y \le i)}) = b_0 + b_1 X_1 \tag{6.3a}$$

In this equation, we model the log of the odds that a case will be at or below a particular threshold (e.g., for the marijuana variable, at or below 0 [never tried MJ], then at or below 1 [tried MJ 1–2 times or never], and then at or below 2 [tried MJ either never, 1–2 times, or 3–19 times]). The intercept is the log of the odds of being in a particular group *or lower* when scores on the other variable(s) are zero. One interesting thing that Grace-Martin (2013; see also Menard, 2010; Norušis, 2012) points out is that SPSS handles ordinal logistic regression differently than in Equation 6.3a (which is how some programs, like SAS, handle it). In SPSS, for example, the equation is slightly different, as in Equation 6.3b:

$$\ln(p_{(Y \le i)} / p_{(1-Y \le i)}) = b_0 - (b_1 X_1 \text{ and any other terms}) \tag{6.3b}$$

SPSS and some other popular statistical packages change the plus sign to a minus between the intercept and the rest of the terms to reverse the meaning of the parameter estimates in order to be more intuitive and in line with interpretations of other types of analyses (e.g., binary or multinomial logistic). Now, with this change, positive slopes indicate higher probabilities of moving into the next higher category. Without the minus sign, positive slopes indicate higher probabilities of moving into a lower category, which is somewhat confusing.

---

6   I thank Karen Grace-Martin for her blog entry "Opposite Results in Ordinal Logistic Regression: Solving a Statistical Mystery" (http://www.theanalysisfactor.com/ordinal-logistic-regression-mystery/), which highlights how different software packages handle ordinal regression differently.

Assumptions of the Ordinal Logistic Model

One important assumption of this model is that the DV is in order. In other words, like the marijuana example, increasing numbers indicate more of something. We often do not know how much more of something, but it is more. If this basic assumption is not met, then multinomial (unordered) analyses are appropriate.

Another important assumption of this model is that of the "proportional odds" assumption, which states that the effect of any (and all) IVs is constant across all groups. Therefore, using our marijuana variable as a continuing example, the effect of student achievement on marijuana use should be the same when comparing groups 0 and 1 (nonusers versus those who tried it 1–2 times) as when comparing groups 2 and 3 (used 3–19 times versus used 20 or more times), and when comparing groups 0 and 3. This assumption is important because, unlike multinomial logistic regression, *the goal of ordinal logistic regression is to create a single estimate* that predicts the probability of being in the next higher group as a function of a change in the IV(s) regardless of which group transition we are talking about. In essence, the ordinal logistic regression model is attempting to model the latent underlying continuous variable rather than a variable that has a series of groups or transitions.

Most statistical packages will test this assumption for you. For example, this is called the "Score test for the proportional odds assumption" in SAS, whereas it is called the "test of parallel lines" in SPSS. Examples of these tests are presented in Table 6.8.

Specifically, the hypotheses being tested by the test of parallel lines are

H_0: Difference in slope coefficients across groups = 0

H_a: Difference in slope coefficients across groups ≠ 0

Thus, in order to meet the assumption, we seek *not* to reject the null hypothesis. As you can see in Table 6.8, this test is not significant ($p < .18$), and we can thus conclude that this analysis would meet the assumption. One problem with this test is that it is very sensitive, especially in large samples in which there are several predictors in the model. You have to be thoughtful in interpreting these and other statistics based on chi-square when large samples are involved.

Table 6.8   Test of Parallel Lines From MJ and zACH Analysis

Test of Parallel Lines[a]

Model	–2LL	Chi-Square	df	p
Null hypothesis	12,995.159			
General	12,991.715	3.443	2	.179

The null hypothesis states that the location parameters (slope coefficients) are the same across response categories.
[a]Link function: Logit.

SOURCE: National Education Longitudinal Study of 1988 (NELS88) from the National Center for Education Statistics (http://nces.ed.gov/surveys/nels88/).

Note the footnote in Table 6.8. In SPSS, you can choose several different link functions for this analysis, including probit, for those of you interested in exploring this link function more thoroughly. The rest of the analysis is relatively straightforward if these assumptions are met. In this case, the effects of each binary comparison discussed above are averaged to provide a single parameter estimate for each IV.

Interpreting the Results of the Ordinal Regression

When these assumptions are met, we get the usual model fit statistics that we have seen with other logistic type analyses as well as individual parameter estimates. In this case, the $-2LL$ is 12,995.16 ($\chi^2_{(1)} = 125.78$, $p < .0001$; note that this is close to the $\chi^2 = 129.49$ from the multinomial analysis when slopes were free to vary). Thus, we know that the addition of zACH and zSES improves model fit (just as we saw in the multinomial analyses earlier in the chapter).

As you can see, there are two pieces of information in the output: the intercepts for each increment and the averaged parameter estimate for zACH, all of which were significant. We can use this information to constitute regression equations for each comparison (Equations 6.4a–6.4c):

Probability of being in the non-MJ group:

$$\ln(p_{(Y \le 0)} / p_{(1-Y \le 0)}) = 1.383 - (-0.236(zACH)) \tag{6.4a}$$

Probability of being in the non-MJ or used 1–2 times groups:

$$\ln(p_{(Y \le 1)} / p_{(1-Y \le 1)}) = 2.120 - (-0.236(zACH)) \tag{6.4b}$$

Probability of being in the non-MJ or used 1–2 times or used 3–19 times groups

$$\ln(p_{(Y \le 2)} / p_{(1-Y \le 2)}) = 3.132 - (-0.236(zACH)) \tag{6.4c}$$

Interpreting the Intercepts/Thresholds

Keep in mind that in this analysis, the comparisons are not between one group and another group; rather, they are between cumulative groups of groups. For each threshold for MJ, the intercept or threshold is the log of the odds of being in that group *or below* given the variables in the equation (estimated at IV = 0, or the mean of zACH). Thus, the first threshold (MJ = 0) is 1.383 in log units, which equates to a probability of about 0.799. This also happens to be within a reasonable rounding error of the percentage of people in the MJ = 0 category (79.8%). The second threshold of MJ = 1 is 2.12, which is the probability in log units of being in group 1 or 0. This converts to a probability of 0.893, which is close to the cumulative percentage of people in groups 0 or 1 (89.1%). The same applies for the other threshold.

The difference between simple counts and predicted probabilities is due to the difference between the two metrics: the conditional probability is estimated for students who are average in achievement, whereas the counts are just everyone in that group.

As you might imagine, it is not necessarily satisfying to be able to assign predicted probabilities to groups of outcomes, as we have above. Each threshold is interpreted as an intercept as in any logistic regression equation, and any threshold can also be converted to cumulative probabilities. As you can see in Table 6.10a, I have taken all of the thresholds from Table 6.9 and converted them to cumulative probabilities (i.e., the probability that an individual with zACH = 0 will be in group 0, or group 0 or 1, or group 0, 1, or 2, etc.). Cumulative probabilities can be broken down into the type of conditional probabilities we are used to seeing in logistic regression models by merely subtracting probabilities. For example, if we have p(MJ = 0, 1) and p(MJ = 0), we can calculate p(MJ = 1) by subtracting the latter from the former. Using this methodology, I calculated the probability of being in any given group when zACH is 0 (average).

Table 6.9   Results of Ordinal Logistic Regression Analysis Predicting MJ From zACH

Model Fitting Information				
Model	−2LL	Chi-Square	df	p
Intercept only	13,120.935			
Final	12,995.159	125.777	1	.000

Link function: Logit.

		Parameter Estimates							
		Estimate	SE	Wald	df	Sig.	95% CI	
							Lower Bound	Upper Bound
Threshold	[MJ = 0]	1.383	0.021	4,470.068	1	.000	1.342	1.423
	[MJ = 1]	2.120	0.027	6,292.968	1	.000	2.068	2.172
	[MJ = 2]	3.132	0.041	5,772.399	1	.000	3.051	3.212
Location	zACH	−0.236	0.021	122.449	1	.000	−0.277	−0.194

Link function: Logit.

SOURCE: National Education Longitudinal Study of 1988 (NELS88) from the National Center for Education Statistics (http://nces.ed.gov/surveys/nels88/).

Table 6.10a   Converting Cumulative Probabilities to Group Probabilities When $b_1 = 0$

	Threshold (Intercept)	Cumulative Probability		Group Probability
$P_{(MJ \leq 0)}$	1.383	0.799	$P_{(MJ = 0)}$	0.799
$P_{(MJ \leq 1)}$	2.12	0.893	$P_{(MJ = 1)}$	0.093
$P_{(MJ \leq 2)}$	3.132	0.958	$P_{(MJ = 2)}$	0.065
$P_{(MJ \leq 3)}$	—	1.00	$P_{(MJ = 3)}$	0.042

Interpreting the Parameter Estimates

The effect of zACH is −0.236 (which is close to the unweighted average of the three regression coefficients in Table 6.4), which means that as achievement increases 1 standard deviation, the probability that a student would be in the next higher group of marijuana use decreases. In other words, as we saw in previous analyses, students with higher achievement test scores are less likely to be in the next highest level of marijuana use. Because the proportional odds assumption is met, this effect is uniform across all levels of the DV and represents an average effect across all groups.

Like any other logistic regression analysis, you can use the regression equation to predict conditional probabilities for each group (Norušis, 2012, pp. 69–89), with some slight modification acknowledging that each threshold is for all groups less than the threshold, rather than just for that group, as discussed above. Thus, we can replicate the type of multinomial analyses we did earlier in the chapter and predict how the probabilities would be affected at each of the three nonreference groups.

To do this, we will use Equations 6.4a–6.4c to predict probabilities for low achievement (e.g., zACH = −2) and high achievement (e.g., zACH = 2) and then use the same subtraction calculations to find the probabilities of being in each of groups 1, 2, or 3 across a range of achievement levels. These results are presented in Table 6.10b and graphically in Figure 6.2.

If you compare Figures 6.1 and 6.2, you should see some marked similarity. This is because when the assumption of proportional odds is met, multinomial and ordinal regression will produce very similar results, as meeting this assumption means that

Table 6.10b   Converting Cumulative Probabilities to Group Probabilities for High and Low Achievement Students

	Predicted Logit	Cumulative Probability		Group Probability
Low achievement				
$P_{(MJ \leq 0)}$	0.911	0.713	$P_{(MJ = 0)}$	0.713
$P_{(MJ \leq 1)}$	1.648	0.839	$P_{(MJ = 1)}$	0.125
$P_{(MJ \leq 2)}$	2.660	0.944	$P_{(MJ = 2)}$	0.096
$P_{(MJ \leq 3)}$	0	1.00	$P_{(MJ = 3)}$	0.065
High achievement				
$P_{(MJ \leq 0)}$	1.855	0.865	$P_{(MJ = 0)}$	0.865
$P_{(MJ \leq 1)}$	2.592	0.930	$P_{(MJ = 1)}$	0.066
$P_{(MJ \leq 2)}$	3.604	0.974	$P_{(MJ = 2)}$	0.043
$P_{(MJ \leq 3)}$	—	1.00	$P_{(MJ = 3)}$	0.026

Figure 6.2 Results of Ordinal Regression Predicting Marijuana Use From Achievement Scores

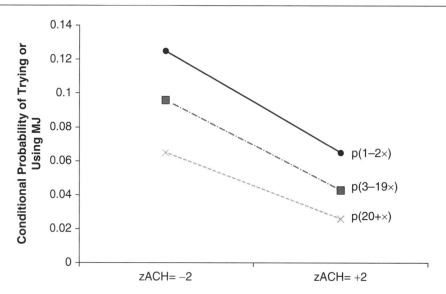

the lines are not significantly different in slope. It is only when this restrictive assumption is not met that the two results will be significantly different (and ordinal regression, of course, should not be used in that case).

Data Cleaning and More Advanced Models in Ordinal Logistic Regression

As with multinomial logistic regression models, those desiring to perform data cleaning (and I hope by now that includes ALL of you) need to perform those data cleaning analyses by performing a series of binary logistic regression analyses, selecting the cases for analysis, and then performing the ordinal logistic regression analysis.[7] In this case, you would constitute three models that compare the cumulative groups: 1 versus 2, 3, and 4; 1 and 2 versus 3 and 4; and 1, 2, and 3 versus 4. Those cases that are inappropriately influential can be removed from the final ordinal logistic regression analysis.

The Measured Variable Is Continuous; Why Not Just Use OLS Regression for This Type of Analysis?

If the latent variable underlying the DV is truly continuous, why not use OLS regression predicting an imperfect outcome variable? The most obvious answer is that having data with nonequal intervals violates some of the most basic assumptions of OLS

---

7   This is true for SPSS at the time this chapter was written. Other software packages may contain more well-developed data cleaning abilities.

regression. For many variables, like rank orderings or Likert-type scales, we cannot assume that the difference between 0 and 1 is the same as between 1 and 2, or 2 and 3. In this specific example of marijuana use, the difference between 0 and 1 is both a small difference in terms of count (1–2 occasions versus 0 occasions) and perhaps monumental in terms of psychology or social environment. The next increment is also different (1–2 versus 3–19, and the difference psychologically or behaviorally between someone who tried marijuana once or twice and someone who is more frequently using marijuana), and so on.

More concretely, the residuals are far from normally distributed, as you can see in Figure 6.3.

A Brief Note on Log-Linear Analyses

Many of you will hear of log-linear analyses and wonder how they relate to logistic regression. Log-linear analyses are part of the generalized linear model (GLM), and they are similar to logistic regression analyses in that they use logarithms of frequency counts to convert unordered categorical DVs for analysis in a more general

Figure 6.3   Residuals From an OLS Regression Predicting MJ From zACH

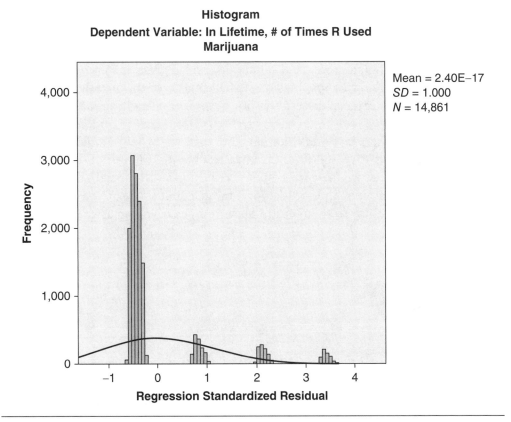

framework than available in other methods of analysis. Log-linear analyses are a special case of the GLM in which all variables (IVs and DV) are unordered categorical variables. Because you are now familiar with linear models in which the DV is either binary or multinomial (unordered), you can imagine a generalization of these models you are familiar with to only categorical IVs. This is essentially a log-linear model. These will be covered in Chapter 12.

SUMMARY AND CONCLUSIONS

Although less common than binary logistic regression, multinomial and ordinal logistic models are available in most commonly used statistical packages and are easily applied when appropriate. Furthermore, they are more appropriate than models often used in the literature. For example, using OLS regression on Likert-type scales is questionable at best and unnecessary with the availability of ordinal logistic models. In addition, simplistic analyses that some authors resort to (i.e., ANOVA analyses rather than multinomial logistic regression) are limited, in that they have difficult assumptions to meet and cannot provide the same level of inference.

In chapters to come, we will see the power of regression models in the ability to examine nonlinear effects, multiple predictors, interactions between multiple predictors, and so forth. All of these valuable characteristics of regression models apply to each of the examples of the GLM we have explored to date (OLS regression, ANOVA or regression with polytomous IVs, binary logistic regression, and multinomial/ordinal logistic regression). Those of you who have mastered the material to this point are well equipped to explore these types of effects within multinomial and ordinal models.

Note also that there are multinomial and ordinal probit models, as briefly alluded to in the narrative and footnotes in this chapter. I have not delved deeper into these models because they are even less common in the social and behavioral sciences than multinomial and ordinal logit models, but those of you interested in exploring them should find them easily mastered given your mastery of logit and OLS models.

ENRICHMENT

1. Download the MJ data used in the examples from this chapter and replicate the results.

2. Using the above NELS88 data, predict EDUCATIONAL PLANS (BYPSEPLN) from student achievement (zACH).

a. Educational plans is coded as follows:

    i. 1 = Won't finish high school
    ii. 2 = Will finish high school
    iii. 3 = Vocational/trade school after high school
    iv. 4 = Will attend college but may not finish
    v. 5 = Will finish college
    vi. 6 = Will seek higher degree college

b. Category 5, completion of college, is the largest group and thus should be set as the reference group.
c. Test whether it is legitimate to combine the first two categories (won't finish high school and will finish high school) for this analysis.
d. Test whether this analysis would be appropriate for ordinal logistic regression (with just zACH in the model). If so, perform an appropriate ordinal logistic regression analysis and interpret the results.
e. Write up conclusions in APA format.
f. Perform the same analyses with socioeconomic status (zSES).

3. Download the NHIS_SM data set and predict the DV SAD ("In the past 30 days, how often have you been so sad that nothing could cheer you up?") from SEX (0 = F, 1 = M), age, ER_USE (number of emergency room visits during the last 12 months), and/or average hours of sleep a night.

a. Is an ordinal logistic regression analysis appropriate for these data? If so, interpret results.
b. If not, perform multinomial logistic regression analysis using the same IVs and DV.
c. Write up results in APA format.

REFERENCES

Grace-Martin, K. (2013). Opposite results in ordinal logistic regression: Solving a statistical mystery. Retrieved from http://www.theanalysisfactor.com/ordinal-logistic-regression-mystery/

Menard, S. W. (2010). *Logistic regression: From introductory concepts to advanced concepts and applications.* Thousand Oaks, CA: SAGE.

Norušis, M. J. (2012). *IBM SPSS Statistics 19 advanced statistical procedures companion.* Upper Saddle River, NJ: Prentice Hall.

SIMPLE CURVILINEAR MODELS

7

Advanced Organizer

In previous chapters, we talked about the assumption in almost any type of linear modeling that the model is, well, *linear*. Whether we are talking about ordinary least squares (OLS) regression or logit or probit models, linear is in the label. However, I think that in many areas of science, this assumption may not be tenable or even desirable. I believe that if we routinely looked for curvilinear relationships, we would find many. In fact, while writing this chapter, I had to explore surprisingly few examples to produce the curvilinear results shown herein. The fact of the matter is that curves are everywhere, and I hope this chapter encourages you to begin looking for them in your data. You will find that this is not terribly painful, and it can produce much more nuanced and interesting results.

In this chapter, we will briefly review the concept of curvilinearity, how to test for curvilinearity more formally, how to account for curvilinearity in your regression analyses, and how to graph curvilinear effects.[1] At the end of the chapter, I will also digress into a brief section on how to have even more fun with curvilinear effects if you know a little calculus (I think everyone should, by the way).

In this chapter, we will cover

- Basic concepts in curvilinear effects
- Curvilinear effects in OLS regression
- Curvilinear effects in logistic regression
- Examples of American Psychological Association (APA)–compliant summaries of analyses

Guidance on how to perform these analyses in various statistical packages will be available online at study.sagepub.com/osbornerlm.

---

1 I believe graphical representations of complex findings like curvilinear effects and interaction effects (where found) are critical to effectively communicating the results of research to the audience of interest.

Zeno's Paradox, a Nerdy Science Joke, and Inherent Curvilinearity in the Universe . . .

My high school science teacher, Larry Josbeno, was not only a brilliant teacher; he was also fond of lousy physics jokes. One of his favorites was related to Zeno's paradoxes and was a variant of what is apparently a classic mathematical joke, which I paraphrase below (although I cannot replicate his spot-on delivery)[2]:

At a high school dance, a group of boys are lined up on one wall of a dance hall, and an equal number of girls are lined up on the opposite wall 10 meters apart. Both groups are then instructed to advance toward each other by one-half the distance separating them every 10 seconds (i.e., if they are distance d apart at time 0, they are d/2 at time = 10, d/4 at time = 20, d/8 at time = 30, and so forth). A mathematician, a physicist, and an engineer are asked when the boys and girls would meet at the center of the dance hall. The mathematician said they would never actually meet because the series is infinite. The physicist said they would meet when time equals infinity. The engineer said that within 1 minute, they would be close enough for all "practical" purposes.

Enthusiastic adolescent laughter ensued, predictably. Thank you, Mr. Josbeno! But what does this have to do with curvilinear effects in regression? Like many things in life, if we were to explore the relationship between time and distance between our girls and boys, the relationship is not linear, as Figure 7.1 shows. And curvilinearity is the topic of this chapter!

Figure 7.1   Zeno's Paradox in the High School Dance

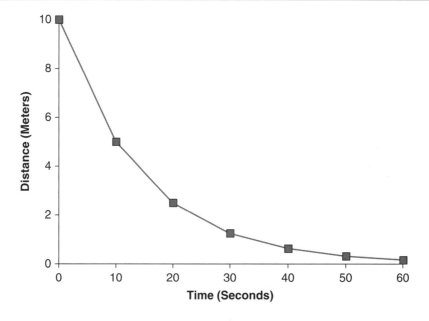

---

2   See "Zeno's paradoxes" by Paul Field and Eric W. Weisstein, available on the Wolfram MathWorld website (http://mathworld.wolfram.com/ZenosParadoxes.html).

A Brief Review of Simple Algebra

Curves in algebra are relatively simple to understand. There are an infinite number of specialized types of curves, but we can advance the field of statistical methods greatly by beginning to apply some simple principles. We have seen many examples of lines in which we have a Y and an X, such as in Equation 7.1, where we have an intercept (when $X = 0$, $Y = 15$) and a slope (for every increment of X, Y will decrease by 8):

$$Y = 15 - 8X \qquad\qquad (7.1)$$

When you don't see a superscript or power next to a variable, that variable is assumed to be raised to the first power (e.g., X^1). Any variable raised to the first power is the same as that variable (i.e., $X^1 = X$), just as multiplying any variable by 1 is the same value ($1X = X$). This is why 1 is called the multiplicative identity in mathematics.[3] You can see the line graphed in Figure 7.2, looking remarkably like many regression graphs we have seen.

When we have a variable raised to a power, we get a curve. For example, if we have a squared term in the equation (this then becomes a quadratic equation), as in Equation 7.2, we get a curve with one inflection point (or change in direction), which is graphed in Figure 7.3.

$$Y = 15 - 8X + 2X^2 \qquad\qquad (7.2)$$

As you can see in Figure 7.3, there is one point, at $X = 2$, where the slope of the curve is 0 (flat), the point where the slope changes from negative to positive. That is the only

Figure 7.2   Graph of Equation 7.1

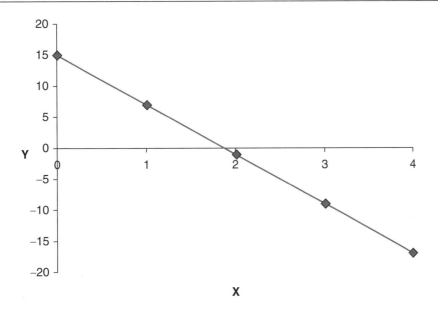

***

3   0 is the additive identity: $X + 0 = X$. Raising anything to the 0 power makes it equal to 1 (i.e., $X^0 = 1$).

inflection point for the curve. If you graph it infinitely in each direction, the curve will have no other inflection point.

If we examine an equation with a cubic term (a variable raised to the third power), we then get two inflection points. For example, Equation 7.3, graphed in Figure 7.4, has X raised to the third power and will therefore have two inflection points.

$$Y = X + X^2 - 0.5X^3 \tag{7.3}$$

Figure 7.3   Graph of Equation 7.2

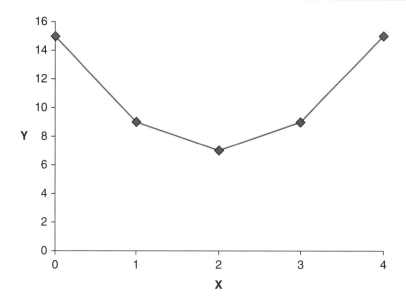

Figure 7.4   Graph of Equation 7.3

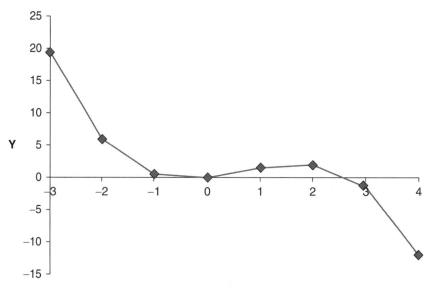

At the end of this chapter, I will review some more fun you can have with curves using calculus, but for now, the general principle is that an equation with a variable raised to a particular power (k) will produce a curve with $k - 1$ inflection points. Keep in mind that lines extend infinitely in both directions, and the inflection points you expect may not occur within the observed range.

One example in this chapter will return to the American Association of University Professors (AAUP) data[4] from earlier chapters and the relationship between size of a university (total number of faculty, NUM_TOT, z-scored) and associate professor salary (SAL_AP, measured in hundreds of dollars). As we saw in Chapters 2 and 3, these data are problematic due to non-normality of residuals, heterogeneity, and so forth. This can sometimes be caused in models that fail to include curvilinearity when it is present in the data.

In Figure 7.5, we see the plot of zPRED versus zRESID (standardized predicted values on the X axis and standardized residuals on the Y axis) from the simple OLS regression predicting SAL_AP from NUM_TOT. Graphs like these can be used to evaluate linearity and homogeneity assumptions, and this one seems to suggest that there might be significant unmodeled curvilinearity in the data.

Figure 7.5   Plot of zPRED Versus zRESID From the AAUP Data Analysis Predicting Salary of Associate Professors From Size of the Institution

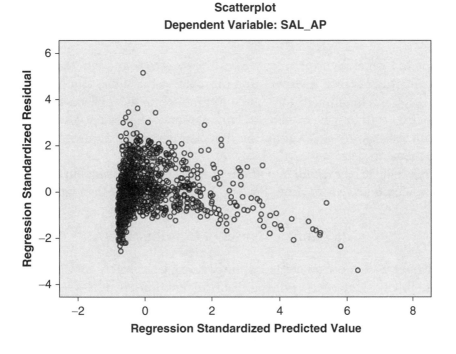

4   Lock, R. (organizer). What's What Among American Colleges and Universities: American Association of University Professors (AAUP) faculty salary data (for the 1995 Data Analysis Exposition, sponsored by the Statistical Graphics Section of the American Statistical Association).

We will explore another example within logistic regression to assess the curvilinear effect of age (AGE) on the probability of being diagnosed with diabetes, using data from the National Health Interview Survey of 2010 (NHIS2010).[5]

Hypotheses to Be Tested

In Equation 7.4, we have an example of a regression equation with a quadratic term in it. In this example, b_2 is the portion of the effect of X_1 on Y that is quadratic, b_3 would be the cubic aspect of the effect, and so on:

$$\hat{Y} = b_0 + b_1 X_1 + b_2 X_1^2 + b_3 X_1^3 + \cdots + b_K X_1^k \tag{7.4}$$

If we fail to test for curvilinearity, we are implicitly asserting that b_2 to b_k are equal to 0. Sometimes they are—there are many linear relationships in the world. However, it is poor practice to simply assert an effect without testing for it. Thus, when we engage in curvilinear analyses, we explicitly test hypotheses about those aspects of the effect that are nonlinear. For example, we can test

$$H_0: b_2 = 0$$

$$H_a: b_2 \neq 0$$

$$H_0: b_3 = 0$$

$$H_a: b_3 \neq 0$$

and so forth. I generally explore curves in the quadratic and cubic range, and not higher, primarily because I suspect that our data in the social and behavioral sciences are not precise enough to confidently model (and replicate) more complex curves. I also begin to have trouble explaining more complex curves rationally. But there is no reason why they cannot be modeled if you can explain them and can replicate them.

If we reject any of these null hypotheses, then we are declaring our assumption of linearity voided, and we should model the curvilinearity rather than ignore it.

Illegitimate Causes of Curvilinearity

In this chapter, I am most concerned with modeling legitimately curvilinear relationships. As I briefly mentioned in Chapter 2 on assumptions, there are several potential sources of curvilinearity that are not, in my mind, legitimate: model misspecification (omission of important variables), converting interval or ratio variables to ordinal variables with unequal intervals, and uncleaned data (i.e., containing influential data points).

---

5   Public data are available from http://www.cdc.gov/nchs/nhis/nhis_2010_data_release.htm.

Model Misspecification: Omission of Important Variables

When discussing the assumption that we have correctly specified the model, we introduced the assumption that we have included all relevant and important variables in the model and have not included extraneous variables. It is possible that omission of important variables can lead to either of these situations. Thus, theory and prior research should help guide you in designing research that accounts for important variables (i.e., prior academic experiences in studying education, prior health events in studying current health status).

Poor Data Cleaning

As also mentioned previously, I have occasionally seen curvilinear effects arise (or masked) merely because of poor data cleaning—a prominent outlier in one range of the data where there are few other cases can pull the regression line in that area out of linearity, leading to the appearance of a curvilinear effect when in fact it is merely poor data cleaning. I suppose it is also possible for highly non-normal data to have the appearance of curvilinearity.

Thus, I would argue that prior to examining data for curvilinear effects, you should be sure that the appropriate variables are modeled in the equation and that you have done due diligence in data cleaning. Once you have satisfied those basic steps (which should probably be part of any analysis regardless of whether curvilinearity is suspected), it is time to explore whether curvilinear effects exist in the data. The end of this chapter includes examples of how data cleaning can reveal an existing curvilinear effect.

Detection of Nonlinear Effects

Theory

First and foremost, theory and common sense is always a good guide. I tend to believe that many things in social science (and health sciences as well) are curvilinear in nature, so I routinely check for these effects. If prior research has indicated curvilinear effects, or if there is good cause to suspect that the effect might not be uniform across the entire range of a variable, it is probably worth taking a few minutes to test.

Ad Hoc Testing

Nonlinear effects are easily tested by entering X, X^2, and X^3 terms into an equation. In my experience, if there is curvilinearity, adding squared and cubed terms tends to capture much of the curvilinearity if there is any.

Box-Tidwell Transformations

Those preferring a more strategic approach to this issue may enjoy exploring Box-Tidwell transformations (Box & Tidwell, 1962) as a more methodical approach to testing and specifying curvilinear effects (and more important, linearizing relationships). Many prominent regression authors and texts (i.e., Cohen, Cohen, West, & Aiken, 2002, pp. 239–240) suggest Box-Tidwell as a method of easily exploring whether any variables have nonlinear effects. However, I have rarely seen studies published in journals or books use it. I personally only used it for the preparation of this book. That is not to say it should not be used.

The essential process for Box-Tidwell is to (a) perform an initial analysis with the independent variables (IVs) of interest in the regression equation, (b) transform all IVs of interest via Box-Tidwell, below, (c) enter them into the regression equation simultaneously along with the original untransformed variables, and (d) see which of the transformed variables (if any) are significant.

The Box-Tidwell transformation is expressed in Equation 7.5:

$$V_i = X(\ln X) \tag{7.5}$$

After computing variable V, it is entered into the equation after X is in the equation, as in Equation 7.6a (OLS regression) or Equation 7.6b (logistic regression):

$$\hat{Y} = b_0 + b_1 X_1 + b_2 V_1 \tag{7.6a}$$

$$\text{Logit}(\hat{Y}) = b_0 + b_1 X_1 + b_2 V_1 \tag{7.6b}$$

The null hypothesis to be tested here is that there is no significant curvilinearity in the data:

$$H_0: b_2 = 0$$

$$H_a: b_2 \neq 0$$

If we reject the null hypothesis, we must recognize that there is curvilinearity in the data. Furthermore, Box-Tidwell provides a way to estimate the nature of the curvilinear effect, as Equation 7.7 shows:

$$\hat{\lambda} = \frac{b_2}{b_1} + 1 \tag{7.7}$$

where b_2 is taken from the second analysis and b_1 is taken from the initial analysis without the V in the equation. You can do successive iterations of this process as well, entering $X^{\hat{\lambda}}$ in place of the original X in both of the original steps and the calculation of V; however, in my opinion, that tends to overfit the model unnecessarily and undermine replicability. Our data in the social sciences are not the same character and nature as in the physical sciences and manufacturing, for example.

Basic Principles of Curvilinear Regression

One issue we will run into with curvilinear regression, and also in the future when we examine interactions, is that X^2 and X^3 are *collinear* with X. In other words, they are highly correlated with the original variable. Thus, we cannot easily evaluate lower-order variables when higher-order variables are in the equation, because high collinearity can distort estimates. We will discuss collinearity in more detail in Chapter 8. For now, we will introduce a couple of important practices to allow you to effectively explore curvilinearity in your data.

Occam's Razor

One general principle in statistical science that we usually work from is that we want to simultaneously find both the best and the simplest explanation for a phenomenon. Thus, when one variable is sufficient to explain a phenomenon, we do not include 10 variables. Likewise, if a linear equation can effectively explain a relationship, we prefer that to a curvilinear relationship, and we prefer a quadratic to cubic (or higher-order) effect if the models are not significantly different in overall fit.

Ordered Entry of Variables

Let us say that we have X, X^2, and X^3. Given our preference for simpler variables over more complex variables, we will enter X into the analysis first, and then enter X^2. If X^2 adds a significant increment to the model, we will keep it. If not, we will not retain it in the model. If X^2 is significant and we enter X^3, it should significantly improve the model over the simpler effects or not be retained.

Each Effect Is One Part of the Entire Effect

Remember that X, X^2, and X^3 are all different aspects of the effect of the single variable X. It is an artifact of how we create linear regression equations that we have to model different aspects of the variable in this way. It is important to remember that X^2 is just one aspect of X. This is partly the reason for the ordered entry that we just discussed. We can only evaluate X^2 when X is already in the equation because that allows us to separate out the linear and quadratic effects. We can only evaluate the effect of X^3 when X and X^2 are already in the equation. Conversely, we cannot evaluate the effect of X when X^2 or X^3 is in the equation because of collinearity.

Centering

We have briefly discussed the benefits of centering (I usually do it through conversion to z-scores, but you can also do it other ways and achieve the same benefit) in prior chapters. In this and future chapters, centering will become mandatory. Aiken and West (1991), in their seminal book on regression (Cohen et al., 2002), defined centering as critical for any analysis in which curvilinear effects (or interactions, as we will discuss in Chapter 9) are found. There are some technical reasons for this. I will refer you to their excellent books if you are interested in that level of detail. The bottom line is that we will center X before calculating X^2 or X^3 or any other curvilinear transform of X.

Curvilinear OLS Regression Example: Size of the University and Faculty Salary

Our first example returns to our example of "problematic" data from Chapter 2: the AAUP data on institutional size (zNUM_TOT) and the salary of associate professors (SAL_AP; salary in hundreds of dollars). Because the residual plot in Figure 7.5 looked so weird, I computed quadratic, cubic, and quartic components of zNUM_TOT (zNUM_TOT^2, zNUM_TOT^3, zNUM_TOT^4) to demonstrate the basic principles of testing for curvilinear effects in OLS regression. Each term was entered on its own step, so that there are four versions of the regression line equation, as you can see in Equations 7.8a–d:

$$SAL\_AP = b_0 + b_1 zNUM\_TOT + e \qquad (7.8a)$$
$$H_0: \Delta R^2 = 0; b_1 = 0$$
$$H_a: \Delta R^2 \neq 0; b_1 \neq 0$$

$$SAL\_AP = b_0 + b_1 zNUM\_TOT + b_2 zNUM\_TOT^2 + e \qquad (7.8b)$$
$$H_0: \Delta R^2 = 0; b_2 = 0$$
$$H_a: \Delta R^2 \neq 0; b_2 \neq 0$$

$$SAL\_AP = b_0 + b_1 zNUM\_TOT + b_2 zNUM\_TOT^2 + b_3 zNUM\_TOT^3 + e \qquad (7.8c)$$
$$H_0: \Delta R^2 = 0; b_3 = 0$$
$$H_a: \Delta R^2 \neq 0; b_3 \neq 0$$

$$SAL\_AP = b_0 + b_1 zNUM\_TOT + b_2 zNUM\_TOT^2 + b_3 zNUM\_TOT^3 +$$
$$b_4 zNUM\_TOT^4 + e \qquad (7.8d)$$
$$H_0: \Delta R^2 = 0; b_4 = 0$$
$$H_a: \Delta R^2 \neq 0; b_4 \neq 0$$

As a point of methodological information, this type of user-controlled entry in which we enter one term at a time as specified by the statistician is traditionally called "hierarchical entry." We will discuss more about different methods of entry in Chapter 8.

The results from this analysis are presented in Table 7.1 in a more expanded form than normal so that we can explore all of the nuances of this analysis. At each step, as a new term is entered into the model, we are testing the null hypothesis that the model is not improved, and that the new term has no relationship to the outcome variable. The alternative hypothesis is that the model is significantly improved, and that the new term is significantly related to the outcome.

As you can see in the first part of Table 7.1, we do indeed have four different models being tested, each one adding a higher-powered version of the z-scored zNUM_ TOT. Having requested the change statistics in the model summary, we can see whether the model significantly improves with the addition of each new term. You can see in the first line of the model summary that the linear relationship between institution size and faculty salary is relatively strong, accounting for about one-quarter of the variance in salary ($R^2 = 0.244$), and that the model has significantly improved (over an empty or intercept-only model; $F_{(1, 1,123)} = 362.90$, $p < .0001$). For this equation, we can reject the null hypothesis for the overall model. Likewise, the model improved significantly with the addition of the second term, $zNUM\_TOT^2$

Table 7.1   Curvilinear Analysis of AAUP Data

Variables Entered/Removed[a]

Model	Variables Entered	Variables Removed	Method
1	zNUM_TOT	.	Enter
2	zNUM_TOT^2	.	Enter
3	zNUM_TOT^3	.	Enter
4	zNUM_TOT^4	.	Enter

[a]Dependent variable: SAL_AP.

Model Summary[a]

Model	R	R^2	Adjusted R^2	SE of the Estimate	Change Statistics				
					R^2 Change	F Change	df1	df2	Sig. F Change
1	0.494[b]	0.244	0.244	62.223	0.244	362.895	1	1,123	.000
2	0.586[c]	0.344	0.342	58.018	0.099	169.688	1	1,122	.000
3	0.618[d]	0.382	0.380	56.318	0.038	69.763	1	1,121	.000
4	0.643[e]	0.413	0.411	54.896	0.031	59.829	1	1,120	.000

[a]Dependent variable: SAL_AP.
[b]Predictors: (constant), zNUM_TOT.
[c]Predictors: (constant), zNUM_TOT, zNUM_TOT^2.
[d]Predictors: (constant), zNUM_TOT, zNUM_TOT^2, zNUM_TOT^3.
[e]Predictors: (constant), zNUM_TOT, zNUM_TOT^2, zNUM_TOT^3, zNUM_TOT^4.

Analysis of Variance[a]

Model		Sum of Squares	df	Mean Square	F	p
1	Regression	1,405,040.683	1	1,405,040.683	362.895	.000[b]
	Residual	4,347,982.294	1,123	3,871.756		
	Total	5,753,022.978	1,124			
2	Regression	1,976,232.584	2	988,116.292	293.547	.000[c]
	Residual	3,776,790.394	1,122	3,366.123		
	Total	5,753,022.978	1,124			
3	Regression	2,197,501.684	3	732,500.561	230.946	.000[d]
	Residual	3,555,521.294	1,121	3,171.741		
	Total	5,753,022.978	1,124			
4	Regression	2,377,801.174	4	594,450.293	197.256	.000[e]
	Residual	3,375,221.804	1,120	3,013.591		
	Total	5,753,022.978	1,124			

[a]Dependent variable: SAL_AP.
[b]Predictors: (constant), zNUM_TOT.
[c]Predictors: (constant), zNUM_TOT, zNUM_TOT^2.
[d]Predictors: (constant), zNUM_TOT, zNUM_TOT^2, zNUM_TOT^3.
[e]Predictors: (constant), zNUM_TOT, zNUM_TOT^2, zNUM_TOT^3, zNUM_TOT^4.

(Continued)

Table 7.1   (Continued)

Coefficients[a]

Model		Unstandardized Coefficients		Standardized Coefficients	t	p	95% CI for B	
		B	SE	Beta			Lower Bound	Upper Bound
1	(Constant)	415.641	1.856		223.997	.000	412.001	419.282
	zNUM_TOT	35.065	1.841	0.494	19.050	.000	31.453	38.676
2	(Constant)	427.239	1.946		219.571	.000	423.421	431.057
	zNUM_TOT	66.442	2.958	0.936	22.465	.000	60.638	72.245
	zNUM_TOT^2	−12.076	0.927	−0.543	−13.026	.000	−13.895	−10.257
3	(Constant)	439.712	2.408		182.617	.000	434.988	444.437
	zNUM_TOT	81.497	3.390	1.149	24.041	.000	74.846	88.149
	zNUM_TOT^2	−35.339	2.927	−1.589	−12.074	.000	−41.081	−29.596
	zNUM_TOT^3	4.075	0.488	0.903	8.352	.000	3.117	5.032
4	(Constant)	451.091	2.770		162.852	.000	445.656	456.526
	zNUM_TOT	79.348	3.316	1.118	23.929	.000	72.842	85.855
	zNUM_TOT^2	−69.306	5.237	−3.116	−13.234	.000	−79.581	−59.031
	zNUM_TOT^3	21.317	2.279	4.726	9.352	.000	16.845	25.789
	zNUM_TOT^4	−1.996	0.258	−2.367	−7.735	.000	−2.502	−1.490

[a]Dependent variable: SAL_AP.

($\Delta R^2 = 0.099$, $F_{(1, 1,122)} = 169.69$, $p < .0001$); the third term, zNUM_TOT^3 ($\Delta R^2 = 0.038$, $F_{(1, 1,121)} = 69.76$, $p < .0001$); and also the fourth term, zNUM_TOT^4 ($\Delta R^2 = 0.031$, $F_{(1, 1,120)} = 59.83$, $p < .0001$). Thus, each term adds to a stronger model. This is not normal in my experience, but it is welcome as an example for this chapter.

The analysis of variance (ANOVA) table summarizes the significance of the overall model (testing the null hypothesis that all coefficients are simultaneously zero). This is not usually of interest but you can see the residual sum of squares (unexplained variance) decreasing with each step, meaning that each additional term added to the model is reducing the unexplained (error) variance and that all models are significant overall. This is generally a good thing. Were the overall model to be nonsignificant, we could not reject the null hypothesis that the regression coefficients are zero and we could not legitimately discuss any effects.

Finally, the last part of the table summarizes the details of each effect, which is what we are usually interested in. I mentioned earlier in this chapter that when exploring complex effects such as curvilinear terms, we generally only interpret the term entered on each step. Because we left the dependent variable (DV) in its original metric (hundreds of dollars), we can interpret this as follows: the average salary of an associate professor at the intercept (salary at an average-sized university) is $41,564.10.[6] The linear effect is significant (which we already saw from the model summary table) and has an unstandardized regression coefficient of 35.07 (with a 95% confidence interval [95% CI] = [31.45, 38.68]). We can interpret this as the average salary of an associate professor increases $3,507.00 (and we are 95% sure the true increase in the population is between $3,145.30 and $3,867.60 based on our 95% CI)

6   Keep in mind that these data are many years out of date. Most professors these days make several dollars more per year than back in the 1990s.

for each increase of 1 standard deviation (*SD*) in size. However, this effect is modified by the subsequent significant curvilinear effects entered on the next steps.

Once the quadratic term is entered, we know that the model improves significantly (accounting for almost 10% more variance), and we now have the regression equation details based on the intercept and the two regression coefficients. Note that the intercept has changed, as has the effect of zNUM_TOT, because now that variable is interpreted as the linear effect controlling for the curvilinear effect. These variables are also highly collinear (correlated over $r = 0.81$) and thus strongly affect each other when both are in the equation.

Of course, this quadratic effect is modified by the cubic effect, which is also significant, and so on. In the service of brevity, I usually report this full table but only interpret the last equation with a significant effect (in this case, the fourth equation, because that effect was also significant).

Note that we always seek to meet assumptions of whatever analysis we are performing. In Figure 7.6, you can see that there are standardized residuals that extend beyond ±3.0, but the normality is not severely violated (skew = 0.69, kurtosis = 0.92). The scatterplot of the standardized predicted values against the standardized residuals in Figure 7.7 shows substantial improvement from the previous linear regression

Figure 7.6   Histogram of the Standardized Residuals From the AAUP Curvilinear Regression Analysis Predicting Salary of Associate Professors From Size of the Institution

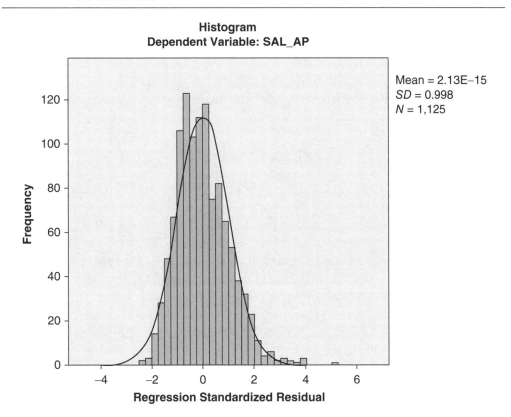

Figure 7.7   Plot of zPRED Versus zRESID From the AAUP Curvilinear Regression Analysis Predicting Salary of Associate Professors From Size of the Institution

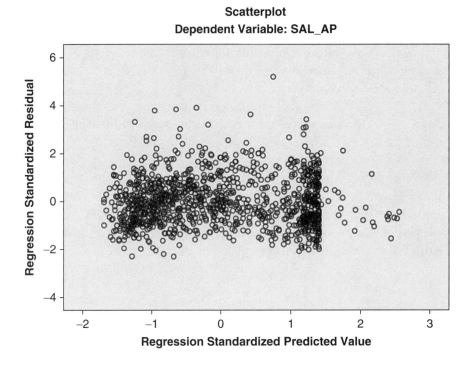

analysis (Figure 7.5). Prior to adding the curvilinear terms to the equation, the plot indicated potential nonlinearity and heterogeneity. Although this is not perfect, it is much improved, particularly in light of the fact that we have not performed data cleaning yet.

Data Cleaning

After all terms were in the equation, I examined both standardized residuals and DfFit statistics. DfBetas would also be appropriate to examine, as would other indicators of influence. There was one case that had an exceptionally large DfFit (it was more than 33 *SD* from the mean DfFit value), and that case was removed. The standardized residuals ranged from −2.27 to 5.14, with 10 cases (out of 1,125) having standardized residuals greater than 3.0. These 11 cases were also removed, resulting in less than 1% of the sample being removed. The results after cleaning are presented (in less detail) below in Table 7.2 and Figures 7.8 and 7.9. The normality of the residuals has improved substantially (skew has now been reduced to 0.38 and the kurtosis is now −0.25). Furthermore, the scatterplot (Figure 7.9) is slightly improved over Figure 7.8, more clearly meeting the assumption of homoscedasticity.

Figure 7.8   Histogram of Standardized Residuals Following Data Cleaning

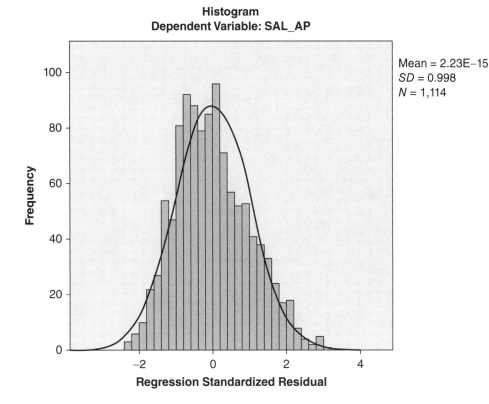

Of course, we would not expect removal of less than 1% of the sample to have massive effects on the results, but we do see incremental improvement in the model. Perhaps more important, we can be somewhat confident that the curvilinear effects are truly in the population and are not the result of a few highly influential cases. You can see that the models are all still significant, meaning that the basic nature of the

Table 7.2   Curvilinear Analysis of AAUP Data After Data Cleaning

					Model Summary[a]				
Model	R	R^2	Adjusted R^2	SE of the Estimate	Change Statistics				
					R^2 Change	F Change	df1	df2	Sig. F Change
1	0.518[b]	0.268	0.268	59.075	.268	408.075	1	1,112	.000
2	0.605[c]	0.366	0.364	55.041	.097	169.940	1	1,111	.000
3	0.643[d]	0.413	0.412	52.958	.048	90.114	1	1,110	.000
4	0.665[e]	0.442	0.440	51.648	.029	58.029	1	1,109	.000

[a]Dependent variable: SAL_AP.
[b]Predictors: (Constant), zNUM_TOT.
[c]Predictors: (Constant), zNUM_TOT, zNUM_TOT^2.
[d]Predictors: (Constant), zNUM_TOT, zNUM_TOT^2, zNUM_TOT^3.
[e]Predictors: (Constant), zNUM_TOT, zNUM_TOT^2, zNUM_TOT^3, zNUM_TOT^4.

(Continued)

Table 7.2 (Continued)

Coefficients[a]

Model		Unstandardized Coefficients		Standardized Coefficients	t	p	95% CI for B	
		B	SE	Beta			Lower Bound	Upper Bound
1	(Constant)	414.025	1.770		233.895	.000	410.552	417.498
	zNUM_TOT	36.113	1.788	0.518	20.201	.000	32.606	39.621
2	(Constant)	425.741	1.878		226.667	.000	422.056	429.427
	zNUM_TOT	66.782	2.883	0.958	23.168	.000	61.126	72.438
	zNUM_TOT^2	−12.402	0.951	−0.539	−13.036	.000	−14.268	−10.535
3	(Constant)	439.687	2.329		188.791	.000	435.117	444.256
	zNUM_TOT	82.261	3.217	1.180	25.568	.000	75.948	88.574
	zNUM_TOT^2	−39.526	3.000	−1.718	−13.174	.000	−45.413	−33.639
	zNUM_TOT^3	5.096	0.537	1.030	9.493	.000	4.043	6.149
4	(Constant)	450.396	2.671		168.609	.000	445.155	455.638
	zNUM_TOT	77.443	3.201	1.111	24.195	.000	71.162	83.723
	zNUM_TOT^2	−73.648	5.350	−3.202	−13.765	.000	−84.146	−63.150
	zNUM_TOT^3	24.375	2.584	4.925	9.431	.000	19.304	29.446
	zNUM_TOT^4	−2.429	0.319	−2.447	−7.618	.000	−3.054	−1.803

[a]Dependent variable: SAL_AP.

Figure 7.9 Scatterplot of Standardized Residuals and Standardized Predicted Values After Data Cleaning

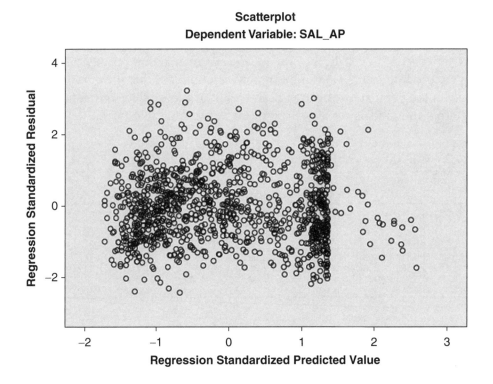

curve has not changed following data cleaning. Second, you can see that the overall model accounts for slightly more variance than it did previously.[7] You can also see that the regression coefficients were modified slightly, which is also to be expected if we removed inappropriately influential cases.

Interpreting Curvilinear Effects Effectively

Looking at the coefficients, is it immediately apparent what the curve will look like? I used to be pretty good at algebra and calculus, but my recommendation to you is to graph complex effects such as curves. I recommend that you graph a reasonable range, and use it to help you (and more important, your reader) intuitively interpret the effect. I provide a graph of all four equations from our analysis in Figure 7.10 only so that you can see the differences between the curves, as well as the value of taking a few minutes of your time to explore curvilinearity. You would normally only graph the final curve and explain that to your reader. I am also graphing the curve across an expanded range (−4 to +4 SD) so that you can see more of the curve. However, graphing an effect beyond the range of the data is difficult to justify and to replicate. I recommend keeping it between −3 and +3 at most.

In addition, because we are talking about salary as our DV, I deleted negative predicted values from the graph, as they did not make sense. As you can see in Figure 7.10,

Figure 7.10   Curvilinear Effects of Institutional Size on Salary of Associate Professors (in Hundreds of Dollars) Following Data Cleaning

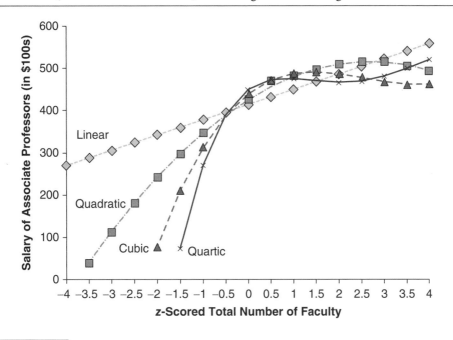

7   This is because we removed the 11 cases out of more than 1,100 that were contributing the most error variance.

the curves add some nuance to the understanding of the relationship between the two variables. For example, there is a good deal of expected change in the lower end of the range and then a flattening out at the upper end. Thus, it might be the case that the most gains are in a particular range, which a curvilinear effect can help you identify more clearly. Looking at the final (quartic) curve, the expected average salary is essentially flat between about the mean (0) and +3.0, indicating that most of the change is in the smaller institutions, and once you reach a certain point, the effect of increased institutional size on professor salary is much weaker. This is a very different conclusion from the model with only the linear effect, which would lead us to include that institutional size has a constant effect on salary across the entire range.

Example Summary of Analysis

If this were my dissertation or study to be published, I would summarize the analyses we have just worked through as follows (eliminating much of the detail you would cover, such as sampling and measurement).

The initial analysis predicting salary from institution size indicated several potential problems. First, there seemed to be substantial heterogeneity, and possible nonlinearity. Second, examination of standardized residuals and DfFit indicators of influence suggested 11 cases that were inappropriately influential (DfFit > 33 *SD* from the mean; ZRE > |3.0|). After they were removed, a curvilinear regression analysis was performed entering the original (linear) effect first, and then the squared, cubed, and quartic terms on separate steps.

All four steps led to significant improvement in model variance accounted for, as you can see in Table 7.2. Ultimately, this model is relatively strong, accounting for over 44% of the variance in salary. Furthermore, after data cleaning and entry of curvilinear terms, normality or residuals was improved and assumption of homoscedasticity was met. Thus, all terms were retained, and the regression line equation from the final model was graphed and is presented in Figure 7.10.

As you can see in Figure 7.10, the salary of associate professors is strongly related to institution size (or our proxy for size, the total number of faculty) when the institutions are below average; however, for those institutions above the mean, there is little added effect of size on faculty salary.

Reality Testing This Effect

Some readers might be skeptical that the data really show this dramatic difference in effects below 0 (mean) and above 0. I heartily encourage researchers to perform reality testing of effects to ensure that some arithmetic mistake is not responsible for the effect observed. Thus, I performed two very quick follow-up analyses. First, I performed a simple linear regression with our two variables, selecting only cases where zNUM_TOT < 0. Next, I performed another identical analysis selecting only cases where zNUM_TOT > 0. If my graphing and interpretation is correct, the first analysis should show a dramatic and strong positive effect, and the second analysis should show a much weaker and flatter effect. I will not waste space with all of the details, but the first regression coefficient was $b_1 = 193.77$ ($\beta = 0.56$, $R^2 = 0.31$). The second analysis contained a regression coefficient of $b_1 = 11.09$ ($\beta = 0.23$, $R^2 = 0.05$), an impressive contrast that strongly supports the conclusion in the text box above (but not analyses you would necessarily report in a paper or dissertation).

Summary of Curvilinear Effects in OLS Regression

I hope at this point you have been persuaded that it is worthwhile to explore your data for curvilinear effects, that it is not difficult to do this exploration, and that the results can be quite interesting. Using these data, we demonstrated an initial linear analysis that did not meet assumptions of OLS regression, and that without modeling the nonlinear effects, the conclusions from this model would have been somewhat misleading. However, this situation improved dramatically once curvilinear terms were added to the model. Not only were the assumptions more clearly met, but the effect was more nuanced and interesting. You may not care about curvilinear effects in professor salaries, but if this were another set of variables, like time spent on homework and student achievement (or exercise and well-being), would it not be helpful to understand the range where the most benefit occurs?

Curvilinear Logistic Regression Example: Diabetes and Age

This second example of curvilinear effects is from NHIS2010, wherein we will predict diagnosis of diabetes from body mass index (BMI) using logistic regression. Most of the same principles from the example of OLS regression will apply here as well, with some slight modifications. We will keep the discussion of model fit and so forth abbreviated in this chapter, as we just covered logistic regression in Chapter 6.

As you can see from Figure 7.11, BMI is positively skewed. Conversion to z-scores would leave the intercept at about 27.69, which is in the "overweight" range according to the US Centers for Disease Control and Prevention.[8] This might not be desirable, so we can center the distribution at 20, which is a healthier BMI.[9] Centering a variable rather than converting to a z-score will leave it in its original metric, with the intercept at a more meaningful point. This will be the variable we will also square and cube to create the curvilinear analysis. These terms will be entered into the analysis one at a time like the previous OLS regression analysis.

Remember that in logistic regression, we are looking at improvement of model fit, not variance accounted for. Therefore, we are looking for a significant change in -2 log likelihood ($-2LL$), evaluated as a chi-square. As you can see in Table 7.3, entry of the first term, BMIc (centered BMI), produced a change in $-2LL$ of 986.98 ($p < .0001$). Entry of the quadratic term produced a reduction in $-2LL$ of 114.91 ($p < .0001$), and

---

8   US Centers for Disease Control and Prevention (http://www.cdc.gov/healthyweight/assessing/bmi/adult_bmi/index.html).

9   Unfortunately, this is a BMI I am not likely to see again for a long while . . .

Figure 7.11   Histogram of Body Mass Index (NHIS2010)

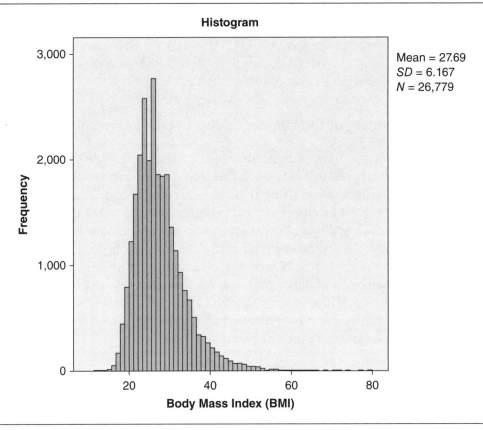

DATA SOURCE: National Health Interview Survey of 2010 (NHIS2010) from the National Center for Health Statistics (http://www.cdc.gov/nchs/nhis/nhis_2010_data_release.htm).

entry of the cubic term failed to significantly improve the model ($\Delta-2LL = 1.20$; $p < .27$). However, this is prior to any data cleaning.

As you can see from Figure 7.12, there are some cases with some rather extreme standardized residuals (105 of 26,779, or 0.39%, had a standardized residual greater than 5.0 and were removed). Following removal of these inappropriately influential cases, the model fit became even better than before, as you can see in Table 7.4. You will also see that the cubic term is now significant, although small in effect, and thus will be retained in the model going forward. You might also see that for the final step, I have expanded the number of decimals reported in the regression coefficient column. When dealing with squared and cubed terms and log-transformed numbers, increased precision is important. If you examine Table 7.3, you will see 0.000 as the coefficient for BMI^3, which is not really the case and is not really helpful if we are trying to predict values using that number in a logistic regression equation. Most statistical software will allow you to get more precision either through setting different preferences or by clicking on the table itself (as in SPSS).

To graph this equation, you would create the logistic regression equation from the last step in Table 7.4, as shown in Equation 7.9:

Table 7.3   Relationship of Body Mass Index and Diabetes

		B	SE	Wald	df	p	Exp(B)	95% CI for Exp(B)	
								Lower	Upper
Step 1	BMIc	0.092	0.003	1,013.633	1	.000	1.096	1.090	1.102
	Constant	−3.071	0.037	6,872.776	1	.000	0.046		
Step 2	BMIc	0.169	0.008	415.291	1	.000	1.185	1.165	1.204
	BMI^2	−0.003	0.000	98.678	1	.000	0.997	0.997	0.998
	Constant	−3.470	0.056	3,789.292	1	.000	0.031		
Step 3	BMIc	0.183	0.015	145.901	1	.000	1.201	1.166	1.238
	BMI^2	−0.004	0.001	15.016	1	.000	0.996	0.995	0.998
	BMI^3	0.000	0.000	1.267	1	.260	1.000	1.000	1.000
	Constant	−3.515	0.070	2,517.835	1	.000	0.030		

Variables in the Equation

Figure 7.12   Standardized Residuals From Body Mass Index and Diabetes Analysis

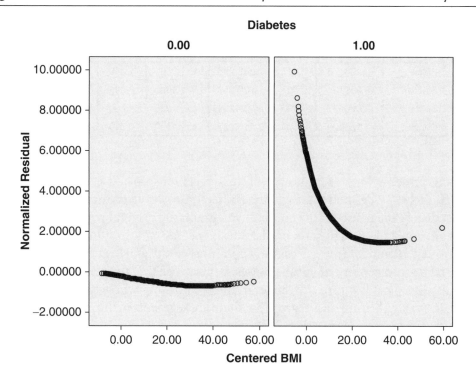

$$\text{Logit}(\hat{Y}) = -4.125529 + 0.298676(\text{BMIc}) - $$
$$0.009354(\text{BMI}^2) + 0.000097(\text{BMI}^3) \tag{7.9}$$

Procedurally, creating predicted logits and conditional probabilities when looking at curvilinear effects is no different than for OLS regression. In this case, recall that we centered BMI at 20, and so 0 will be a BMI of 20, and any increment of ±1 will be a change in BMI of 1.0 (i.e., we did not convert to z-scores). Furthermore, as discussed in Chapter 6, I encourage readers to convert predicted scores from logit to conditional probability for easier interpretation. I will demonstrate why now.

Table 7.4   Relationship of Body Mass Index and Diabetes After Data Cleaning

	Chi-Square (Δ–2LL)	df	p
Step 1	1,161.731	1	.000
Step 2	218.891	1	.000
Step 3	34.140	1	.000

Variables in the Equation

		B	SE	Wald	df	Sig.	Exp(B)	95% CI for Exp(B)	
								Lower	Upper
Step 1	BMIc	0.101	0.003	1,174.386	1	.000	1.106	1.100	1.112
	Constant	−3.214	0.039	6,922.436	1	.000	0.040		
Step 2	BMIc	0.218	0.009	540.553	1	.000	1.244	1.221	1.267
	BMI^2	−0.004	0.000	168.583	1	.000	0.996	0.996	0.997
	Constant	−3.840	0.064	3,614.639	1	.000	0.021		
Step 3	BMIc	0.298676	0.017	306.258	1	.000	1.348	1.304	1.394
	BMI^2	−0.009354	0.001	90.563	1	.000	0.991	0.989	0.993
	BMI^3	0.000097	0.000	37.201	1	.000	1.000	1.000	1.000
	Constant	−4.125529	0.084	2,416.437	1	.000	0.016		

As you can see in Figures 7.13a and 7.13b, the relationship between BMI and diabetes is definitely best defined as a curvilinear effect.[10] However, the nature of logarithms tends to distort the nature of the curve. If you graph the original regression equation, you would conclude that the risk of diabetes is increasing fastest for those with BMI between, say, 10 and 20 (the very skinny and most healthy). However, once these are converted back to conditional probabilities, you can see that this is exactly the wrong interpretation. That is one of the ranges in the relationship where the slope is flattest. As with the example from OLS regression, there are segments of the relationship where there is almost no increase in risk as BMI increases (e.g., between 10 and 18 and from 46 to 60) and some areas where each increment in BMI increases the risk of being diagnosed with diabetes dramatically (e.g., 26–42).

Curvilinear Effects in Multinomial Logistic Regression

Thus far we have explored some basic principles around curvilinearity where we have continuous IVs and DVs, as well as continuous IVs and binary DVs. If you have followed to this point, you should be wondering whether we can apply curvilinear

---

10   Also note that despite centering BMI at 20, for the convenience of the reader, I converted BMI back to the original scale when presenting it. Simple touches such as this make the reader's job much easier and reduce the chance of misinterpretation.

Figure 7.13a   Curvilinear Relationship Between Body Mass Index and Diabetes
Graphed in Logits

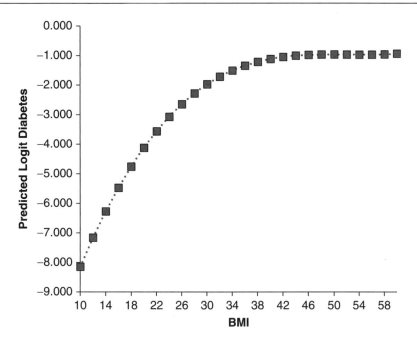

Figure 7.13b   Curvilinear Relationship Between Body Mass Index and Diabetes
Graphed in Conditional Probabilities

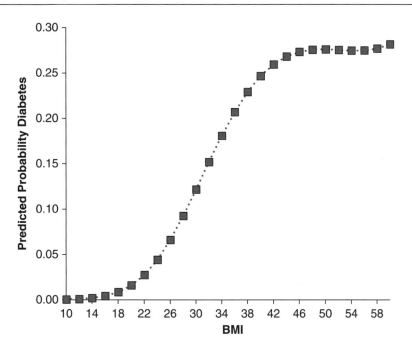

An Example Summary of This Analysis

In order to explore the curvilinear relationship between diabetes and BMI, the IV was centered at 20, and then squared and cubed versions of the centered BMI variable were created and entered sequentially (on individual steps) into the analysis. A small percentage (less than 0.4%) of the sample had inappropriate levels of influence by virtue of having standardized residuals greater than $|5.0|$. After these cases were removed, the entry of each term (linear, quadratic, cubic) contributed to a significant improvement in model fit, as Table 7.4 shows. Because all three terms were significant, the final logistic regression equation was used to create predicted values across a broad range of BMI (10–60). These predicted values were converted from logits to conditional probabilities for ease of interpretation.

As you can see in Figure 7.13b, the probability of being diagnosed with diabetes is relatively low and slow to accelerate in adults with low BMI, but it begins to rise more rapidly as BMI moves toward the high 20s and continues to increase rapidly until the high 40s, where it levels off at a high prevalence.

effects to other parts of the generalized linear model: polytomous IVs and DVs. The answer is that because we look for curvilinearity in the IVs, any analysis with a continuous IV is a candidate for potential curvilinearity. This rules out looking for curvilinearity in simple ANOVA-type analyses (but not repeated measures!). Let us expand our exploration to multinomial logistic regression, and return to our example from Chapter 6 and the National Education Longitudinal Study of 1988 (NELS88) data involving student achievement (zACH, our z-scored version of the student achievement variable) and marijuana use (MJ; coded 0 = never tried it, 1 = tried it 1–2 times, 2 = tried it 3–19 times, and 3 = tried it 20 or more times). In the previous analysis, we observed that higher achievement test scores tended to relate to lower probabilities of trying marijuana at each level.

Let us expand that analysis to add a quadratic term ($zACH^2$) to the equation already described above. The initial model had a likelihood ratio test of $\chi^2_{(3)} = 129.49$, $p < .0001$ when the first term was entered, with a final $-2LL$ of 12,991.445. Adding zACH and $zACH^2$ significantly improves the model $-2LL$ to 12,969.033 (for a likelihood ratio test of $\chi^2_{(3)} = 22.41$, $p < .0001$). The cubic effect did not add a significant improvement to model fit and thus was disregarded (Table 7.5).

Binary logistic regression models were created to determine whether there were any inappropriately influential outliers with the curvilinear effect in the analysis. Examining the deviance residuals, for example, revealed many cases with values over 2.00, which would be potential candidates for removal. However, none exceeded 3.00, and thus all were retained. As with the logistic regression analysis, the predicted values were restricted to reasonable ranges (in this case, zACH between -2 and $+2$, which captured most of the sample) and values were converted from logits to conditional probabilities. As you can see in Figure 7.14, the probability of trying marijuana 1–2 times, 3–19 times, or 20 or more times (compared with 0 times) remains relatively flat while achievement is below average and then tends to drop more steeply. The exception is the second group (tried marijuana 1–2 times), which

Table 7.5   Final Multinomial Model With Curvilinear Achievement Effect Predicting Marijuana Use

Model Fitting Information

Model	Model Fitting Criteria	Likelihood Ratio Tests		
	−2LL	Chi-Square	df	p
Intercept only	13,120.935			
Final	12,969.033	151.902	6	.000

Parameter Estimates

MJ[a]		B	SE	Wald	df	p	Exp(B)	95% CI for Exp(B)	
								Lower Bound	Upper Bound
1	Intercept	−2.104	0.040	2,813.651	1	.000			
	zACH	−0.221	0.032	48.273	1	.000	0.801	0.753	0.853
	zACH²	−0.049	0.030	2.635	1	.105	0.952	0.897	1.010
2	Intercept	−2.404	0.046	2,736.645	1	.000			
	zACH	−0.159	0.038	17.068	1	.000	0.853	0.791	0.920
	zACH²	−0.110	0.037	9.030	1	.003	0.896	0.834	0.962
3	Intercept	−2.806	0.057	2,457.751	1	.000			
	zACH	−0.247	0.048	26.506	1	.000	0.781	0.711	0.858
	zACH²	−0.165	0.048	12.033	1	.001	0.848	0.772	0.931

[a]The reference category is: 0.

Figure 7.14   Curvilinear Effect of Student Achievement on Marijuana Use

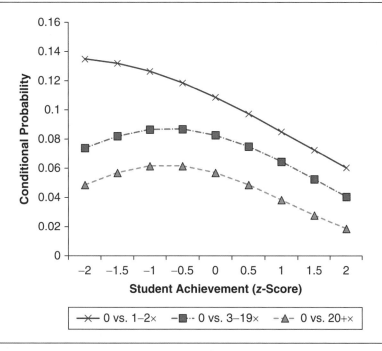

DATA SOURCE: National Education Longitudinal Study of 1988 (NELS88) from the National Center for Education Statistics (http://nces.ed.gov/surveys/nels88/).

is more linear of an effect, with the probability decreasing more monotonically across the entire range of achievement. This is expected because the quadratic effect was not significant for this group.

Replication Becomes Important

All effects should be replicated in independent data sets. This is one of the most basic aspects of the endeavor we call science. Without replicability (the ability to produce similar results by performing similar tasks on similar objects), we do not have science. Yet few pursue exploration of replicability. The more complex analyses get, the more important it becomes to replicate our results to ensure that the effects are not merely taking advantage of a peculiar sample. This is particularly important with the smaller samples that are common even in top-tier journals in the behavioral and social (and often, health) sciences. In future chapters, we will explore this issue in depth with a variety of effects. For now, let me demonstrate the volatility of curvilinear effects with two random, small samples from the AAUP data we opened the chapter with. Two random samples of approximately 15% ($N = 176$) will be analyzed in the same way, and the curves will be compared to see how closely the effect would replicate in two independent samples.

As you can see from Table 7.6, the model summaries show similar patterns, in that all steps are significant; however, the details, including the variance accounted for, vary substantially (54% versus 44%). In addition, in the second sample, the last two steps are only marginally significant, meaning that had the details been slightly different, the conclusions about the nature of the curve might have been different (quadratic only versus quartic).

As you can see in Figure 7.15, when graphed across a more reasonable range that includes only positive predicted salaries, the curves look similar, although the details differ, particularly in the lower ranges. In this case, the basic conclusions from the analysis were replicated generally. The lesson we will explore in more detail during later chapters is that in many sciences, replication helps establish whether an effect is likely to be found in subsequent samples or whether it was largely an artifact of a particular sample. In this case, a new sample from the same population is likely to give us the same general result, within a certain range.

More Fun With Curves: Estimating Minima and Maxima as Well as Slope at Any Point on the Curve

Although we will explicitly discuss logistic regression in this section because that is the focus of this book, these principles should work with any type of regression. In fact, Aiken and West (1991, see pp. 72–76) explicitly discuss this issue in their excellent treatise on interactions in OLS regression.

Any equation can be manipulated with calculus according to simple rules to allow post hoc probing of regression line equations. This can be particularly fun in complex

Table 7.6   Replication of Two Small(er) Samples Predicting Faculty Salary From Institution Size

Model Summary[a]

Model	R	R^2	Adjusted R^2	SE of the Estimate	Change Statistics				
					R^2 Change	F Change	df1	df2	Sig. F Change
Sample 1 (N = 176)									
1	0.530	0.281	0.277	60.016	0.281	67.991	1	174	.000
2	0.660	0.436	0.430	53.295	0.155	47.653	1	173	.000
3	0.712	0.506	0.498	50.018	0.070	24.408	1	172	.000
4	0.732	0.536	0.525	48.636	0.030	10.913	1	171	.001
Sample 2 (N = 176)									
1	0.553	0.306	0.302	62.393	0.306	73.577	1	167	.000
2	0.641	0.411	0.404	57.662	0.105	29.526	1	166	.000
3	0.651	0.424	0.414	57.161	0.014	3.925	1	165	.049
4	0.663	0.440	0.427	56.541	0.016	4.638	1	164	.033

Sample 1: Final model (Step 4)

Model		Unstandardized Coefficients		Standardized Coefficients	t	p	95% CI for B	
		B	SE	Beta			Lower Bound	Upper Bound
	(Constant)	462.213	6.504		71.063	.000	449.373	475.052
	zNUM_TOT	89.000	8.495	1.283	10.476	.000	72.231	105.769
4	zNUM_TOT^2	−87.862	13.265	−4.136	−6.624	.000	−114.046	−61.678
	zNUM_TOT^3	29.678	7.094	6.536	4.183	.000	15.675	43.682
	zNUM_TOT^4	−3.072	0.930	−3.297	−3.303	.001	−4.908	−1.236

Sample 2: Final model (Step 4)

Model		Unstandardized Coefficients		Standardized Coefficients	t	p	95% CI for B	
		B	SE	Beta			Lower Bound	Upper Bound
	(Constant)	449.811	7.933		56.698	.000	434.146	465.476
	zNUM_TOT	75.028	8.780	1.123	8.545	.000	57.691	92.365
4	zNUM_TOT^2	−56.205	15.277	−2.635	−3.679	.000	−86.370	−26.040
	zNUM_TOT^3	19.617	7.840	4.240	2.502	.013	4.136	35.098
	zNUM_TOT^4	−2.217	1.029	−2.303	−2.154	.033	−4.250	−0.184

[a]Dependent variable: SAL_AP.

curvilinear equations, because you can estimate where the curve reaches a minimum or maximum, or you can estimate the slope at any particular point on the curve to estimate how fast the probabilities are changing.[11]

Those of you who have taken (and remember) basic calculus will remember that taking the first derivative of any equation allows you to estimate slope. For example, taking a simple linear equation from the NHIS2010 database, we can look at the

---

11   As many authors have pointed out (Aiken & West, 1991, pp. 73–75; DeMaris, 1993), technically what you are estimating is the slope of a line *tangent to* the point where we are estimating the value for the first derivative. For our purposes, these concepts are identical.

Figure 7.15 Replication of AAUP Curvilinear Analysis in Two Smaller Samples

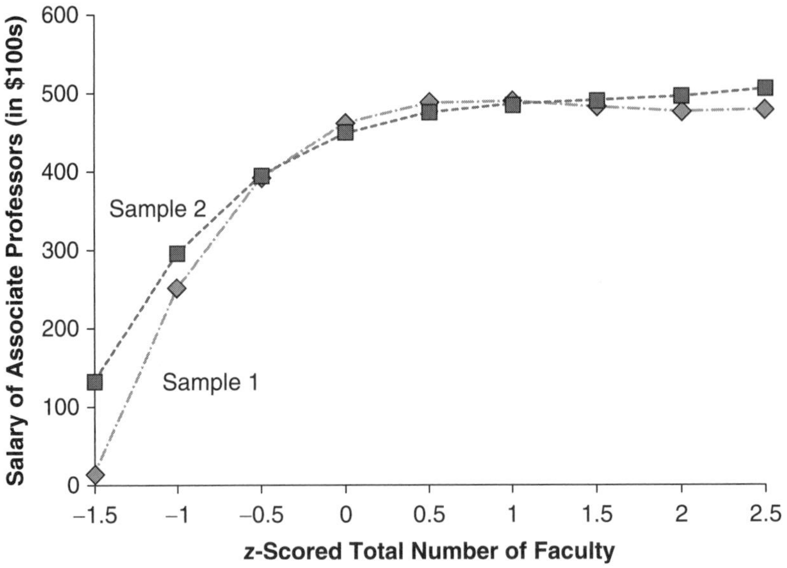

logistic regression equation relating AGE and DIABETES (you will be performing these analyses in this chapter's enrichment exercises). The original equation was as follows (Equation 7.10a):

$$\text{Logit}(\hat{Y}) = -4.631 + 0.45X \qquad (7.10a)$$

or expressed more fully (Equation 7.10b):

$$\text{Logit}(\hat{Y}) = -4.631X^0 + 0.45X^1 \qquad (7.10b)$$

Being more specific, the intercept has an X raised to the 0 power, which is 1 (anything raised to the 0 power is 1); thus, it is often eliminated from the regression equation by convention. Furthermore, the X is raised to the first power, and anything raised to the first power is itself. This might seem like more detail than is needed, but once we start adding quadratic and cubic terms, or taking derivatives, this starts to make some sense. For example, the quadratic equation for AGE and DIABETES is

$$\text{Logit}(\hat{Y}) = -8.56625X^0 + 0.19402X^1 - 0.001301X^2 \qquad (7.11a)$$

The simple rules for taking a derivative are that you multiply each term by the exponent of the X, reducing that exponent by 1.[12] The first term will drop out, because anything multiplied by 0 is 0. Thus, taking the derivative of Equation 7.10b (or 7.10a), we get Equation 7.10c:

---

12 Unfortunately, we cannot include an entire course in calculus here. Please refer to good calculus references if you are not familiar with this concept.

$$\frac{d\left(logit\left(\hat{Y}\right)\right)}{dX} = (1)0.45X^0 \qquad (7.10c)$$

which simplifies to

$$\frac{d\left(logit\left(\hat{Y}\right)\right)}{dX} = 0.45$$

In other words, because this is a *linear* equation, not a curvilinear equation, the slope is constant across the entire regression: 0.45 logits per increase in X of 1.0. This is perhaps not the most surprising or illuminating outcome, but it is a simple example of a derivative.

Let's move to the curvilinear example. The derivative for the quadratic formula (Equation 7.11a) is (dropping the constant and simplifying)

$$\frac{d\left(logit\left(\hat{Y}\right)\right)}{dX} = 0.19402 - 2(0.001301X) \qquad (7.11b)$$

or

$$\frac{d\left(logit\left(\hat{Y}\right)\right)}{dX} = 0.19402 - 0.002602X$$

Once we have this first derivative, we can look for the point where the slope is 0 (the minimum or maximum) by setting $\frac{d\left(logit\left(\hat{Y}\right)\right)}{dX}$ equal to 0 and solving for X. We get Equation 7.11c:

$$0 = 0.19402 - 0.002602X \qquad (7.11c)$$

By adding 0.002602X to both sides in Equation 7.11c, we get

$$0.002602X = 0.19402$$

Solving for X, we get

$$X = 74.57 \text{ years}$$

Looking at the curve, this makes sense because visually we can see that the curve levels off around that point and then curves downward (Figure 7.16).

Figure 7.16   Calculating the Inflection Point of a Curve

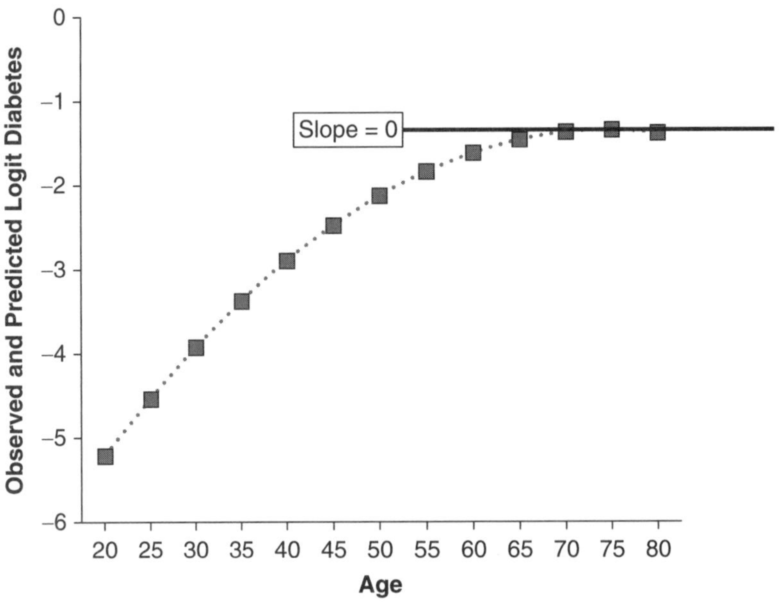

DATA SOURCE: National Health Interview Survey of 2010 (NHIS2010) from the National Center for Health Statistics (http://www.cdc.gov/nchs/nhis/nhis_2010_data_release.htm).

Note that we are predicting the slope of logit($\hat{Y}$), and when it reaches zero, that is where the curve has a minimum or maximum and change in direction.[13] However, with the first derivative, we can do more. We can also estimate slopes (in logits) at particular values of X. For example, let us look again at the first derivative of the quadratic equation and estimate the slope at two other time points (we already know the slope around AGE = 75): AGE = 25 and AGE = 50. By substituting these into the equation, we get slopes of 0.12897 for AGE = 25 and 0.06392 for AGE = 50. This suggests that the log odds of having diabetes are increasing faster at age 25 years than at 50 years. Looking at the graph of logits (Figure 7.16), this seems to hold.

However, looking at the graph of the predicted conditional probabilities (Figure 7.17), it does not. The change in probabilities seems to be much slower at age 25 than age 50 years. Thus, we must be careful to be clear when reporting post hoc probes of these types of analyses, but they can be useful at times. However, note that the extrema are calculated to be identical for both graphs.

Let us examine a more interesting curve, that predicting EVERMJ from student achievement. We performed the linear logistic regression analysis for these data in Chapter 5, but if you work through all of the enrichment examples at the end of this

---

13   For you calculus nerds out there, technically we are estimating the slope of a line tangent to the point, but proofs can show that the slope of that line is also the instantaneous slope of our curve at that point.

Figure 7.17 The Same Curve Graphed as Conditional Probabilities

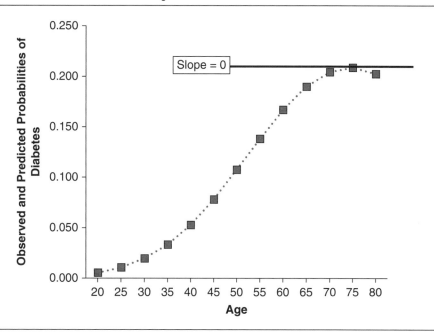

DATA SOURCE: National Health Interview Survey of 2010 (NHIS2010) from the National Center for Health Statistics (http://www.cdc.gov/nchs/nhis/nhis_2010_data_release.htm).

chapter, you will find there is a cubic curve, meaning it has two points where the slope is equal to 0. The original equation (after data cleaning) is shown below as a spoiler in Equation 7.12a:

$$\text{Logit}(\hat{Y}) = -1.1514 + 0.2683\,(z\text{ACH}) - 0.0214\,(z\text{ACH}^2)$$
$$- 0.3168\,(z\text{ACH}^3) \tag{7.12a}$$

which gives us a first derivative, shown in Equation 7.12b:

$$\frac{d\left(logit\left(\hat{Y}\right)\right)}{dX} = 0.2683 - 0.0428\,(z\text{ACH}) - 0.9504(z\text{ACH}^2) \tag{7.12b}$$

If you set Equation 7.12b equal to zero and solve, you find two *extrema* at −0.55 and at 0.51, both of which seem reasonable given the graph shown in Figure 7.18 (graphed in logits rather than predicted probabilities).

We could again predict slopes at particular points using the first derivative. With this example, let us examine the following three points: −1.75, 0, and 1.75. Substituting these points into the equation, we get slopes of −2.57, 0.27, and −2.72, respectively. This tells us that the logits are decreasing relatively steeply in the

Figure 7.18 Two Inflection Points for a Cubic Curve

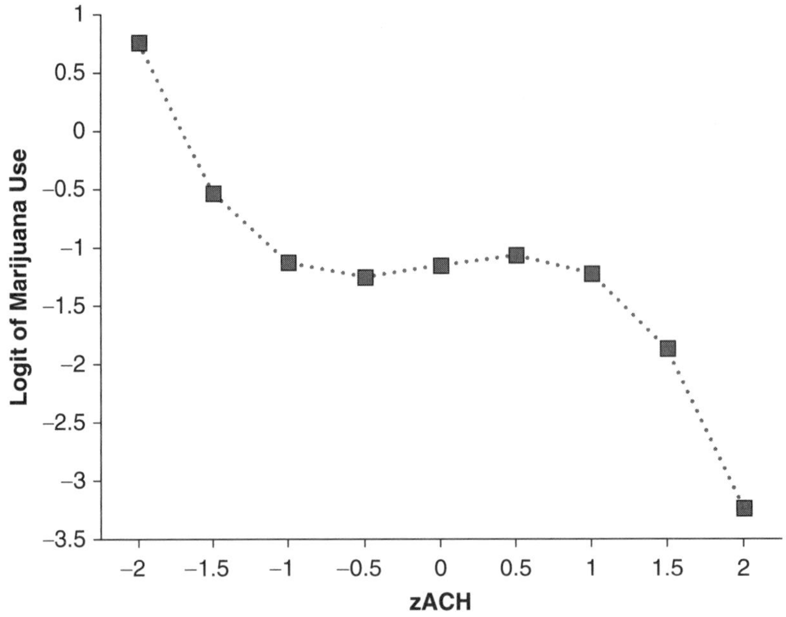

DATA SOURCE: National Education Longitudinal Study of 1988 (NELS88) from the National Center for Education Statistics (http://nces.ed.gov/surveys/nels88/).

extremes of the distribution and are relatively flat in the center of the distribution of achievement scores.[14]

In general, this procedure should allow reasonable estimation of extrema (minima and maxima) for curves expressed either as logit or probability. Note that although these calculations give us very exact estimates (our diabetes equation has an inflection point at AGE = 74.75 years), the precision of estimates through this method is only as good as the data. In curves that replicate well, this might be useful. Where curves are not able to be replicated, this process is really not terribly useful. This is a warning all statisticians using regression or linear modeling need to keep in mind! One can model complex, beautiful curves with poor-quality, biased, or error-filled data, and the results are only as good as the ingredients.

Furthermore, there have been discussions of how to test whether individual point estimates for slope are significantly different from 0. For example, Aiken and West (1991, pp. 77–78) discuss this in regard to OLS regression. I have some reservations about probing the data too much because (a) that increases the risk of overinterpreting

---

14 There are interesting examples of application of this technique throughout various literatures in science. For example, Boyce and Perrins (1987) used this type of technique of locating extrema to understand and estimate the optimal clutch size for Great Tits (*Parus major*, the bird, although I could see how this particular phrasing could lead to confusion) in varying environmental conditions. Apparently there is a curvilinear relationship between clutch size (number of eggs laid) and the number of chicks that survive to breed as adults, and this curve is also influenced by whether the year was "bad" or "good" for the birds.

the data, unless it is a very large and representative sample, and (b) this too is beyond the scope of this chapter. Perhaps if you encourage all of your colleagues and friends to buy this book, I will add more of these advanced topics in a second edition!

SUMMARY

This chapter explored how to model curvilinear effects in OLS regression as well as logistic and multinomial regression. It is relatively simple to find curvilinear effects if you look for them. There are several more examples in the Enrichment section, below.

I had intended to include an example of a curvilinear effect that was due to extreme scores (certainly a possibility!) but I was unable to find one in the data sets I was working with. One of the reasons there are so many examples at the end of the chapter relative to other chapters is that as I kept searching for a counterexample (removing inappropriately influential scores removed a curvilinear effect), I repeatedly came across relatively powerful and interesting examples of how data cleaning enhanced curvilinear effects. After trying many different modes of data cleaning, I failed to find a reasonable example that used appropriate data cleaning to remove a curvilinear effect due solely to inappropriately influential cases. Of course I could manufacture an artificial example, and perhaps I will in the future. At this point, there are two main messages from this chapter.

First, checking analyses for curvilinear effects is not terribly difficult nor is it particularly time-consuming. In a few minutes you can create quadratic and cubic terms for important variables, and in a few seconds an analysis can demonstrate whether there might be a nonlinear effect. A few minutes more spent data cleaning may amplify or attenuate the effect, and you may end up with a very interesting result.

Second, if you are familiar with simple calculus concepts, you can extract interesting details from well-modeled (and replicable) curvilinear equations (such as where the curve flattens out and turns the opposite direction). If you enjoyed this chapter, you will enjoy the chapters to come, in which we have fun with multiple predictors, interactions, and even curvilinear interactions![15]

ENRICHMENT

1. Download the AAUP, NELS88, and NHIS2010 data used in the examples from the chapter and replicate the results from the chapter.

---

15 You may think we were performing analyses that included multiple predictors in this chapter—and in a sense we did, because there were multiple terms being entered as predictors. However, technically, BMI, BMI^2, and BMI^3 are all different aspects of the same variable. So, in my mind, we were still performing univariate analyses.

2. Download the EVERMJ data (NELS88) (similar to the data that we explored in Chapter 5) and explore whether there is a curvilinear effect of zACH on EVERMJ.

    a. Perform appropriate data cleaning and tests of assumptions.
    b. Summarize results in APA format.
    c. Graph effect in conditional probabilities.
    d. *Bonus:* Calculate extrema as in the calculus section above to see if your results match mine.

3. Within the NHIS2010 data, explore whether AGE predicts DIABETES, whether there are curvilinear effects, and so forth.

    a. Perform appropriate data cleaning and tests of assumptions.
    b. Summarize results in APA format.
    c. Graph effect in conditional probabilities.
    d. *Bonus:* Calculate extrema as in the calculus section above to see if your results match mine.

4. OLS regression example: Within the NHIS2010 data on BMI and AGE, complete the following:

    a. Perform appropriate data cleaning.

        i. Both variables have problematic values.

    b. Center age at the median (46) and explore whether age predicts BMI.
    c. Graph.

        ii. Convert centered AGE to actual age to make it easier on the reader, but be sure to do the calculations so that you use the centered age.

    d. Summarize in APA format.
    e. *Bonus:* Calculate the age at which average BMI peaks in this population.

5. Using the Natality 2013 data (from the Centers for Disease Control and Prevention data on births in the United States) on the book website, predict birth weight (BIRTHWT; measured in grams) from gestational age (GESTWEEK38; number of weeks gestation centered at 38, which is generally considered full term).

    a. Perform appropriate data cleaning.
    b. Graph the curve.
    c. Summarize in APA format.

6. Using the Natality 2013 data, determine if the amount of weight the mother gains (MOM_WTGAIN) is a significant and important predictor of the birth weight of the infant (BIRTHWT).

    a. Perform appropriate centering of the predictor variable, and then perform appropriate analyses.
    b. Graph the curve.
    c. Summarize in APA format, particularly focusing on effect size (R^2 in particular).

REFERENCES

Aiken, L. S., & West, S. (1991). *Multiple regression: Testing and interpreting interactions*. Thousand Oaks, CA: SAGE.

Box, G. E. P., & Tidwell, P. W. (1962). Transformation of the independent variables. *Technometrics, 4*(4), 531–550. doi: 10.1080/00401706.1962.10490038

Boyce, M. S., & Perrins, C. M. (1987). Optimizing Great Tit clutch size in a fluctuating environment. *Ecology, 68*(1), 142–153. doi: 10.2307/1938814

Cohen, J., Cohen, P., West, S., & Aiken, L. S. (2002). *Applied multiple regression/correlation analysis for the behavioral sciences*. Mahwah, NJ: Lawrence Erlbaum.

DeMaris, A. (1993). Odds versus probabilities in logit equations: A reply to Roncek. *Social Forces, 71*(4), 1057–1065. doi: 10.2307/2580130

MULTIPLE INDEPENDENT VARIABLES

8

Advance Organizer

Those individuals who visit doctors are more likely to die than those who do not—a significant and replicable correlation in the literature. Do we have an epidemic of homicidal doctors rampaging through our society? Or is it perhaps an artifact of another variable? Perhaps people who are more ill, and therefore more likely to die, are those most likely to visit doctors. Or perhaps there is a curvilinear effect, such that those most likely to visit doctors are either very health conscious or very ill, and individuals in the former group are much less likely to die and the latter group much more likely to die.

Students who come from families with lower socioeconomic status (SES) are less likely to thrive in school. Does that mean poorer students are just dumber? Or perhaps is it another set of factors that explain the link between affluence and academic performance? I am sure you have some ideas about this already.

Regardless of what is truly going on, we cannot begin to think about answering these questions with the techniques we have explored thus far. In this chapter, we will explore regression with multiple independent variables (IVs), and in Chapter 9, we will explore how to model interaction effects between IVs. Having multiple variables in a single analysis allows us to examine the *unique* contributions of each variable.

In this chapter, we will cover

- Basics of multiple predictors in a single equation
- Zero-order versus semipartial correlation
- Different methods of entering variables into the analysis
- Real data examples of multiple regression (ordinary least squares [OLS] and logistic)
- Interpreting the intercept when there are multiple IVs
- Issues with collinearity
- Using multiple regression for theory testing (and analysis of covariance [ANCOVA])
- Examples of American Psychological Association (APA)–compliant summaries of analyses

Guidance on how to perform these analyses in various statistical packages will be available online at study.sagepub.com/osbornerlm.

The Basics of Multiple Predictors

In simple regression analyses, we have thus far modeled a single predictor as in Equation 8.1a:

$$\text{Logit}(\hat{Y}) = b_0 + b_1 X_1 \tag{8.1a}$$

or as in Equation 8.1b:

$$\hat{Y} = b_0 + b_1 X_1 \tag{8.1b}$$

We have also modeled curvilinear effects, which look a little more complicated but are still *univariate* analyses in that there is only one dependent variable (DV), although we model different aspects of the DV, as you can see in Equations 8.2a and 8.2b:

$$\text{Logit}(\hat{Y}) = b_0 + b_1 X_1 + b_2 X_1^2 + b_3 X_1^3 + \cdots + b_K X_1^K \tag{8.2a}$$

$$\hat{Y} = b_0 + b_1 X_1 + b_2 X_1^2 + b_3 X_1^3 + \cdots + b_K X_1^k \tag{8.2b}$$

In fact, even in Chapter 4 we had multiple terms in our equation, as you can see in Equations 8.3a and 8.3b, although again, these were different aspects of the same polytomous IV with k levels or groups:

$$\text{Logit}(\hat{Y}) = b_0 + b_1 \text{DUM1} + b_2 \text{DUM2} + b_3 \text{DUM3} + \ldots + b_{k-1} \text{DUM}(k-1) \tag{8.3a}$$

$$\hat{Y} = b_0 + b_1 \text{DUM1} + b_2 \text{DUM2} + b_3 \text{DUM3} + \ldots + b_{k-1} \text{DUM}(k-1) \tag{8.3b}$$

In this chapter, we will add a second variable to the equation, as you can see in Equations 8.4a and 8.4b:

$$\text{Logit}(\hat{Y}) = b_0 + b_1 X_1 + b_2 X_2 \tag{8.4a}$$

$$\hat{Y} = b_0 + b_1 X_1 + b_2 X_2 \tag{8.4b}$$

These could be generalized to more than two IVs, of course.

Procedurally, when we talked about dummy coding variables in Chapter 4 and curvilinear effects in Chapter 7, we already introduced the procedural aspects of adding more than one term to the equation, although in those examples we were adding multiple things to an equation in order to represent different aspects of a *single variable*. In this case, let us assume that we are truly adding a second, conceptually distinct variable to the equation.

What Are the Implications of This Act?

When we only have one variable in a regression equation, we can answer the question "Does this variable have a significant/important relationship to the outcome?" The simple act of including a second variable in the equation allows us to answer a more important and interesting question: "Which is the *most important* predictor of the outcome?"

When looking at simple (one-predictor) equations, it is difficult to assess relative importance of variables. Sure, family SES is an important predictor of academic outcomes, but is it *more important or less important* than prior academic achievement? And perhaps even more intriguing, are SES and achievement independent predictors of academic outcomes, or do they interact in some way such that the effect of achievement is dependent on level of SES, or vice versa? Unless we get both variables in the same equation, we cannot answer these more interesting questions.

When we have only one variable in the equation, what we get is a *zero-order* relationship (e.g., a zero-order correlation), which is another, more technical term for a simple relationship between two variables. This does not take into account the effects of other variables. When we calculate the relationship between two variables controlling for, or taking into account, a third variable, we get *first-order* or *semipartial* relationships (e.g., a semipartial correlation).[1] With two or more variables in the regression equation, the effect of each variable is assessed *controlling for all other variables* in the equation. In other words, what we are getting is an estimate of the *unique* effect of a particular variable above and beyond the effects of all other variables in the equation, which can be very informative and useful. By taking any semipartial correlations and squaring them, we can describe and quantify the unique relationship between two variables, controlling for other specific variables as unique variance accounted for.

In Figure 8.1, I have tried to represent this graphically. If you have two IVs (X_1 and X_2) and one DV (Y), you could perform two simple regression analyses, describing the simple (zero-order) relationship between each IV and Y. These relationships (percent variance accounted for) are represented by a and b in the left side of Figure 8.1.

But what neither of these analyses will tell us is whether one variable is *more important* than the other. If X_1 and X_2 are completely uncorrelated, then and only then can we compare which is most important via simple regression, as entering both analyses into a multiple regression analysis will produce identical results. More likely, the two IVs are at least modestly correlated. In this case, when we enter both variables into the equation, their unique relationships (unique variance accounted for, or their semipartial correlations) will change, as you can see in the right side of Figure 8.1. The unique relationships are still represented by a and b, but now we also have a representation of

---

1   Nerdly trivia: These relationships are called "zero order" and "first order" because there are zero variables partialed out or covaried in the simple correlation. With one variable covaried or partialed out, that is a first-order relationship. Technically, we can represent a correlation between two variables as r_{xy} and a semipartial correlation between X and Y controlling for Z as $sr_{xy.z}$. If we had multiple variables being covaried, we could have terms like $sr_{xy.zabc}$. Hopefully you get the idea.

Figure 8.1   Graphical Representation of Zero-Order and Semipartial Correlations

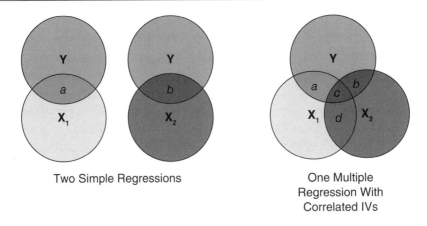

Two Simple Regressions

One Multiple
Regression With
Correlated IVs

Figure 8.2   Permutations of Multiple Regression and Relationships Between
Independent Variables

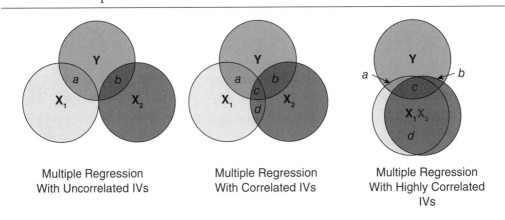

Multiple Regression
With Uncorrelated IVs

Multiple Regression
With Correlated IVs

Multiple Regression
With Highly Correlated
IVs

the correlation between the two variables $(c + d)$. As you can see, the unique relationships are each diminished to an extent—the stronger the correlation between the two
IVs, the more the unique variance accounted for is likely to change. I have represented
three possible permutations of this in Figure 8.2.

As you can see in Figure 8.2, we can have a wide range of correlation between IVs
from two variables that are perfectly uncorrelated (left), moderately correlated (center), or very highly correlated (right). In multiple regression, one of our effect sizes
is R (and R^2), which is the multiple correlation between all IVs and the DV. When
IVs are uncorrelated, R^2 will be the sum of the two squared semipartial correlations
($sr_{Y1.2}^2$ and $sr_{Y2.1}^2$), simply adding the variance accounted for in Y by X_1 (controlling
for X_2, which has no effect) and the variance accounted for in Y by X_2 (controlling
for X_1, which again has no effect). This is represented in Equation 8.5:

$$R^2 = sr_{Y1.2}^2 + sr_{Y2.1}^2 \qquad (8.5)$$

Once we move to a set of IVs that are at least minimally correlated, the sum of these two semipartial correlations, these unique variances, will always be less than R^2 (as Equation 8.6 shows) because the shared (nonunique) variance (depicted by c in Figures 8.1 and 8.2) is included in the total variance accounted for (R^2) but is not included in either semipartial (a, b).

$$R^2 > sr^2_{Y1.2} + sr^2_{Y2.1} \tag{8.6}$$

As the correlation between IVs grows, the disparity between R^2 and the sum of the semipartials will increase. At some upper limit, the correlation between the two variables gets so high that we have issues with collinearity, and the results will become chaotic or perhaps the analysis will result in an error message. We will discuss collinearity more later in this chapter.

Hypotheses to Be Tested in Multiple Regression

Because of the nuances and possible permutations in the various scenarios presented above, which you might encounter with data in the wild, we need to be methodical in our approach to modeling as complexity increases. In simple OLS regression, the overall model statistics were identical to the statistics for the single IV in the analysis. In several other chapters, you saw that the overall model statistics were the sum of the effects of several dummy variables, or of linear and nonlinear effects related to the same variable. At this point, we need to clearly distinguish the overall model statistics from the individual variable statistics. Thus, from this point forward, we will test one set of hypotheses for the overall model, which is represented by a multiple correlation (R) or the overall variance accounted for (R^2):

$$H_0: R = 0; R^2 = 0$$

$$H_a: R \neq 0; R^2 \neq 0$$

Generally, this omnibus null hypothesis is testing whether the overall model explains any variance in the DV. Another way of thinking about this hypothesis is that it is testing whether all coefficients are simultaneously zero. If we do not reject this omnibus null hypothesis, we should declare the model "not significant" and move on to something different. However, if we reject the null hypothesis, we assert that there is *some relationship somewhere* within the data that is explaining some variance in the DV. It is then appropriate for us to examine individual terms in the analysis for significant and important effects. Thus, we have hypotheses about individual effects, such as the intercept:

$$H_0: b_0 = 0$$

$$H_a: b_0 \neq 0$$

As I have mentioned in previous chapters, my experience is that many researchers overlook the intercept out of habit because it is too often not a meaningful data point. However, I encourage you to keep it in mind because it may be valuable or interesting, as we have seen already in this book. In addition to this hypothesis test concerning the intercept, we have hypotheses about each individual variable in the equation, each of which is independent. Thus, if we have k variables in the equation, we will have

$$H_0: b_1 = 0$$

$$H_a: b_1 \neq 0$$

$$H_0: b_2 = 0$$

$$H_a: b_2 \neq 0$$

.

.

.

$$H_0: b_k = 0$$

$$H_a: b_k \neq 0$$

Assumptions of Multiple Regression and Data Cleaning

When we have two or more IVs in the equation, things can get slightly more complex but not much more than for simple OLS regression. We still have the assumptions we discussed already, including completeness of the model, independence of observations, linearity, homoscedasticity, normality of residuals, and so forth.

There are a few wrinkles and nuances that are worth noting, however. First, because we have multiple IVs, we have the potential for issues of collinearity,[2] which can cause problems for the model. Collinearity issues can happen when predictor variables are highly correlated with each other. Variables that are very highly correlated with each other (e.g., over $r = 0.90$, usually) can substantially distort the parameter estimates for each variable, because each variable is estimated controlling for all other variables in the equation. When there is such substantial overlap in variables, there is little unique variance, and as this goes toward zero, the ability for our algorithms to accurately estimate parameters diminishes in either OLS or logistic regression. For example, in the analysis in Table 8.1a, I created a version of zSES that is highly correlated ($r = 0.98$) but not redundant with the original zSES.

As you can see in Table 8.1a, both versions of SES individually account for substantial improvement in model fit ($\chi^2_{(1)} = 1{,}167.51$ and $1{,}112.53$, respectively). However,

---

2   Some folks refer to this issue as "multicollinearity" and some simply "collinearity." I have not been able to figure out if there is truly a difference between the two terms . . . so I prefer the latter just because it requires less typing. This reminds me of the "inflammable" versus "flammable" issue we have in this beautiful language of ours. Or, if you grew up in upstate New York, you might be familiar with the inaccurate vernacular interchangeability of "unthaw" versus "thaw."

Table 8.1a   Example of High Collinearity Among Two Predictors

			Score	df	p
Step 0	Variables	zSES	1,167.513	1	.000
		zSES_v2	1,112.529	1	.000
	Overall statistics		1,321.861	2	.000

Omnibus Tests of Model Coefficients

		Chi-Square	df	p
Step 1	Step	1,247.025	2	.000
	Block	1,247.025	2	.000
	Model	1,247.025	2	.000

Variables in the Equation

		B	SE	Wald	df	p	Exp(B)	95% CI for Exp(B)	
								Lower	Upper
Step 1	zSES	−1.407	0.568	6.126	1	.013	0.245	0.080	0.746
	zSES_v2	0.130	0.030	18.473	1	.000	1.139	1.074	1.209
	Constant	−10.356	3.044	11.571	1	.001	0.000		

SOURCE: National Education Longitudinal Study of 1988 (NELS88) from the National Center for Education Statistics (http://nces.ed.gov/surveys/nels88/).

when both are entered, they improve model fit by only 1,247.03 collectively. This is a significant clue that high collinearity is present. Furthermore, we know that zSES should have an odds ratio (OR) of around 2.0, but these ORs are far from 2.00; furthermore, the 95% confidence intervals (95% CIs) do not include that value. zSES has an OR of 0.245, which is strong and in a direction opposite from what it should be, and zSES_v2 has a much weaker OR than expected. This is because each is evaluated taking into account the other. Another clue to problems is when a large effect (the OR of 0.245 is roughly equivalent to OR = 4.0) is only significant at $p < .013$ with a very large sample, whereas a much smaller effect (OR = 1.14) is significant at $p < .0001$ at the same time. This doesn't make sense, and when things that simple don't make sense, odds are you have issues with collinearity.

Although highly correlated variables are problematic, linearly dependent variables are particularly devastating to an analysis. Linearly dependent variables are those that are either identical, inverses, or sums of other variables. For example, if I had three subscales (SUB1, SUB2, and SUB3) and a total score that was the sum of those three subscales, entering all three subscales *and* the total score can result in failure of the model to converge or in the failure of one or more of the variables to enter the equation.

Another example of this sort of issue is entering a variable and then a transformation of that same variable (e.g., SES and zSES08). When I attempted this in SPSS, I received the output shown in Table 8.1b, indicating a collinearity issue.

Furthermore, one of the redundant variables is removed from the analysis. When you do this in logistic regression, you get the same result with a slightly different message (but the same outcome), as shown in Table 8.1c.

Linear dependency is usually simple to modify. In this case, SPSS refused to enter both variables, and I can choose which version I want and perform the analysis again. When variables are highly correlated but not linearly dependent, you can request collinearity statistics. When variables are perfectly uncorrelated, both the variance

Table 8.1b When Linearly Dependent Variables Are Entered Into the Same Analysis in Ordinary Least Squares Regression

Excluded Variables[a]

Model		Beta In	t	p	Partial Correlation	Collinearity Statistics		
						Tolerance	VIF	Minimum Tolerance
1	SES	.[b]	.	.	.	−1.169E-013	−5.926E+13	−1.169E-013

[a]Dependent variable: ACH.
[b]Predictors in the model: (constant), zSES08.

Table 8.1c When Linearly Dependent Variables Are Entered Into the Same Analysis in Logistic Regression

Warnings
Due to redundancies, degrees of freedom have been reduced for one or more variables.

inflation factor (VIF) and tolerance will be close to 1.0. VIF is calculated as 1/tolerance, so as one increases, the other will decrease accordingly. When these numbers get appreciably far from 1.0, there might be concern for collinearity issues, and variables should be examined for redundancy or high correlation.

When I find two variables that are so highly correlated that they might be causing collinearity issues, there are two possible methods I resort to for dealing with this. First is simply to eliminate one or the other. Some researchers feel queasy about this option, but if two variables are this highly correlated, they are almost entirely redundant and thus removing one is not harming the model. The other option I have occasionally used is to combine the two variables (e.g., by averaging) so that both pieces of (virtually redundant) information are in the model but are not causing problems.

The other issue we will face immediately in terms of data cleaning is that we have more terms in the analysis; so if you are looking at individual indicators of influence (e.g., DfBetas), you will now have more to examine. Remember, we have a DfBeta for each term in the equation. Therefore, we will have to examine each one, and we also must be very judicious and conservative in our data cleaning so that we do not remove all of our data.

Predicting Student Achievement From Real Data

In prior chapters, we examined how family SES can influence student achievement (correlation of about 0.51 between eighth-grade SES and ACH). Prior achievement is also a strong predictor of future achievement. Using our National Education Longitudinal Study of 1988 (NELS88) data, let us change the question slightly. We will now ask which eighth-grade variable, achievement test score (zACH08; converted to z-scores) or family affluence in eighth grade (zSES08; converted to z-scores), predicts achievement test scores in twelfth grade (zACH12; converted to z-scores). In our data set, the zero-order correlation between achievement in eighth grade and twelfth grade is $r_{(12,571)} = 0.846$, meaning 71.57% of the variance in twelfth-grade achievement test scores is explained by eighth-grade scores. The zero-order correlation between eighth-grade family SES and twelfth-grade achievement is $r_{(12,571)} = 0.502$, accounting for 25.20% of the variance in twelfth-grade achievement. Does this mean that if we know both scores, we can explain 96.77% of the variance in twelfth-grade achievement? Probably not, because zACH08 and zSES08 are also correlated ($r_{(12,571)} = 0.498$), meaning they share about 25% of their respective variances.[3] Let us see how the multiple regression analysis allows us to better understand these dynamics. First, we will examine the overall model statistics (Table 8.2).

As you can see in Table 8.2, we would reject the omnibus hypothesis because the model is clearly accounting for variance in the DV ($R^2 = 0.724$, $F_{(2, 12,568)} = 16,478.67$, $p < .0001$). Thus, we conclude the model is explaining a good portion of the variance in the outcome (but not nearly as much as the two zero-order correlations squared). Because we have a significant model, we can examine the individual parameters (also in Table 8.2). All three are significant (all $p < .0001$, meaning we would reject the null hypothesis for each). Thus, the constant is different from zero, as are the slopes of both zACH08 and zSES08. Interestingly, if you look at the standardized regression coefficients, which are what we must use to compare the effects of different IVs, we see a substantial difference between the magnitudes of the two effects. The standardized regression coefficient (β) for zACH08 is 0.793, but it is only 0.107 for zSES08. Thus, the unique effect of prior achievement is *much* stronger than the unique effect

Table 8.2   Multiple Regression Predicting Twelfth-Grade Achievement From Eighth-Grade Achievement and Family Socioeconomic Status

Model Summary[a]									
Model	R	R^2	Adjusted R^2	SE of the Estimate	Change Statistics				
					R^2 Change	F Change	df1	df2	Sig. F Change
1	0.851[b]	0.724	0.724	0.51654307	0.724	16,478.671	2	12,568	.000

[a]Dependent variable: zACH12.
[b]Predictors: (constant), zSES08, zACH08.

Coefficients[a]										
Model	Unstandardized Coefficients		Standardized Coefficients	t	p	95% CI for B		Correlations		
	B	SE	Beta			Lower Bound	Upper Bound	Zero-Order	Partial	Part
(Constant)	−0.074	0.005		−15.904	.000	−0.077	−0.059			
1 zACH08	0.785	0.005	0.793	146.648	.000	0.775	0.796	0.846	0.794	0.687
zSES08	0.107	0.005	0.107	19.736	.000	0.270	0.330	0.502	0.173	0.092

[a]Dependent variable: zACH12.

---

3   Some of you might be wondering why the correlation is not 0.505, as in a previous chapter. In this case, if you are paying attention to the degrees of freedom, we have a different sample because far fewer students (approximately half) completed the twelfth-grade achievement test than the eighth-grade achievement test. Therefore, the sample has changed, and the parameter estimate has thus changed slightly. At this point, I also have not done any data cleaning.

of SES. We can also quantify these effects in a slightly different way, by examining the squared semipartial correlations for each (sr^2). These are labeled "PART" correlations by SPSS.[4] If you square each semipartial correlation (0.687 and 0.092, respectively), you can estimate the unique variance accounted for (47.20% and 0.85%, respectively). Thus, we can say that after controlling for family SES, eighth-grade achievement accounts for almost half of the variance in twelfth-grade achievement. However, after controlling for achievement, eighth-grade family SES accounts for almost no variance in twelfth-grade achievement.

Where Is the Missing Variance?

Note that if you add these two squared semipartials together, you can see about 48.04% variance accounted for (these are conceptually a and b in Figures 8.1 and 8.2); however, in the overall model, there is about 72.4% accounted for overall. The difference between these two numbers represents the shared (nonunique) variance accounted for in the DV by both IVs—we represented this as c in Figures 8.1 and 8.2. It is still variance explained in the DV, but it does not get assigned to a particular IV because it is not unique to one or the other.

Testing Assumptions and Data Cleaning in the NELS88 Data

Let us briefly explore some of the typical data cleaning one would perform with this type of analysis. We still get residuals in this type of analysis—now the residuals reflect the unexplained portions of the data after accounting for all IVs. We still want them normal. As you see in Figure 8.3, these are relatively symmetrical in distribution, but somewhat kurtotic (skew = −0.34, kurtosis = 1.33). Removal of cases with standardized residuals more than |3| in magnitude ($N = 109$, new sample of 12,462) brought the residuals into closer alignment with the standard normal distribution (skew = −0.13, kurtosis = 0.16). As you can see in Figure 8.4, after removal of these inappropriate cases, the zPRED versus zRESID plot is somewhat egg shaped, indicating that we could argue for not meeting the assumption of homoscedasticity and recommend further transformations or data cleaning. However, the effect is so overwhelming in magnitude that it is not likely to change substantially regardless. Thus, at this point, I would note the final results with the caveat that there was some modest heteroscedasticity. These final results are presented in Table 8.3, and as you can see, they do not change substantially from Table 8.2 (results prior to data cleaning). With relatively narrow confidence intervals around the effect sizes, we can feel relatively confident in generalizing these results.[5]

---

4 I have this vague feeling that I knew the history of this odd naming convention at one time and have now forgotten it. There are various theories on various websites but none explain them satisfactorily. We will have to live in ambiguity.

5 Of course, I did not do the necessary weighting and adjustments for sampling in this complex data set and thus we would not draw *any* conclusions from these results without first taking proper actions in that area. This is true of all analyses in this book.

Figure 8.3   Distribution of Standardized Residuals Prior to Data Cleaning

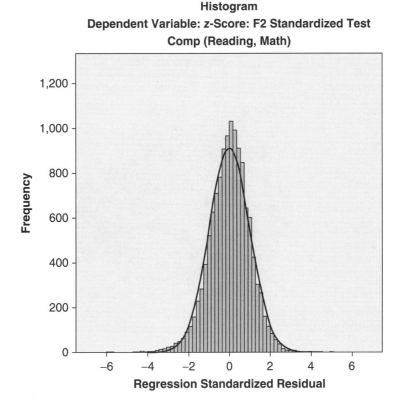

What Does the Intercept Mean When There Are Multiple IVs?

As you may remember from previous chapters, the intercept is the value when all other terms are 0. When there are multiple IVs in the equation, the intercept is the expected value of Y when all other variables are zero. Depending on your variables, this will have different meanings. In this example, because all variables were centered through conversion to z-scores, the intercept is the expected twelfth-grade achievement when both eighth-grade achievement and family SES are zero (or the mean). If all variables are centered, then the meaning of the intercept is the predicted value of the DV when one is average on all other variables. In this example, the intercept is −0.068, meaning that if a student is average on achievement and family SES at eighth grade, we can expect that student to be slightly below average at twelfth grade.

Methods of Entering Variables

There are several methods of entering variables into the equation, depending on your goals for the analysis. In general, the methods fall into two groups: analyst controlled and software controlled. I will begin this section by telling you that some fields have strong biases against software-controlled entry (which includes procedures such as

Figure 8.4 Standardized Predicted Values Versus Standardized Residuals Following Minimal Data Cleaning

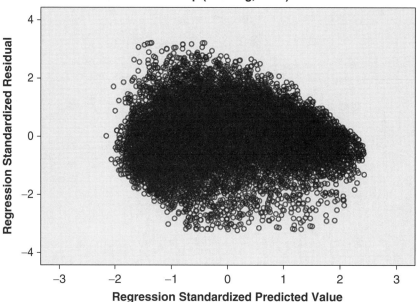

Scatterplot
Dependent Variable: z-Score: F2 Standardized Test
Comp (Reading, Math)

Table 8.3 Multiple Regression Predicting Twelfth-Grade Achievement From Eighth-Grade Achievement and Family Socioeconomic Status, Following Data Cleaning

Model Summary[a]

Model	R	R^2	Adjusted R^2	SE of the Estimate	Change Statistics				
					R^2 Change	F Change	df1	df2	Sig. F Change
1	0.868[b]	0.753	0.753	0.48644641	0.753	18,982.192	2	12,459	.000

[a]Dependent variable: zACH12.
[b]Predictors: (constant), zSES08, zACH08.

Coefficients[a]

Model		Unstandardized Coefficients		Standardized Coefficients	t	p	95% CI for B		Correlations		
		B	SE	Beta			Lower Bound	Upper Bound	Zero-Order	Partial	Part
1	(Constant)	−0.068	0.004		−15.420	.000	−0.077	−0.059			
	zACH08	0.801	0.005	0.812	157.789	.000	0.791	0.810	0.863	0.816	0.703
	zSES08	0.102	0.005	0.103	19.929	.000	0.092	0.112	0.509	0.176	0.089

[a]Dependent variable: zACH12.

stepwise, forward, or backward entry). All forms of entry were developed to address specific needs, and software-controlled entry procedures were developed with legitimate goals in mind.

User-Controlled Methods of Entry

The most widely used and most widely accepted is simultaneous or forced entry. In this entry method, the analyst tells the software what variables to enter, they are all entered at once (simultaneously), and that is the end of it. This is one of the most common and most accepted methods of entry because it is often theory driven—meaning that the analysis is proceeding based on prior theory or research and is thus supposedly more defensible. This is the method of entry I use most often.

Hierarchical Entry

A variant of the user-controlled simultaneous entry is hierarchical (forced-order) entry, where the user specifies which variables enter the equation in which order. Each variable represents a unique step, and each time a variable is entered into the equation. In OLS regression, we can compare the increment in R and R^2 that is attributable to that particular variable, in addition to examining how the individual effects of other variables already in the equation change as a result of entry of that variable. This often allows us to draw conclusions that are theoretically important, such as whether a particular variable's effect is spurious, or whether a particular variable adds any unique predictive effect over that of those variables already in the equation. Of course, in logistic regression, we do not look at R and R^2 (nor do I advocate the use of their *pseudo*-variants); rather, we can look at overall model fit in -2 log likelihood ($-2LL$) increments. One thing I do not like about the labeling of this procedure as "hierarchical" entry is the potential for confusion with *hierarchical linear modeling*, which is a completely different procedure with completely different goals, methods, and so forth and will be discussed in Chapter 13.

Blockwise Entry

A variant of hierarchical entry, this entry method allows us to enter blocks of variables in groups, rather than individually, assessing change in model fit and individual variable effects after each *group* of variables is entered. For example, we could have a group of variables that represent covariates (variables we need to have in the equation but are not substantively interested in), a group of variables that represent a theoretically coherent group, and another group of variables that represent another coherent group. To illustrate this example, let's refer to Figure 8.5, which I adapted from an article by my colleague Brett Jones and me (Osborne & Jones, 2011). As you can see in Figure 8.5, we hypothesize a rather complicated model to explain motivation and academic outcomes for students.

In a blockwise-entry scheme, we could enter all antecedent variables (e.g., race, sex, family environment, etc.) as one group of variables, and we could then enter identification with academics, goals, beliefs, and so forth as a group. After entering that second block, we could look at change in model fit and so forth as a function of the entire block of variables in addition to the individual variables. We could then enter a third block of variables representing the group of variables labeled choices, persistence, and effort and again assess the contribution of this group of variables to the overall model.

Figure 8.5   Precursors and Consequences of Identification With an Academic Domain

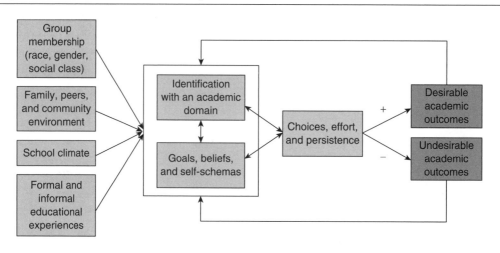

SOURCE: Osborne, J. W., & Jones, B. D. (2011). Identification with Academics and Motivation to Achieve in School: How the Structure of the Self Influences Academic Outcomes. *Educational Psychology Review,* 1-28. © Springer Science+Business Media, LLC 2011. With permission of Springer.

This is particularly useful when you have dummy- or effects-coded variables but want to assess them as a single effect. As we covered previously, three dummy-coded variables that represent smoking status are not three separate variables; rather, they are three terms that collectively account for one variable. Using blockwise entry, we could enter the three dummy-coded variables representing smoking on a single block, assessing the overall effect of the variable in the individual model.

One important thing to keep in mind is that it does not matter how a variable gets into a regression equation. Once a particular set of variables are in the equation, their parameters are all estimated simultaneously controlling for all other variables in the equation; thus, the results are identical regardless of the order in which they were entered, whether they were entered all at once or one at a time or in groups or blocks. What will be different is the information you get in looking at incremental change in the model.

Software-Controlled Entry

Stepwise entry is an innocuous-sounding method of entry that has long been part of statistical modeling. Stepwise methods were developed to let researchers automatically comb through large data sets to empirically select an optimal set of variables that best predict an outcome. For example, if you are analyzing a database (e.g., our National Center for Education Statistics database examining dropout/graduation) with many thousands of data points and more than 2,000 variables, *and* if you do not have a theoretical basis for choosing variables, then you could imagine that stepwise regression methods might be valuable to help you identify important predictor variables. I also suspect that stepwise methods can help identify *unexpected* results, which are often more interesting than expected results.[6] So if I were only interested in best predicting who will drop out or graduate from among the hundreds of variables available to me, I can run a stepwise regression analysis and the software will identify variables with the largest predictive effects and enter them into the equation for you. I might discover that variables that make sense are the best predictors (e.g., indicators of academic achievement), but it might be the case that something unusual is a strong predictor (e.g., shoe size). If you only care about creating the best prediction equation, this might be a good path for you. An unexpected result such as shoe size predicting dropout risk could also lead to new discoveries about the nature of the phenomenon (it could get you laughed out of your profession; see Fleischmann, Pons, & Hawkins, 1989).

There are probably important applications for stepwise methods in many fields from medicine and business to the social sciences, and particularly in the "big data" movement that has recently emerged. But you should know that in the academic literature, there has been significant pushback against these methods in recent decades (e.g., Cohen, Cohen, West, & Aiken, 2002; Thompson, 1995), particularly within the social sciences. Specifically, many authors argue that atheoretical analyses lead to less valuable analyses than theory-driven analyses (although I might disagree, depending in particular on the application). One of the biggest criticisms of this approach is that stepwise techniques will capitalize on chance quirks in your data, which may lead to nonreproducible results (but see my article on prediction equations and cross-validation of regression analyses for a method to test reproducibility; Osborne, 2000, 2008). Thompson (1995) outlines some other serious challenges to the use of these methods, including issues with incorrect degrees of freedom (df) and inability to identify a best variable set of a given size. However, Cohen et al. (2002) do concede that stepwise methods are less problematic when (a) the research goal is entirely predictive (as discussed above), (b) the sample size is very large (to reduce odds of spurious results due to capitalization on chance), (c) the number of variables is reasonable (they suggest a ratio of 40:1 for sample size to variable number), and (d) the results are cross-validated (as I discuss in my articles). The easy availability of resampling techniques (which we will discuss in later chapters) also makes this potentially less of an issue, because researchers can test whether a particular solution is routine and generalizable.

There are two basic approaches to stepwise regression: forward entry and backward elimination (and often a "hybrid" version that leverages both approaches). In forward entry, we start with a blank slate, and all variables are assessed for their potential predictive power. The variable with the single greatest relationship is entered into the equation, and then all remaining variables are assessed as to whether they add significant predictive power to the equation above that variable in the equation. The next strongest predictor is added, and the process continues until no variable rises above a particular threshold. In backward elim ination, all variables are put in the equation and the nonsignificant ones are removed from the equation one at a time based on

---

6   Unless you are analyzing your dissertation data. In this case, you definitely want only your expected results, or more precisely, the results expected by your dissertation chair and committee.

criteria similar to that used in forward entry. Once all variables in the equation are above the threshold for elimination, the equation is finalized. Depending on the software used, there might be various hybrid techniques available. For example, if forward entry is used, it is possible that a variable entered early in the process has become a nonsignificant predictor once other variables are in the equation. In some procedures, those variables could be dropped from the analysis and reassessed. Likewise, in backward elimination, it is possible that a variable eliminated from the equation early on would be significant at a later point after other variables are removed. Hybrid processes can handle these situations, but again have benefits and drawbacks.

One significant objection I have to these processes is that they cannot take into account nonlinear relationships (curvilinear relationships) or interactions. We have already seen examples of curvilinear effects that would not enter an equation. Later in this chapter, we will see interaction terms that are important and not modeled in these methods. Finally, data cleaning is not straightforward in these situations. It is entirely possible that a variable has a strong effect but is not recognized as important without data cleaning. Because stepwise methods cannot easily handle these issues (which eliminates many important effects from being examined), I do not endorse these methods except for use in an exploratory manner.

Using Multiple Regression for Theory Testing

One of the reasons why we use multiple regression is because we can do more than describe simple relationships. We can determine the strongest predictors of a particular outcome, and we can also do some modest theory testing. For example, early in my career, I was exploring racial disparities in educational outcomes. Back in the 1980s, when the NELS88 data were gathered, there were predictable disparities between racial groups. Usually, African American, Latino, and Native American students were more likely to underperform on academic tasks, particularly high-stakes tests such as the standardized test(s) in NELS88. One narrative that evolved in this line of research was that it was not race itself that was the primary issue, but rather poverty. In fact, there is a strong relationship between race and poverty, so this seems like one plausible explanation for at least part of the pattern.

Using regression, we can explore hypotheses like this (we will examine this methodology in more detail in Chapter 13). Using the same data set, I dummy coded the RACE variable as shown in Table 8.4.

We will now perform a generalized linear model (GLM) version of ANCOVA, where we have a polytomous IV (RACE) and zSES08 covaried, predicting eighth-grade achievement (zACH08). First, to get a baseline of the size of the effect, we will enter the dummy-coded variables that represent the effects of race alone. Then we will enter the dummy variables that represent the effect of RACE *after* we enter zSES08 to see how much variance the combined effect of RACE accounts for after SES is taken into account. As you can see in Table 8.5, RACE accounts for about 10.9% of the variance in student achievement (before any data cleaning or testing of assumptions). As you can also see, consistent with the expectations from that era,

Table 8.4   Dummy Coding of RACE Variable in NELS88

RACE	DUM1	DUM2	DUM3	DUM4
White	0	0	0	0
African American	1	0	0	0
Latino	0	1	0	0
Asian/Pacific Islander	0	0	1	0
Native American	0	0	0	1

Table 8.5   Effect of RACE on Eighth-Grade Student Achievement (zACH08)

Model Summary

Model	R	R^2	Adjusted R^2	SE of the Estimate	R^2 Change	F Change	df1	df2	Sig. F Change
1	0.331[a]	0.109	0.109	0.94291558	0.109	719.017	4	23,427	.000

[a]Predictors: (constant), DUM4, DUM3, DUM1, DUM2.

Coefficients[a]

Model		Unstandardized Coefficients		Standardized Coefficients			95% CI for B	
		B	SE	Beta	t	p	Lower Bound	Upper Bound
1	(Constant)	0.192	0.008		25.580	.000	0.177	0.207
	DUM1	−0.796	0.019	−0.261	−41.597	.000	−0.833	−0.758
	DUM2	−0.666	0.019	−0.223	−35.529	.000	−0.703	−0.630
	DUM3	0.119	0.026	0.029	4.656	.000	0.069	0.169
	DUM4	−0.802	0.056	−0.089	−14.408	.000	−0.912	−0.693

[a]Dependent variable: zACH08.

students from traditionally disadvantaged minority groups tend to score significantly and substantially lower than White students, who also score a bit lower than Asian/Pacific Islander students.

If my assertion is correct (that much of this effect is due to poverty and affluence, rather than racial/ethnic background), the effect of RACE should be substantially diminished once SES is in the equation.

In the second analysis, I entered zSES08 first and then entered RACE on a second step so we would get change in R^2 and other statistics to allow us to evaluate the collective effect of the four dummy variables. The results of this new analysis are presented in Table 8.6. By requesting statistics related to change in model fit and entering the dummy variables as a separate block, we can get valuable information that informs questions such as this. As you can see in Table 8.6, after controlling for the effect of SES on achievement, RACE has a substantially diminished effect (change in R^2 of 0.032 versus 0.109 prior to controlling for SES). Although still significant, you can see by comparing the unstandardized regression coefficients that the adjusted mean differences between groups are now also much diminished. Thus, we have a

Table 8.6 Effect of RACE After Controlling for the Effect of Socioeconomic Status

Model Summary									
Model	R	R^2	Adjusted R^2	SE of the Estimate	Change Statistics				
					R^2 Change	F Change	df1	df2	Sig. F Change
1	0.502[a]	0.252	0.252	0.86422380	0.252	7,876.619	1	23,423	.000
2	0.533[b]	0.284	0.284	0.84534774	0.032	265.430	4	23,419	.000

[a]Predictors: (constant), zSES08.
[b]Predictors: (constant), zSES08, DUM4, DUM3, DUM1, DUM2.

Coefficients[a]							
Model	Unstandardized Coefficients		Standardized Coefficients	t	p	95% CI for B	
	B	SE	Beta			Lower Bound	Upper Bound
1 (Constant)	0.001	0.006		0.124	.902	−0.010	0.012
zSES08	0.502	0.006	0.502	88.750	.000	0.491	0.514
2 (Constant)	0.101	0.007		14.775	.000	0.088	0.114
zSES08	0.444	0.006	0.444	75.648	.000	0.433	0.456
DUM1	−0.494	0.018	−0.162	−28.023	.000	−0.528	−0.459
DUM2	−0.301	0.017	−0.101	−17.214	.000	−0.336	−0.267
DUM3	0.098	0.023	0.024	4.259	.000	0.053	0.143
DUM4	−0.536	0.050	−0.060	−10.716	.000	−0.634	−0.438

[a]Dependent variable: zACH08.

simple example of how we can test ideas and explanations for phenomena via multiple regression (and also a relatively simple example of ANCOVA within GLM).

What Is the Meaning of This Intercept?

When we have dummy variables in a regression analysis, the meaning of the intercept is technically the same—the expected value when all other variables are zero. In this case, when all dummy variables are zero, we have the expected value of White students; because SES is centered via conversion to z-score, the intercept is the expected value for White students who are average in eighth-grade family SES. In this case, it is slightly (0.10 standard deviations [SD]) above the mean.

Logistic Regression With Multiple IVs

Now that you have already mastered logistic (and multinomial, ordinal, and Poisson) regression, in addition to OLS regression, much of the rest of this book will provide examples of these new and fun techniques in the context of different approaches. Let us explore an example of multiple regression in a logistic framework. We will examine an abbreviated example of binary logistic regression predicting high school graduation (GRADUATED) that you have seen in previous chapters.

When looking at family zSES,[7] we see that variable provides for significant improvement in model fit (analogue to R^2) when entered into the equation. The initial −2LL for the model was 9,967.24, and once zSES entered the equation, it was reduced to 8,739.91, a reduction of 1,227.33 ($\chi^2_{(1)} = 1{,}227.33$, $p < .0001$). The variable statistics are represented in Table 8.7.

---

7 In this example, we are using all eighth-grade data from NELS, so we will simply call the variables zACH and zSES.

Table 8.7   Simple Logistic Regression Equation Predicting GRADUATED With Only zSES in the Equation

		B	SE	Wald	df	p	Exp(B)	95% CI for Exp(B)	
								Lower	Upper
Step 1[a]	zSES	1.041	0.033	1,018.355	1	.000	2.831	2.656	3.018
	Constant	2.740	0.037	5,409.564	1	.000	15.480		

SOURCE: National Education Longitudinal Study of 1988 (NELS88) from the National Center for Education Statistics (http://nces.ed.gov/surveys/nels88/).

It should not be surprising that SES is a significant predictor of graduation, with an OR of 2.83, and a 95% CI of [2.66, 3.02]. We have seen this information in previous chapters. Similarly, we can look at whether eighth-grade achievement test scores (zACH) predict graduation. Again, starting with a −2LL of 9,967.24, once ACH is in the equation, this is reduced to 8,475.86, a reduction of 1,491.38 ($\chi^2_{(1)}$ = 1,491.38, $p < .0001$). At first glance, both models seem strong, with zACH having a bit larger of an effect (i.e., the model was improved more by adding that variable than the other variable with only the intercept in the equation). The variable statistics reflect this as well, as you can see in Table 8.8.

In this case, the OR is a bit higher (3.89 [95% CI = 3.58, 4.23]). Of course, this tells us important information—that both variables are strongly related to graduation. However, SES and achievement also tend to be strongly correlated—in this case, the simple correlation between the two variables exceeds $r = 0.50$, meaning there is substantial overlap in the two variables. It looks as though achievement is the more important predictor from these first simple analyses, but we can do better. As we just did in the earlier part of this chapter, we can evaluate the unique, independent predictive effects of these two variables in the context of multiple regression. Note also that combining multiple predictors into a single equation eliminates thorny issues such as the increased chance of Type I errors inherent in performing several simple logistic analyses. Thus, combining variables into a single equation has many benefits from controlling Type I error rate to being able to ask more nuanced questions of the data.

Entering both variables in the same equation yields a significant overall model. With the initial −2LL of 9,967.24, the model summary shows a final −2LL of 8,062.99, a reduction of 1,904.25 ($\chi^2_{(2)}$ = 1,904.52, $p < .0001$). This overall model fit improvement has two degrees of freedom because we entered two variables. Note also that

Table 8.8   Simple Logistic Regression Equation Predicting GRADUATED From zACH

		B	SE	Wald	df	p	Exp(B)	95% CI for Exp(B)	
								Lower	Upper
Step 1	zACH	1.359	0.043	1,003.838	1	.000	3.891	3.577	4.232
	Constant	2.955	0.045	4,326.004	1	.000	19.194		

SOURCE: National Education Longitudinal Study of 1988 (NELS88) from the National Center for Education Statistics (http://nces.ed.gov/surveys/nels88/).

entering the two variables does *not* produce the same change in −2LL as simply adding the effects from the simple logistic regression analyses. The two effects mentioned above (1,491.38 and 1,227.33) add to 2,718.71, but the change when both are entered simultaneously was 1,904.25, a difference of 814.46. This reflects the fact that the two predictors are correlated, and thus account for overlapping variance, in the same way we saw previously that the squared semipartial correlations in OLS regression do not sum to the overall variance accounted for in the model if the predictors are correlated. In logistic regression (as with any other regression analysis), if a set of predictors were perfectly orthogonal (uncorrelated), then there would be no difference between the added effects of the separate analyses and the model change when both are entered into the regression equation.[8]

As you can see in Table 8.9, the individual parameter estimates are also attenuated somewhat over what they were in the simple regression analyses. The point estimates for the ORs, for example, dropped about 0.8 and 0.9, respectively. Again, as with OLS regression, these are similarly part or corrected effects—effects corrected for (or controlling for) other variables. Thus, when student achievement is covaried, the OR for SES is 2.00, $p < .0001$. Likewise, when SES is covaried, the OR of ACH is 2.90, $p < .0001$.

We are not limited to two predictor variables. We can add student sex (coded 0 = female, 1 = male) to the equation as shown in Table 8.10.

Table 8.9   Logistic Regression Predicting GRADUATED With Both zSES and zACH in the Equation

| | | B | SE | Wald | df | p | Exp(B) | 95% CI for Exp(B) | |
								Lower	Upper
	zSES [a]	0.695	0.036	381.240	1	.000	2.004	1.869	2.149
Step 1	zACH	1.066	0.046	543.383	1	.000	2.904	2.655	3.176
	Constant	3.119	0.048	4,145.332	1	.000	22.613		

SOURCE: National Education Longitudinal Study of 1988 (NELS88) from the National Center for Education Statistics (http://nces.ed.gov/surveys/nels88/).

Table 8.10   Logistic Regression Predicting GRADUATED From Socioeconomic Status, Achievement, and Student Sex

| | | B | SE | Wald | df | p | Exp(B) | 95% CI for Exp(B) | |
								Lower	Upper
	zSES	0.693	0.036	376.976	1	.000	1.999	1.864	2.144
	zACH	1.069	0.046	542.835	1	.000	2.913	2.663	3.187
Step 1	SEX	−0.055	0.059	0.875	1	.350	0.947	0.844	1.062
	Constant	3.147	0.057	3,005.791	1	.000	23.265		

SOURCE: National Education Longitudinal Study of 1988 (NELS88) from the National Center for Education Statistics (http://nces.ed.gov/surveys/nels88/).

---

8   Except for the increased probability of Type I error.

Example Summary of Previous Analysis

Prior to analysis, both SES and student achievement (ACH) were converted to z-scores (standard normal distribution) and entered simultaneously into the equation. With an initial $-2LL$ of 9,967.24 and final $-2LL$ of 8,062.99, the model represented a significant improvement in fit ($\chi^2_{(2)} = 1,904.52$, $p < .0001$). As you can see in Table 8.10, both SES and ACH were significant, unique predictors of graduation (student sex was not found to be significant after controlling for these variables and will not be discussed further). Specifically, after controlling for all other variables in the analysis, as SES increased, the probability of graduation increased (OR = 2.00 [95% CI = 1.87, 2.15]). Similarly, after controlling for all other variables in the analysis, increases in student achievement were associated with increased probability of graduation (OR = 2.90 [95% CI = 2.66, 3.18]). Because these variables were standardized, each OR represents the increase in odds of graduation for every increase of 1 *SD* in either SES or ACH.

To put these effects into perspective, students who come from families with SES 2 *SD* below the mean have an 84.93% chance of graduating high school, whereas students who come from families that are 2 *SD* above the mean in SES have a 98.91% chance of graduating (assuming average achievement, held constant). Likewise, holding SES constant at the mean, a student with achievement scores 2 *SD* below the mean would have a 72.85% chance of graduation, whereas a student with achievement scores of 2 *SD* above the mean would have a 99.48% chance of graduation.

In this case, sex is not a significant predictor of graduation, once the other variables are in the equation.

Assessing the Overall Logistic Regression Model:
Why There Is No R^2 for Logistic Regression

We are used to exploring overall model statistics in OLS regression, such as the F statistic that tests whether the proposed model explains an amount of variance that is significantly different than 0. We also get an effect size, R (and R^2), which tells us the proportion of variance in the DV that might be explained from our model.[9] In this way, we get our first look at the goodness of our model—a significance test and an effect size. When predictors are entered on more than one step, you can ask for (or calculate) the change in R (and R^2) and get a significance test of ΔR (which is also the test for ΔR^2).

It seems as though it should be simple to design analogous indicators for logistic regression. Indeed, we do have some similar pieces of information. Some of this was

---

9   R^2 is easily calculated. We can calculate the total sum of squares (SST) as the sum of all squared differences between each individual score on the DV (y_i) and the mean of the DV, $\overline{Y}$. The sum of squares for the error term (SSE) is the sum of the squared differences between each predicted value ($\hat{y}_i$) and the actual score on the DV (y_i). The regression sum of squares is merely the difference between SST and SSE, and R^2 is SSR/SST (conceptually and literally, the amount of variance accounted for by the IVs divided by the total possible variance in the DV).

already covered in previous chapters, so I will briefly review the basics. The overall summary for a logistic regression analysis has a −2LL statistic, which tests that all predictors' coefficients are 0.00. We also get a model chi-square. This is the difference between the −2LL for the initial *independence* model (no predictors in the equation) and the model being evaluated (which may be an intermediate step in the analysis or the final analysis), with degrees of freedom equal to the number of predictors added to the equation. If this chi-square statistic is significant, we can conclude that the model has improved significantly over either the independence model (no predictors) or the previous model. This is conceptually analogous to the *F* test in OLS regression. The difference is that there is no direct measure of effect size. Of course, larger chi-square changes are better than smaller changes, but this is not measured on an absolute scale. As such, it is difficult to compare non-nested models, whereas we can directly compare non-nested models in OLS regression: an $R^2 = 0.25$ is better than $R^2 = 0.10$ and is substantively interpretable. In the former analysis, we are explaining 15% more variance in the outcome than the second. Because the actual values of the −2LLs vary widely from analysis to analysis, is a change from −2LL = 17,000 to 16,000 (chi-square of 1,000) equal in magnitude and change in prediction as a change from −2LL = 5,000 to 4,000 or from −2LL = 1,500 to 500? Probably not.

It is definitely legitimate to compare nested models, and lower −2LL indicates better fit within that particular set of nested models, but it is not legitimate to compare −2LL across non-nested models (different samples, or a different DV within the same sample, for example).

Those of us with roots in OLS regression yearn to know the answer to the question "How much variance am I accounting for?" (or "How good is my model?"). Unfortunately, this is a question that is more challenging to answer in logistic regression. Conceptually, the difference between the −2LL for the independence model and the saturated model (which is perfect fit) is analogous to the SST—the goodness or lack of fit in the model. This is also sometimes referred to as *deviance*. Once predictor variables are entered into the model, the −2LL is a reduced amount of lack of fit and is now conceptually similar to SSE. The difference between the two can be thought of as SSR—the amount of the model goodness of fit or deviance that is reduced by the IVs. So it should be simple to construct an R^2 analogue, by dividing SSR/SST (model chi-square by deviance [0 minus the −2LL for the independence model]. This is essentially the approach taken in calculating the Nagelkerke R^2.

Other variants (e.g., Cox and Snell R^2, the contingency coefficient R^2, the Wald R^2, and the Brier index, to name merely a few of the proposed options) have been proposed, but none seem to be universally accepted or without serious conceptual or mathematical problems. Menard (2010, pp. 45–57) has an extensive and thorough examination of the options, and DeMaris (2002) also reviews many of the common options through Monte Carlo study. Based on several studies, conceptual ambiguities, and personal experience, my conclusion is that none of them should be used because they tend to be contradictory and problematic. For example, one analysis I performed for another purpose produced the information in Table 8.11 from SPSS.

Table 8.11 The Random Frustrations of R^2 Equivalents in Logistic Regression

Step	−2LL	Cox and Snell R^2	Nagelkerke R^2
1	507.465	0.050	0.438

SOURCE: Author's previous work.

These statistics (which are not altered in any way from the SPSS output I saw) encapsulate, in an extreme way, the issue with R^2 analogues in logistic regression. Which of the two R^2 analogues do I report? The one I *want* to report is the Nagelkerke R^2 = 0.44, rather than the Cox and Snell R^2 = 0.05, but there is no clear way for me to determine which is most accurate. In fact, throughout my writing of this book, I routinely saw these two differ, often substantially, and not always in predictable ways.

More to the point, I don't think we need an R^2 analogue at all. In structural equation modeling, we have a similar issue and seemingly endless possible indicators of goodness of fit (Akaike information criterion, normed fit index, comparative fit index, root mean square error of approximation, chi-square, and so on). That literature has not reached consensus as to what model fit statistics to report and how to evaluate them after several decades of development. Thus, I recommend against focusing on one specific index of overall model fit, and either report many (as I often do in structural equation modeling analyses) or report only the basic −2LL values and chi-square change statistics. The reality is that we have quantitative indicators of model goodness, but we also have qualitative indicators as well, such as the classification tables. Depending on your goal, it might be more important to have good classification than strong R^2 analogues.

SUMMARY AND CONCLUSIONS

Now that you have mastered simple linear models with continuous and categorical IVs and DVs, we can start having some real fun. Our first step is to understand how to add more than one IV to the equation. The benefits of this are that we not only control Type I error, but we can also answer interesting questions, such as identifying which variables are the strongest unique predictors of an outcome. We also began playing with the idea that we can ask more sophisticated, theory-driven questions of the data, as in our example of RACE and SES predicting student achievement. In that example, we were able to test an idea that racial disparities in achievement are at least partly explained by the concentration of poverty in groups who are traditionally disadvantaged.

ENRICHMENT

1. Download the data from the book's website and replicate the results in the chapter.

2. Explain what a semipartial (part) relationship is and how it is different from a zero-order correlation.

   a. Be able to explain why the sum of the squared semipartials does not equal the squared multiple R when IVs are correlated.

3. Let's return to our multinomial logistic example from Chapter 6, predicting marijuana use (MJ) from student achievement. Let's add zSES to the analysis and see which is the strongest predictor of MJ use.

4. Once you have done Exercise 3, do the same for MJ using ordinal logistic regression.

   a. See if the analysis with zSES and zACH in the equation meets the assumptions for ordinal regression.
   b. If so, summarize the results in APA format (briefly) describing which variable is the strongest predictor of MJ use.

5. Download the NHIS2010 data predicting diabetes as the DV. In Chapters 4 and 5, we explored whether smoking status (never smoked, former smoker, occasional smoker, daily smoker) was related to body mass index (BMI) or diabetes diagnosis, which turns out to be an unexpectedly interesting question, although I know almost nothing about any of the literatures.[10] In this chapter, let us begin to play with these findings. We previously saw that there was a small but significant effect of smoking status on BMI, and that smoking status (converted to dummy-coded variables) has a significant effect on DIABETES. This raises interesting questions. Let's see if BMI (centered at a reasonable number or converted to z-scores) helps explain part of the relationship between smoking and diabetes by entering BMI into the equation.

   a. Using this data set, which has only cases with valid data on all variables, replicate the analysis from Chapter 5 (predict diabetes from smoking) with only the smoking dummy variables in the equation, noting $\Delta-2LL$.[11]
   b. Using the same data, run another analysis entering zBMI before the dummy-coded variables.

---

10 Well, it is interesting to me, anyway. Knowing nothing about any of these literatures has not heretofore stopped me from using them as examples. I offer apologies to those whose life work is studying these phenomena. I hope I do not undermine your efforts!

11 You cannot compare –2LL between analyses when the sample size changes because –2LL is a function of sample size and model fit.

c. Compare Δ–2LL with that from Exercise 5a when only smoking was in the equation. How much is the effect diminished? Can we conclude that BMI helps explain the effect of smoking on diabetes?

REFERENCES

Cohen, J., Cohen, P., West, S., & Aiken, L. S. (2002). *Applied multiple regression/correlation analysis for the behavioral sciences*. Mahwah, NJ: Lawrence Erlbaum.

DeMaris, A. (2002). Explained variance in logistic regression: A Monte Carlo study of proposed measures. *Sociological Methods & Research, 31*(1), 27–74.

Fleischmann, M., Pons, S., & Hawkins, M. (1989). Electrochemically induced nuclear fusion of deuterium. *Journal of Electroanalytical Chemistry, 261*(2), 301–308. doi: 10.1016/0022-0728(89)80006-3

Menard, S. W. (2010). *Logistic regression: From introductory concepts to advanced concepts and applications*. Thousand Oaks, CA: SAGE.

Osborne, J. W. (2000). Prediction in multiple regression. *Practical Assessment, Research & Evaluation, 7*(2).

Osborne, J. W. (2008). Creating valid prediction equations in multiple regression: Shrinkage, double cross-validation, and confidence intervals around prediction. In J. W. Osborne (Ed.), *Best practices in quantitative methods* (pp. 299–305). Thousand Oaks, CA: SAGE.

Osborne, J. W., & Jones, B. D. (2011). Identification with academics and motivation to achieve in school: How the structure of the self influences academic outcomes. *Educational Psychology Review, 23*(1), 131–158. doi: 10.1007/s10648-011-9151-1

Thompson, B. (1995). Stepwise regression and stepwise discriminant analysis need not apply here: A guidelines editorial. *Educational and Psychological Measurement, 55*(4), 525–534. doi: 10.1177/0013164495055004001

INTERACTIONS BETWEEN INDEPENDENT VARIABLES

Simple Moderation

Advance Organizer

Thus far in the book, you have mastered the art of examining simple relationships between many different combinations of independent variables (IVs) and dependent variables (DVs). In Chapter 8, we began asking more nuanced questions by entering multiple predictors into our regression equations. For example, we can now explore whether particular variables are stronger predictors than others, as well as issues of unique variance accounted for when other variables are controlled for.

However, we are merely beginning to scratch the surface in terms of interesting questions we can ask of our data. Even in multiple regression, we assume that all variables have the same effect across all other groups. In this chapter, we begin to ask questions such as whether the effect of a variable differs for different groups or at different levels of another variable. For example, we can ask questions of our data such as the following:

- Does the effect of body mass index (BMI) on diabetes remain constant for the young as well as the old, or for males and females?
- Does the effect of achievement on graduation differ depending on socioeconomic status (SES) or race/ethnicity?
- Is the effect of a drug or intervention the same for all subgroups (i.e., for whom is the intervention most or least effective)?

These are, to my mind, more interesting questions than simply describing a relationship between two variables. Technically, what we are referring to is the concept of *moderation,* the idea that one variable could moderate, or influence, the effect of another. Throughout this chapter, we will be exploring examples of how one variable can moderate, or influence, the effect of another variable in predicting an outcome. However, at this point, I will not get overly involved in using the term *moderator* or moderation language. Researchers get passionate about that term, as well as its cousin, *mediation*, and the inevitable discussion of *mediated moderation.* We do not want to get bogged down in this quagmire at this time. We will explore mediation and moderation more fully in Chapter 11.

In this chapter, we will introduce linear interactions for the various models you have become familiar with (ordinary least squares [OLS], logistic, multinomial, etc.). In Chapter 10, we will increase the fun even further to explore what happens when different groups have different *curvilinear* effects (i.e., curvilinear interactions for various models).

In this chapter, we will cover

- Conceptualizing an interaction
- General procedures for testing interactions
- Testing interactions between different types of variables
- Testing interactions in various types of linear models
- Real data examples
- Interpreting and graphing results for your reader
- Post hoc probing of interactions (simple slopes analysis)
- Examples of American Psychological Association (APA)–compliant summaries of analyses

Guidance on how to perform these analyses in various statistical packages will be available online at study.sagepub.com/osbornerlm.

What Is an Interaction?

In simple regression analyses, we have thus far modeled a single predictor as in Equation 9.1a:

$$\text{Logit}(\hat{Y}) = b_0 + b_1 X_1 \tag{9.1a}$$

or as in Equation 9.1b:

$$\hat{Y} = b_0 + b_1 X_1 \tag{9.1b}$$

In Chapter 8, we added more variables to the equation, as you can see in Equations 9.2a and 9.2b:

$$\text{Logit}(\hat{Y}) = b_0 + b_1 X_1 + b_2 X_2 \tag{9.2a}$$

$$\hat{Y} = b_0 + b_1 X_1 + b_2 X_2 \tag{9.2b}$$

When we create any regression model, we are assuming that model is appropriately specified. When we have multiple IVs in the model, we should test whether the additive model (the effects of each variable add together) is the appropriate model, or we cannot be certain we are meeting this assumption. If you have a regression model with multiple IVs, you are making an explicit statement (although you may not realize it) that the effects of each IV are additive.

An interaction is a nonadditive (multiplicative) effect. By not modeling an interaction as part of a routine regression analysis, we are asserting that the slope of X_1 (e.g., SES) is constant regardless of where a person is on X_2 (e.g., achievement).

Let's look at a regression equation containing an interaction term (X_1 multiplied by X_2, which is represented by X_1X_2):

$$\text{Logit}(\hat{Y}) = b_0 + b_1X_1 + b_2X_2 + b_3X_1X_2 \tag{9.3a}$$

$$\hat{Y} = b_0 + b_1X_1 + b_2X_2 + b_3X_1X_2 \tag{9.3b}$$

In analysis of variance (ANOVA) types of analyses, interactions are routinely checked because most statistical software packages model and test them by default. In most implementations of regression that I am aware of, interactions must be created by the researcher and entered manually. Thus, many researchers fail to routinely check for nonadditive effects (e.g., interactions) in their regression analyses.

If you as a researcher do not enter this term, you are asserting that the coefficient b_3 is 0: that there is no interaction, or that the slope b_1 is constant across the range of b_2. Although this is sometimes (or perhaps even usually) true, it is impossible to know whether it is true in your particular analysis unless you actually test whether b_3 is different from 0. My experience with interactions is similar to that of curvilinear effects. They are often found in real data, if you have enough power, and particularly if you have performed your due diligence with respect to data cleaning. When they are present, we often have much more interesting and nuanced results, answering questions such as those we started the chapter with.

Procedural and Conceptual Issues in Testing for Interactions Between Continuous Variables

Interactions are tested by entering the cross-product of two other variables (e.g., X_1 and X_2, represented by X_1X_2). When we are dealing with continuous variables, those variables should already be centered at 0 by virtue of conversion to z-scores (or simply centering) as a best practice (Aiken & West, 1991) prior to the creation of the interaction term.

Once we have the term X_1X_2 created, it contains three ingredients: the simple effect of X_1, the simple effect of X_2, and the nonadditive effect (if any) of the combination of the two variables. The procedure for testing whether there is a significant nonadditive effect flows from this concept. In order to isolate the interaction from the simple effects of each variable from this term, we must covary X_1 and X_2. Any significant effect of X_1X_2 present after we covary the other terms represents the nonadditive or interaction effect. If no effect remains, no interaction exists in our data and it is safe to remove X_1X_2 from the model. If, however, a significant effect remains, you now have a much more interesting model than before. Congratulations!

To test whether this is the case, we will enter variables in blocks or groups (blockwise entry, discussed in the previous chapter) and examine the overall

Figure 9.1 Fictitious Interaction Graphed Without the Origin

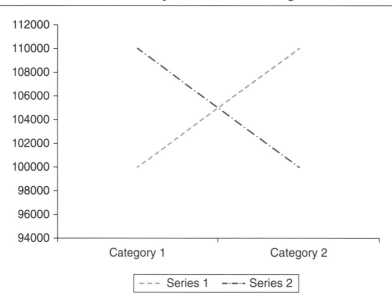

change in model fit, such as change in R^2 (ΔR^2) or change in −2 log likelihood (Δ−2LL), to see whether the addition of the interaction terms to the model improves model fit (or increases variance accounted for) beyond that of the individual variables. If there are significant interactions between the IVs, you should not substantively interpret their simple effects, as those simple effects are modified by the interaction term.[1]

As you might expect, it is often the case that the cross-product of the two variables ($X_1 X_2$) is highly collinear with the individual variables. Thus, it is sometimes the case that the interaction term looks relatively small once the other variables are in the equation—similar to what we sometimes saw with curvilinear effects. This does not necessarily mean the interaction is not important.

Furthermore, as with curvilinear analyses, it is not necessary for the simple effects to be significant in order to examine the higher-order effects. Thus, if X_1 and X_2 are not significant predictors of the DV, it is still possible that $X_1 X_2$ is. Just remember that regardless of whether they are significant, the main effects of the variables must *always* be in the model prior to evaluating the effect of the nonadditive effect. For example, the effect in Figure 9.1 would result in no significant main effects for either IV, but there would obviously be a highly significant interaction effect.

---

1   Again, this is like asserting that Drug X will cure you, or Drug Y will cure you, but taking both drugs simultaneously will kill you. The simple effect of either Drug X or Y is immaterial when both are present in the equation. Feel free to substitute your own random romantic relationship analogy here if you desire (i.e., life is wonderful if only romantic partner X is present in the room, and life is fabulous if only romantic partner Y is present, but if both were to be present in the same room at the same time, things might not go so well for you . . .).

Procedural and Conceptual Issues in Testing for Interactions Containing Categorical Variables

Dichotomous predictors coded as 0 and 1 can be directly multiplied with continuous variables to create interactions between binary and continuous variables (e.g., SEX × SES). Dichotomous variables not coded 0/1 must be recoded prior to examining interactions. In this example, our original SEX variable was coded 1 for boys and 2 for girls. I recoded the variable so girls were 0 and boys were 1, and I labeled the new variable SEXr.[2] Polytomous categorical variables that are unordered are more complex and must first be converted to dummy-coded variables before multiplying each dummy variable by the other continuous variable. It sounds complicated but is really not that much more complicated than any other interaction. We will explore this type of analysis later in the chapter. Interactions between two categorical (dummy-coded) variables become more complicated as the number of dummy variables increases. They are certainly doable within the general linear modeling framework; however, if you only have categorical IVs in your analysis, using the ANOVA routines in your statistical software will make life a bit easier in many cases (and as we demonstrated earlier in the book, it will produce identical results to analyses using regression models).

Hypotheses to Be Tested in Multiple Regression With Interactions Present

In this type of analysis, we will have a multistep process in evaluating nonadditive (interaction) effects similar to that we used in curvilinear analysis. We first start with the overall model, examine simple effects, and then examine interaction effects. We will first test one set of hypotheses for the initial model containing the intercept and simple effects all entered simultaneously on the first block, which is represented by a multiple correlation (R) or the overall variance accounted for (R^2):

$$H_0: R = 0; R^2 = 0$$

$$H_a: R \neq 0; R^2 \neq 0$$

Regardless of the outcome of the hypothesis test for this first block, we will enter the interaction term(s) on a separate block, and most statistical software packages will calculate ΔR^2 and the significance tests associated with the change in the model to answer the question as to whether the addition of the interaction term(s) significantly improves the model:

$$H_0: \Delta R = 0; \Delta R^2 = 0$$

$$H_a: \Delta R \neq 0; \Delta R^2 \neq 0$$

---

2   I always recode original variables into new variables, so that if there is ever any question, I have the original variables in my data. I recommend this policy for you also.

Or, if you are using logistic regression, the hypotheses would be

$$H_0: \Delta-2LL = 0$$

$$H_a: \Delta-2LL \neq 0$$

If the addition of the interaction terms results in significant improvement in the model, we can examine the individual effects and prepare to interpret them. If not, the block of interaction terms can be removed from the analysis and the simple/main effects can be interpreted with full confidence that an additive model is the appropriate model to test.

An OLS Regression Example: Predicting Student Achievement From Real Data

In prior chapters, we examined how family SES and prior achievement can predict future achievement, and we showed that prior achievement is the strongest predictor of future achievement. SES was still significant but contributed a relatively small effect after controlling for prior achievement. But what if that was not the end of the story? We can ask other questions, like whether the effect of achievement is similar across all of the ranges of SES, or across all racial/ethnic groups. In other words, we assumed that an additive model was appropriate, and we neglected to test whether that assumption was supported. We will use this example to demonstrate the procedural steps in testing for interactions.

By the end of Chapter 8, we already had eighth-grade SES and achievement converted to z-scores. Once variables are centered in this way, we create a new variable (in this case, I call it "INT" so it is obvious to me that it is an interaction; you will develop your own naming conventions for variables) by multiplying zACH08 by zSES08. You can easily do that through point-and-click interfaces in most statistical software packages, or you can easily perform these types of variable manipulations via SPSS syntax[3] as follows:

```
compute int=zach08*zses08.

execute.
```

To set up the model, we should enter the zSES08 and zACH08 variables on the first block of variables and the interaction on the second block, and request R^2 change statistics (see the Chapter 9 SPSS guide on the book website for a visual tutorial). An initial analysis showed some influential cases (in this case, about 0.7% scores on either Cook's Distance [or Cook's D] or standardized residuals more than 4 SD from the mean; $N = 92$ of 12,571), which left us with non-normally distributed residuals (skew = −0.41, kurtosis = 1.45) and heteroscedasticity. Following removal of these cases, the

---

3 SAS and other software packages have similarly simple syntax.

residuals were more normally distributed (skew = −0.276, kurtosis = 0.62) and had much improved homoscedasticity. Thus, these results will be reported below.

First, we will examine the overall model statistics in Table 9.1. As you can see, the initial block of variables (achievement and SES) have a very strong effect overall on twelfth-grade achievement ($R = 0.86$, $R^2 = 0.75$, $F_{(2, 12,476)} = 18,251.22$, $p < .0001$). This is a massive and unusually large effect in the social sciences. What is more relevant to this example are the change statistics for Model 2, wherein we add the interaction term. The change to the model is significant ($\Delta R^2 = 0.004$, $F_{(1, 12,475)} = 188.33$, $p < .0001$). Although the effect size is *small* (improvement of less than 1% in variance accounted for), there is not much variance left over from the first model, and thus it is not surprising. As I mentioned above, small effects are not uncommon in interaction effects.

Interpreting the Results From a Significant Interaction

You might be interested to note that if you request the part and partial correlations in this analysis, the interaction term has a $sr = -0.062$. Recall that the semipartial correlation is the unique relationship between a predictor and the DV, and the squared semipartial is the unique variance accounted for. In this case, squaring −0.062 produces a squared semipartial correlation (sr^2) of 0.003844, which rounds to 0.004 (the same value as the ΔR^2 from the first part of the table).

Once we know that we have a significant interaction (meaning that the effects of the variables are nonadditive), we focus on the results represented by the final model.

Table 9.1 Multiple Regression Predicting Twelfth-Grade Achievement From Eighth-Grade Achievement and Family Socioeconomic Status, Including Interaction Term

Model Summary[a]

Model	*R*	R^2	Adjusted R^2	*SE* of the Estimate	Change Statistics				
					R^2 Change	*F* Change	*df*1	*df*2	Sig. *F* Change
1	0.863[b]	0.745	0.745	0.49516637	0.745	18,251.216	2	12,476	.000
2	0.865[c]	0.749	0.749	0.49149027	0.004	188.326	1	12,475	.000

[a]Dependent variable: zACH12.
[b]Predictors: (constant), zSES08, zACH08.
[c]Predictors: (constant), zSES08, zACH08, INT.

Coefficients[a]

Model		Unstandardized Coefficients		Standardized Coefficients	*t*	*p*	Correlations		
		B	*SE*	Beta			Zero-Order	Partial	Part
1	(Constant)	−0.072	0.004		−16.026	.000			
	zACH08	0.799	0.005	0.807	153.986	.000	0.859	0.809	0.696
	zSES08	0.103	0.005	0.103	19.639	.000	0.511	0.173	0.089
2	(Constant)	−0.043	0.005		−8.665	.000			
	zACH08	0.812	0.005	0.820	155.095	.000	0.859	0.811	0.696
	zSES08	0.111	0.005	0.111	21.188	.000	0.511	0.186	0.095
	INT	−0.063	0.005	−0.064	−13.723	.000	0.178	−0.122	−0.062

[a]Dependent variable: zACH12.

Once we know that the simple effects of the IVs are modified by an interaction, I find them to be no longer of much interest.

Here, the constant is still meaningful despite the existence of an interaction term. Because all variables were centered prior to the creation of that term, the constant is still the expected achievement score in twelfth grade if a student had average scores on both SES and achievement test scores in eighth grade. In this case, the expected score for that student is slightly below average. I tend not to focus much on the simple effects of the IVs (SES, achievement) except as a tool for discussing the nature of the interaction. Before I do this, however, I find that the easiest way to communicate the nature of an interaction effect is to graph the results.

Graphing Interaction Effects

There is actually a good deal of controversy in the statistical literature over how to honestly and accurately graph these types of results.[4] I will not review all of the controversies here, but I will present a few general principles that have served me well:

a. You should graph interaction effects within reasonable, realistic boundaries, ideally within a reasonable range observed within the data that produced the effect;

b. You should focus the range of values in the graph so that it communicates the effect to the reader in the clearest possible way; and

c. You should interpret the effect for the reader, including both nature and implications of the effect as well as magnitude of the effect.

Staying Out of Trouble on the X Axis

I will often use values at ± 2 SD (z-score of -2 and $+2$ representing "low" and "high," sometimes also including 0, or average as well). The goal is to emphasize the nature of the effect while at the same time remaining within reasonable boundaries. Although using ± 5 SD would make for a much more dramatic representation of the effect, projecting beyond the observed data is risky and usually not advisable. Furthermore, our goal is to *accurately* represent the nature of the effect in the population, not inappropriately inflate the size of the effect.

However, this guidance is not a rigid rule; rather, it is meant to be general guidance. In your research, it might be desirable to project beyond the observed data. Regression lines were made for prediction, and thus prediction of scores outside the observed data is defensible, if done carefully and with good, clean data to support it. I would suggest that effects be graphed only within reasonable ranges, however. While graphing an interaction with age might produce spectacularly impressive effects if you use 1 and 99 years of age, you should probably stick to reasonable adult ages (e.g., 18–80) if you are dealing with the adult population. If your sample included

---

4   I find it fascinating that statisticians can find controversy in almost any issue, no matter how trivial or unrealistically extreme the example.

adults across a broad age range, this is probably defensible. If your sample was from an undergraduate psychology class in which most participants were between 18 and 22 years, then I would recommend against projecting far outside the observed range. Following these simple rules helps keep you out of trouble on the X axis.

Staying Out of Trouble on the Y Axis

There is even more controversy over how to think about the Y axis. Some statisticians believe you should always represent the 0 (origin) so that readers can get a realistic picture of where the interaction falls in relation to that 0 point. Other authors argue that the Y axis should always include the entire theoretical range of the variable. Although this might make sense in the abstract, there are cases where this is not ideal. Let us take a fictitious example of a simple crossover interaction that involves annual income of currently employed professionals. This is the fictitious example in Figure 9.1 already presented. This nice, strong, interesting effect could be important to someone, but if we heed the dogmatic advice of always including 0 in the graph, or always graphing the entire theoretical range of the variable, we are representing an income range that is not relevant or interesting (or even sensible) perhaps. As you can see in Figures 9.2a and 9.2b, this obfuscates[5] the nature of the effect when the point of graphical representation is to *clarify* communication. Thus, I would argue that it is acceptable to focus the Y axis on a reasonable interval around the observed (or predicted) range of the DV so as to communicate the nature of the effect clearly. It is not defensible to artificially restrict the range of the interaction to make it appear to be larger or more important than it should be, and I think it is silly to dogmatically

Figure 9.2a   Fictitious Interaction Graphed With the Origin

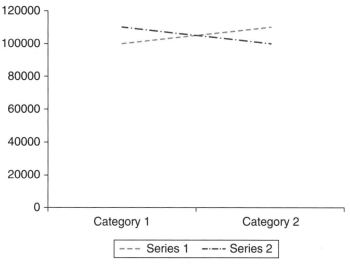

---

5   I have been a fan of National Public Radio's *Car Talk* program for longer than I care to admit. I am not sure if they invented this word or merely popularized it.

Figure 9.2b   Fictitious Interaction Graphed With the Origin and Entire
Theoretical Range

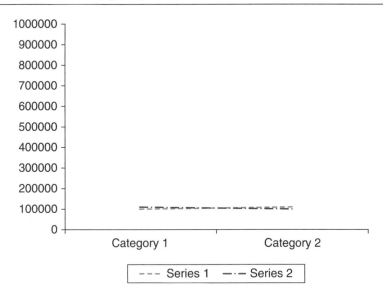

extend the Y axis far outside the observed range of the DV. Thus, we have a decision rule for the Y axis.

Procedural Issues With Graphing

The first thing we will do when setting out to graph the interaction effect(s) is calculate some predicted values. To facilitate this, we need to remove effects that are irrelevant to the interaction, if there are other variables in the model. In this case, there are not. If there were, we would hold these effects constant at 0, which should be the mean, and thus remove them from calculations.

The next thing is to capture the regression line equation from the last step in the analysis, as in Equation 9.4 (extracted from the final step in the analysis presented in Table 9.1). It is often desirable (particularly in logistic regression) to extract more than three decimal points' worth of precision from your statistical output.

$$\hat{Y} = -0.043 + 0.812(zACH08) + 0.111(zSES08) - 0.063(zACH08*zSES08) \qquad (9.4)$$

When we have two continuous variables as part of an interaction, we have an almost infinite number of points that we could plot. However, remember that our goal is clarity concerning the general nature of the interaction. Thus, I usually plot points that represent high/low combinations from each variable (or high/average/low combinations) to communicate the general effect. Thus, the first step is to decide what values will represent "high" and "low" when making these calculations. I often use ± 2 *SD* (z-score of -2 and $+2$ representing "low" and "high," and sometimes 0 for "average") for each variable, giving me values that I can put into an Excel spreadsheet to create predicted values that can then be plotted graphically (see Table 9.2).

Table 9.2   Excel Spreadsheet Using Regression Equation to Create Predicted Scores

	zACH08	zSES08	INT	Predicted
LOW-LOW	−2	−2	4	−2.141
LOW-HIGH	−2	2	−4	−1.193
HIGH-LOW	2	−2	−4	1.611
HIGH-HIGH	2	2	4	1.551

I have graphed these values two different ways. In Figure 9.3a, I used eighth-grade achievement as the X axis and SES as different lines. In Figure 9.3b, I used eighth-grade SES as the X axis and eighth-grade achievement as the different lines. They contain the same information, but depending on the way you wanted to discuss the results, it might make sense to have one or the other as the X axis.

When I am discussing an interaction effect, I usually start by discussing the overall model statistics, such as the following:

As you can see in Table 9.1, the simple effects of SES and achievement in eighth grade were both significant predictors ($R = 0.86$, $R^2 = 0.75$, $F_{(2, 12,476)} = 18,251.22$, $p < .0001$). In addition, once the interaction term was entered into the model, there was a significant increase in variance accounted for ($\Delta R^2 = 0.004$, $F_{(1, 12,475)} = 188.326$, $p < .0001$). This indicates that the interaction is significant. To explore the nature of the interaction, values of −2 and +2 were entered into the regression equation from the last step to create four predicted values for all possible combinations of high and low achievement and SES.

Once the overall model statistics and methods are briefly reviewed, I usually interpret the interaction effect for the reader by starting with the general effect of the variable I put on the X axis, and then discuss how that general effect is different depending on the value of the other IV represented by different lines. For example, if I was focusing on the effect of achievement, I would show the reader Figure 9.3a and proceed as follows:

In general, increasing achievement in eighth grade predicts increased achievement in twelfth grade. This effect is modified by the effect of eighth-grade SES. As you can see in Figure 9.3a, for students coming from low-SES homes, the effect of low achievement is more dramatic than for students coming from high-SES homes. Specifically, low-achieving students from low-SES homes are expected to have twelfth-grade achievement almost a full standard deviation below low-achieving students coming from more affluent homes. Conversely, SES does not seem to matter much for students who are high achieving.

If I wanted to focus on the effect of SES, I would present the reader with Figure 9.3b and discuss as follows:

> In general, increasing affluence (SES) in eighth grade predicts increased achievement in twelfth grade. However, as you can see in Figure 9.3b, this is primarily true for low-achieving students. For low-achieving students, there is approximately a full standard deviation difference in expected twelfth-grade achievement between students coming from a low-SES home and those coming from a high-SES home. For high-achieving students, SES does not seem to make a significant difference in future achievement.

Figure 9.3a   Interaction Between Eighth-Grade Achievement and Socioeconomic Status Predicting Twelfth-Grade Achievement

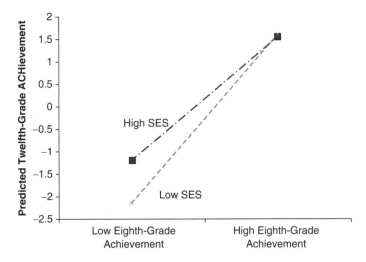

Figure 9.3b   Interaction Between Eighth-Grade Achievement and Socioeconomic Status Predicting Twelfth-Grade Achievement

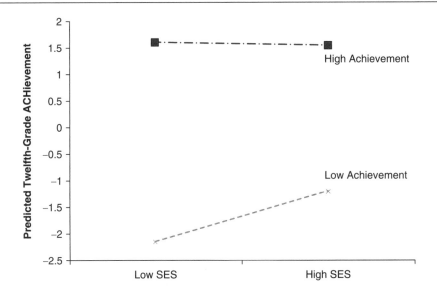

An Interaction Between a Continuous and a Categorical Variable in OLS Regression

Conceptually, there is no difference when considering interactions between categorical and continuous variables compared with interactions between two continuous variables. Procedurally, things can get a little more complex because we have multiple dummy variables to represent the categorical variable, but otherwise we progress similarly as in the last example. In this example, we will look at the interaction of SES (zSES08) and race (dummy coded as in previous chapters, and as shown again for your convenience in Table 9.3) in predicting eighth-grade achievement (zACH08).

Conceptually, an interaction between a categorical and a continuous variable simply means that the slope of the continuous variable is significantly different across groups. The only difference between this and the previous example with two continuous variables is that the groups we will compare in this example are predefined by the categorical variable.

Procedurally, testing an interaction of this type requires that you first dummy code the categorical variable and then create an interaction term for each dummy variable. This literally means multiplying each dummy variable (DUM1–DUM4) by the continuous variable (zSES08). In SPSS, I do this simply as

```
compute int1= dum1*zses08.
compute int2= dum2*zses08.
compute int3= dum3*zses08.
compute int4= dum4*zses08.
execute.
```

Then, we will use blockwise entry to place the dummy variables, representing the categorical variable, into the equation. We will then add the continuous variable, and

Table 9.3 Dummy Coding of the RACE Variable in NELS88

RACE	DUM1	DUM2	DUM3	DUM4
White	0	0	0	0
African American	1	0	0	0
Latino	0	1	0	0
Asian/Pacific Islander	0	0	1	0
Native American	0	0	0	1

SOURCE: National Education Longitudinal Study of 1988 (NELS88) from the National Center for Education Statistics (http://nces.ed.gov/surveys/nels88/).

then we will add the block of interaction terms that represent the interaction between the two variables. If the last block of variables (the collective interaction term) accounts for significant variance above that of the individual variables, then we have an interaction and a slightly more complex process to graph predicted values compared with the last analysis.

Note that I have tended to skim over data cleaning and testing of assumptions as the models have gotten more complex and interesting, but we still have to test and meet all assumptions even in the most complex models. So for this analysis, I examined Cook's D (just to provide a change of pace) and then standardized the values. Approximately 0.6% (134 out of an original N of 23,424) of the sample had values above 4 SD from the mean, and thus, they were removed. Following this basic data cleaning, the residuals were close to normally distributed (skew = 0.17, kurtosis = −0.40), as you can see in Figure 9.4a, below, and the assumption of homoscedasticity is reasonable, as you can see in Figure 9.4b.

Once we have established a reasonable case that we have met assumptions, we can examine the overall model information, presented in Table 9.4. As you can see, I entered all of these terms in three blocks (race dummy variables, zSES08, and then the interactions between zSES08 and each dummy variable), each of which accounted

Figure 9.4a   Reasonably Normal Residuals From NELS88 RACE × zSES08 Analysis

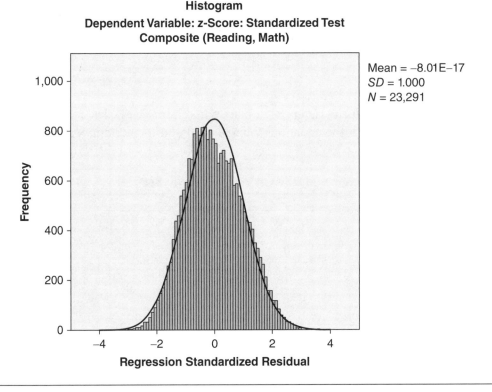

SOURCE: National Education Longitudinal Study of 1988 (NELS88) from the National Center for Education Statistics (http://nces.ed.gov/surveys/nels88/).

Figure 9.4b   Plot of Standardized Predicted Values Versus Residuals From NELS88
RACE × zSES08 Analysis

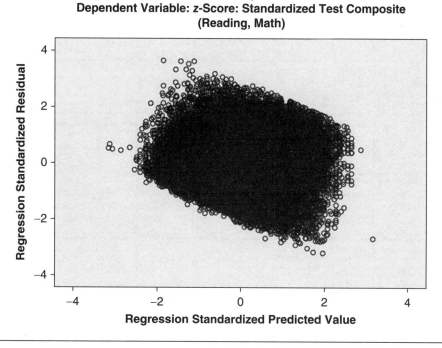

Scatterplot
Dependent Variable: z-Score: Standardized Test Composite
(Reading, Math)

SOURCE: National Education Longitudinal Study of 1988 (NELS88) from the National Center for Education Statistics (http://nces.ed.gov/surveys/nels88/).

for a significant amount of variance above the prior step(s). Of most relevance, the block of terms representing the interactions contributed $\Delta R^2 = 0.006$ ($F_{(4, 23,381)} = 49.82$, $p < .0001$) to a model accounting for almost 30% of the variance in eighth-grade achievement test scores. Thus, we will examine the final model for the regression line equation and create predicted values for each group in order to facilitate graphing and interpretation of this effect.

As you can see in the final model, having two variables (race, SES) and the interaction term quickly starts to look complicated due to the fact that four dummy variables represent the effect of a single variable. However, if you work through the analysis methodically and carefully, it really is not much more of a challenge than any other analysis. Let us look at the regression line equation, presented in Equation 9.5:

$$\hat{Y} = 0.091 - 0.561(\text{DUM1}) - 0.376(\text{DUM2}) + 0.088(\text{DUM3})$$
$$- 0.735(\text{DUM4}) + 0.494(\text{zSES08}) - 0.185(\text{INT1}) - 0.176(\text{INT2})$$
$$+ 0.028(\text{INT3}) - 0.270(\text{INT4}) \tag{9.5}$$

My strategy in this situation is to make the X axis the continuous variable (zSES08) and predict low and high (−2, +2) values for each group, making a graph that contrasts the effect of SES for each subgroup. Thus, we will turn to Excel (or your

Table 9.4   Summary of NELS88 Analysis Predicting zACH08 From RACE and zSES08

Model Summary[a]

Model	R	R^2	Adjusted R^2	SE of the Estimate	Change Statistics				
					R^2 Change	F Change	df1	df2	Sig. F Change
1	0.339[b]	0.115	0.115	0.93682529	0.115	757.977	4	23,286	.000
2	0.541[c]	0.292	0.292	0.83788837	0.177	5,824.838	1	23,285	.000
3	0.546[d]	0.298	0.298	0.83439719	0.006	49.815	4	23,281	.000

[a]Dependent variable: zACH08.
[b]Predictors: (constant), DUM4, DUM3, DUM1, DUM2.
[c]Predictors: (constant), DUM4, DUM3, DUM1, DUM2, zSES08.
[d]Predictors: (constant), DUM4, DUM3, DUM1, DUM2, zSES08, INT3, INT4, INT1, INT2.

Coefficients[a]

Model		Unstandardized Coefficients		Standardized Coefficients	t	p	Correlations		
		B	SE	Beta			Zero-Order	Partial	Part
1	(Constant)	0.192	0.007		25.746	.000			
	DUM1	−0.812	0.019	−0.267	−42.548	.000	−0.234	−0.269	−0.262
	DUM2	−0.678	0.019	−0.228	−36.273	.000	−0.189	−0.231	−0.224
	DUM3	0.122	0.026	0.030	4.755	.000	0.080	0.031	0.029
	DUM4	−0.954	0.061	−0.096	−15.566	.000	−0.078	−0.101	−0.096
2	(Constant)	0.100	0.007		14.770	.000			
	DUM1	−0.503	0.018	−0.166	−28.715	.000	−0.234	−0.185	−0.158
	DUM2	−0.304	0.017	−0.102	−17.415	.000	−0.189	−0.113	−0.096
	DUM3	0.097	0.023	0.024	4.234	.000	0.080	0.028	0.023
	DUM4	−0.626	0.055	−0.063	−11.389	.000	−0.078	−0.074	−0.063
	zSES08	0.449	0.006	0.448	76.321	.000	0.508	0.447	0.421
3	(Constant)	0.091	0.007		13.341	.000			
	DUM1	−0.561	0.019	−0.185	−29.812	.000	−0.234	−0.192	−0.164
	DUM2	−0.376	0.020	−0.126	−19.146	.000	−0.189	−0.125	−0.105
	DUM3	0.088	0.024	0.021	3.711	.000	0.080	0.024	0.020
	DUM4	−0.735	0.064	−0.074	−11.479	.000	−0.078	−0.075	−0.063
	zSES08	0.494	0.007	0.493	68.645	.000	0.508	0.410	0.377
	INT1	−0.185	0.018	−0.068	−10.255	.000	0.193	−0.067	−0.056
	INT2	−0.176	0.018	−0.070	−9.822	.000	0.191	−0.064	−0.054
	INT3	0.028	0.022	0.008	1.275	.202	0.152	0.008	0.007
	INT4	−0.270	0.064	−0.027	−4.222	.000	0.057	−0.028	−0.023

[a]Dependent variable: zACH08.

spreadsheet of choice) to methodically work through this task. Each dummy variable is coded either 0 or 1, as you know, so most terms will drop out of any given estimate. For example, if we want to estimate two values for African American students, those students have a value of 1 for DUM1 but zeros for DUM2–DUM4; thus, those variables (as well as INT2–INT4) will drop out of the equation. In Table 9.5, I show how I set up the spreadsheet I used to calculate predicted values.

As you can see in Figure 9.5, the effect of SES is not uniform across all groups.[6] In general, as SES increases, student achievement increases, but for some groups, the

---

6   As an aside, if you were to attempt to perform an analysis of covariance predicting student achievement from race controlling for SES, one of the assumptions of that analysis is that the effect of the covariate is uniform across all groups. The presence of a significant interaction between race and SES violates that assumption. Instead of viewing this as a failed test of assumptions, we can examine the differences substantively. This is a much more interesting way to look at the data, don't you think?

Table 9.5   Spreadsheet Used to Predict Values for Each Group

	zSES08	DUM1	DUM2	DUM3	DUM4	INT1	INT2	INT3	INT4	Predicted
Low zSES08										
Caucasian	−2	0	0	0	0	0	0	0	0	−0.897
African American	−2	1	0	0	0	−2	0	0	0	−1.088
Hispanic	−2	0	1	0	0	0	−2	0	0	−0.921
Asian/ Pacific Islander	−2	0	0	1	0	0	0	−2	0	−0.865
Native American	−2	0	0	0	1	0	0	0	−2	−1.092
High zSES08										
Caucasian	2	0	0	0	0	0	0	0	0	1.079
African American	2	1	0	0	0	2	0	0	0	0.148
Hispanic	2	0	1	0	0	0	2	0	0	0.351
Asian/ Pacific Islander	2	0	0	1	0	0	0	2	0	1.223
Native American	2	0	0	0	1	0	0	0	2	−0.196

Figure 9.5   Interaction Between Socioeconomic Status and Race Predicting Eighth-Grade Student Achievement

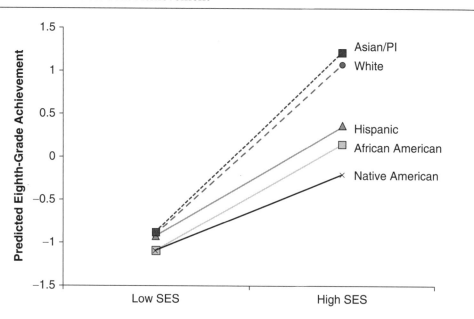

effect is stronger than others. For example, students from Asian or Pacific Islander backgrounds[7] and students who identify as White show strong effects of SES and a relatively steep slope. Students from other groups (e.g., African American, Hispanic, and Native American) show less strong effects (flatter slopes). This may or may not be substantively interesting to those who have scholarly interests in this area, but this type of analysis within the context of whatever variables you are interested in might yield similarly interesting results.

Interactions With Logistic Regression

Procedurally and conceptually, testing and reporting interactions in logistic regression is very similar to that of OLS regression. The primary differences include (a) assumptions that must be tested, (b) conversion of predicted values in logit metrics to probabilities, and (c) evaluation of change in model fit as change in $-2LL$ and evaluated as a chi-square, rather than as change in R^2 and evaluated as an F statistic. Because we have been through many of the procedural issues already, I will move more quickly through these examples.

Let us return to the example from Chapter 8, in which we predicted graduation based on eighth-grade achievement (we will return to a simpler naming convention because all data are from eighth grade in this example: zACH), socioeconomic status (zSES), and student SEX (SEXr, coded 0 = female, 1 = male). You might recall from the previous chapter that both zSES and zACH were significant predictors of graduation and that sex was not. However, it might be the case that these variables interact in interesting ways, and so let us test for all possible "two-way" interactions (SEX × SES, SEX × ACH, ACH × SES) and, as long as we have three predictors in the equation, we can also test for the "three-way" interaction of SEX × ACH × SES.

In this example, I labeled the interactions with names that would make them immediately identifiable as follows:

```
compute sexSES=SEXr*zSES.
compute sexACH=SEXr*zACH.
compute SESACH=zSES*zACH.
compute INT3=SEXr*zSES*zACH.
execute.
```

Prior to data cleaning, the three-way interaction was nearly significant ($p < .066$), which many would be tempted to report and interpret as a "near significant trend." However, after some simple data cleaning (removal of cases with Cook's D values above 5 SD: 86 cases out of 16,608, or approximately 0.5% of the cases, distribution of

---

7   I know some might find these labels from the 1980s outdated. Apologies. These are the government data labels from these data.

these distances presented in Figure 9.6),[8] the three-way interaction was clearly *not significant*. This step was disregarded and the prior step was considered the final model. These results are summarized in Tables 9.6a and 9.6b.

As you can see in Table 9.6b, neither interaction with SEX is significant and the main effect of SEX was not significant. The only significant effect was the interaction between zSES and zACH, which is significant at $p < .0001$. The full regression line equation for this model is presented in Equation 9.6a.

$$\text{Logit}(\hat{Y}) = 3.593 + 0.955(\text{zSES}) + 1.411(\text{zACH}) - 0.140$$
$$(\text{SEXr}) + 0.293(\text{zSES*zACH}) + 0.082(\text{SEX*zACH}) +$$
$$0.034(\text{SEX*zSES}) \tag{9.6a}$$

However, we want to focus on the significant interaction, and the variables that are related to that interaction. Thus, we want the constant, zSES, zACH, and the interaction in the equation, and the rest of the terms to be held constant. If we enter zero for

Figure 9.6   Cook's Distance (Converted to *z*-Scores) From Logistic Regression

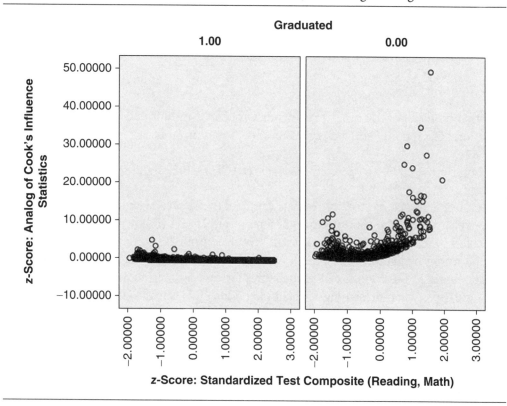

8   I performed the analysis using a cutoff of 4 also, which removed about 0.2% more cases but did not produce a meaningful difference, so I opted for the more conservative cutoff of 5. I also experimented with deviance residuals and various cutoff points, which produced similar results.

Table 9.6a   Summary of Model Fit as Each Block of Variables Entered

Variables Entered	−2LL	Chi-square for Change	df	p
Initial model	9,545.990			
Model 1: only variables	7,399.348	2,146.642	3	.0001
Model 2: 2-way interactions	7,374.180	25.168	3	.0001
Model 3: 3-way interactions	7,373.494	0.686	1	.408

Table 9.6b   Summary of Final Logistic Regression: Model 2

Variables in the Equation[a]							95% CI for Exp(B)	
	B	SE	Wald	df	p	Exp(B)	Lower	Upper
zSES	0.955	0.070	185.325	1	.000	2.600	2.265	2.983
zACH	1.411	0.083	289.511	1	.000	4.099	3.484	4.822
SEX	−0.140	0.114	1.522	1	.217	0.869	0.696	1.086
SESACH	0.293	0.061	22.793	1	.000	1.341	1.189	1.512
SEXACH	0.082	0.104	0.622	1	.430	1.085	0.885	1.331
SEXSES	0.034	0.075	0.206	1	.650	1.035	0.893	1.200
Constant	3.593	0.187	367.865	1	.000	36.325		

[a]Variable(s) entered on step 1: SESACH, SEXACH, SEXSES.

sex, that holds all of the other extraneous variables constant and they drop out of the equation,[9] leaving the reduced regression line Equation 9.6b.

$$\text{Logit}(\hat{Y}) = 3.593 + 0.955(\text{zSES}) + 1.411(\text{zACH}) + 0.293(\text{zSES*zACH}) \quad (9.6b)$$

Because I have converted continuous variables to *z*-scores, we can use −2 to represent LOW regardless of the variable and +2 to represent HIGH (following the same logic, as in Table 9.7). Zero is also fine to use as an "average" group if you desire that.

One recommendation I have made and continue to make is to convert the results of logistic regression to conditional probabilities for ease of interpretation. I also think that there are times when presenting results in log odds units is misleading. For example, the results of this analysis are presented graphed in log odds in Figure 9.7a.

If you were to interpret this graph, you would conclude that family SES has the strongest effect on those students who have the highest achievement and the least effect on those students with low achievement. However, this does not make sense given what we know of how these variables influence graduation rates. For comparison, the actual graduation rates are presented in Figure 9.7b.

As you can see from the actual observed probabilities, the interaction graphed in logits is misleading—it leads researchers to conclude the biggest gap between high- and low- achieving students is found in high-SES students. In fact, the actual

---

9   Taking advantage of the mathematical rule that anything multiplied by zero is zero.

Table 9.7 Graphing the Interaction Between zSES and zACH Predicting Graduation

	zSES	zACH	INT	Predicted	Odds	Probability
LOW-LOW	−2	−2	4	0.033	1.03355	0.508249
LOW-AVG	−2	0	0	1.683	5.38167	0.843301
LOW-HIGH	−2	2	−4	3.333	28.02228	0.965543
HIGH-LOW	2	−2	−4	1.509	4.52220	0.818912
HIGH-AVG	2	0	0	5.503	245.42711	0.995942
HIGH-HIGH	2	2	4	9.497	13,319.70771	0.999924

SOURCE: National Education Longitudinal Study of 1988 (NELS88) from the National Center for Education Statistics (http://nces.ed.gov/surveys/nels88/).

Figure 9.7a Interaction of Socioeconomic Status and Achievement Predicting Graduation Graphed in Logits

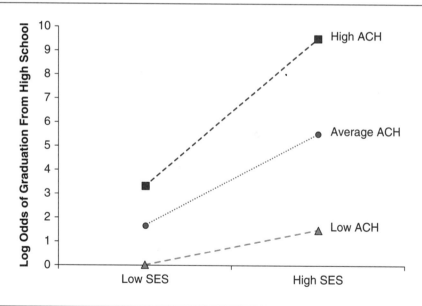

SOURCE: National Education Longitudinal Study of 1988 (NELS88) from the National Center for Education Statistics (http://nces.ed.gov/surveys/nels88/).

probabilities reflect the opposite—that at the higher SES ranges there is a much smaller gap than at the low-SES range. Now, when we convert logits to conditional probabilities and graph the interaction in that metric, we get something much closer to what we expect, as presented in Figure 9.7c.

Figure 9.7b   Observed Probabilities for Three Groups (Groups With $N < 10$ Are Not Represented)

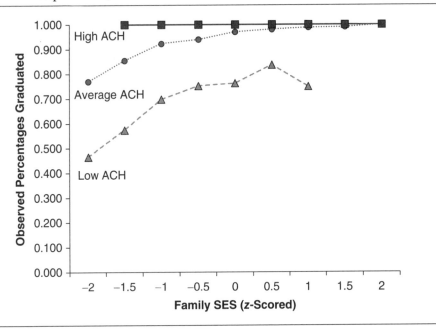

SOURCE: National Education Longitudinal Study of 1988 (NELS88) from the National Center for Education Statistics (http://nces.ed.gov/surveys/nels88/).

Figure 9.7c   Interaction Between Socioeconomic Status and Achievement Predicting Graduation From High School, Graphed in Conditional Probabilities

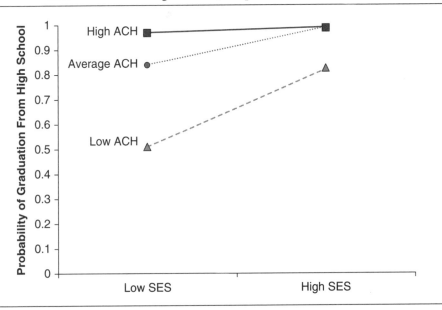

SOURCE: National Education Longitudinal Study of 1988 (NELS88) from the National Center for Education Statistics (http://nces.ed.gov/surveys/nels88/).

As you can see in Figure 9.7c, we see a very different (and more accurate) picture of graduation when the results were converted to predicted probabilities compared with when they were graphed in logits. As this final graph indicates, family affluence has the strongest effect on graduation for students with the lowest achievement test scores, whereas family SES has almost no observable effect on graduation for the highest-achieving students.

You may still note that this is not a strong representation of the actual probabilities, but it is a more accurate representation of the observed probabilities than the graph of the interaction in logits. That is because the relationships in Figure 9.7b seem to be nonlinear, but this logistic regression equation does not account for that nonlinearity. Nonlinear interactions are complex creatures, and we will explore their applications in the next chapter.

Example Summary of Interaction Analysis

The effects of SES and achievement (ACH) seem to be modified by a significant interaction. In this case, the direct effects should be downplayed because they are modified by the interaction.

Following creation of the cross-product terms for ACH, SES, and sex, all three interaction terms were entered together on a separate block following entry of the main effects. Conservative data cleaning was performed. Cook's D values were converted to z-scores, and 86 cases (out of 16,608, or approximately 0.5%) had values more than 5 SD from the mean and thus were removed. Following removal, the analysis was performed on the cleaned data.

Addition of the simple variables significantly improved the model fit, as you can see in Table 9.6a ($\chi^2_{(3)} = 2,146.64$, $p < .0001$). Entry of the two-way interactions also significantly improved the model fit ($\chi^2_{(3)} = 25.17$, $p < .0001$). Entry of the three-way interaction did not improve model fit ($\chi^2_{(1)} < 1$, $p < .41$) and thus was disregarded.

As you can see from Table 9.6b, only one interaction was significant: the interaction of ACH and SES. To explore the nature of the interaction more fully, we graphed the regression line equation assigning a value of –2 for "low," 0 for "average," and +2 for "high." All values were converted from logits to conditional probabilities and graphed in Figure 9.7c. As you can see in this figure, increased achievement in eighth grade tends to increase the probability of graduation. However, there is less effect of achievement on graduation for students from families with high SES, whereas the effect of achievement is much stronger on students from families with lower SES. Thus, the effect of either achievement or SES must be considered in the context of the other variable.

Interactions and Multinomial Logistic Regression

We can do everything in multinomial logistic regression (and other models we have covered, such as ordinal regression models) that we could do in OLS or binary logistic regression. We explored curvilinear effects in multinomial logistic regression in

Chapter 7 and multiple predictors in Chapter 8; now we can add interactions to our tool box with this model as well. Of course, multinomial logistic regression brings some added complexity, but if you approach the process methodically, you can certainly manage the complexity and use this tool to uncover interesting results.

Let us return to the example of student achievement (zACH) and family socio-economic status (zSES) in predicting level of marijuana use (MJ; 0 = never tried marijuana, 1 = tried it 1–2 times, 2 = tried it 3–19 times, and 3 = tried it 20 or more times). We will also add SEX to the equation, because it was found to be a significant predictor of marijuana use in some earlier analyses. As in binary logistic regression, we create the interaction term as the cross-product of the various predictors and then enter the variables on one block with the interaction term on the second step. I did test the three-way SEX × zACH × zSES interaction just for fun, but like the last analysis, it was not significant and thus was not included in our analyses below.

Data Cleaning

Data cleaning for multinomial logistic regression has been described previously, and thus we will not dwell on this somewhat cumbersome but important process. I performed three binary logistic regression analyses, each comparing group "0" (never tried marijuana) with each of the other three groups in order, examining standardized Cook's D statistics for each analysis. The three analyses produced results such as presented in Figure 9.8. I tend to be more conservative with multinomial analyses because we are removing cases from three different analyses, which can quickly reduce your overall sample size. In the first analysis, standardized values of Cook's D ranged up to over 15; in the second, they ranged as high as 25; and in the third, they ranged to over 29. Removing cases with values greater than 5.0 resulted in 324 of 14,861 cases being removed (2.18%), which is more than I routinely like to remove but is still relatively small in the context of this large a sample.[10]

Calculation of Overall Model Statistics

Multinomial in SPSS does not allow for automated blockwise entry; therefore, you must manually enter the main effects once you clean your data and arrive at your final sample (zSES, zACH, SEX), noting the overall model fit and change in model fit, as I have noted in Table 9.8a. Then, entering the interaction terms, you can calculate the change in −2LL from the first analysis to the second and you can calculate the chi-squared statistic and the significance level manually.[11]

---

10 I experimented with using a cutoff of 6.0 but some of the interaction effects were closer to $p < .05$ in this analysis while only retaining a small number of cases more.

11 As with many things, a spreadsheet program like Excel is your friend here, because the =CHIDIST(χ^2, df) function will return the exact probability. In this case, it is $p < 0.000000000000000019$, rounded to two significant figures.

Figure 9.8a   Cook's Distance, Standardized, From a Binary Logistic Regression
Analysis Comparing MJ = 0 With MJ = 1

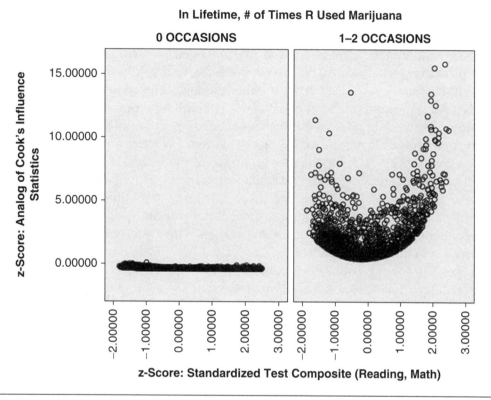

SOURCE: National Education Longitudinal Study of 1988 (NELS88) from the National Center for Education Statistics (http://nces.ed.gov/surveys/nels88/).

Table 9.8a   Summary of Model Fit as Each Block of Variables Entered

Variables Entered	−2LL	Model *df*	Chi-Square for Change	*df* for Change	*p*
Initial model	19,104.044				
Model 1: only variables	18,823.435	9	280.608	9	< .0001
Model 2: 2-way interactions	18,723.839	18	99.596	9	< .0001

Example Summary of Findings

These findings are complex, but walking through them systematically will aid you (and your audience) in making sense. Thus, I will walk you through these findings as I would if this were my dissertation or publication. First, let us start with the overall model summaries. As you can see in Table 9.8a, addition of the simple effects produces a significant improvement in model fit ($\chi^2_{(9)} = 280.61, p < .0001$), and addition of the interaction terms improved the model again ($\chi^2_{(9)} = 99.60, p < .0001$).

Because both models significantly improve model fit, we can examine the individual parameter estimates. As you can see in Table 9.8b, each category (1–3) is compared with the first category (0), and for all, student achievement (zACH) is a significant predictor. In each case, increasing achievement is associated with lower probabilities of using marijuana (odds ratios [ORs] of 0.730, 0.763, and 0.661, respectively). In addition, for categories 1 and 3, SEX (coded 0 = girls, 1 = boys) is a significant predictor. In both cases, boys are more likely to admit having used marijuana (ORs of 1.22 and 1.62, respectively), and interestingly, as noted in previous examples, sex is not a significant predictor of using marijuana 3–19 times (after controlling for all other variables). In none of the analyses is SES significant, although it is nearly so in the first group. Moving to Table 9.8c, we can see that there are several significant interactions. Specifically, for the first group (tried marijuana 1–2 times), there is a significant interaction of SES and student achievement, as well as an interaction of sex and achievement. For the second (used marijuana 3–19 times) and third (used marijuana 20 or more times) groups, there is a significant interaction of SES and achievement, but no other significant interactions. These effects will be graphed and presented in Figures 9.8b and 9.8c.

As you can see in Figure 9.8b, increasing achievement tends to be associated with decreasing marijuana use. As this figure shows, in low-achieving students, boys are more likely to admit using marijuana than girls, while at higher achievement levels, both are approximately equal and the probabilities are much lower. As you can see in Figure 9.8c, the effect of SES on the probability of using marijuana is very different for students who are lower in achievement compared with those with higher achievement. In general, for those with high achievement scores, increasing SES decreases the probability of reporting having used marijuana at each of the three levels (1–2, 3–19, and 20 or more times). However, for those students who have relatively low achievement test scores, increasing family SES is associated with strong increases in the probability of reporting use of marijuana at all three levels. In addition, within each group (low achievement or high achievement), SES can

Table 9.8b Parameter Estimates of Model 1, With Only Simple Effects in the Model

Parameter Estimates[a]

MJ[a]		B	SE	Wald	df	p	Exp(B)	95% CI for Exp(B)	
								Lower Bound	Upper Bound
1 1–2 OCCASIONS	Intercept	-2.341	0.043	2,926.095	1	.000			
	zACH	-0.315	0.036	75.755	1	.000	0.730	0.680	0.783
	zSES	-0.063	0.036	3.118	1	.077	0.939	0.875	1.007
	SEXr	0.197	0.060	10.842	1	.001	1.218	1.083	1.370
2 3–19 OCCASIONS	Intercept	-2.653	0.050	2,821.322	1	.000			
	zACH	-0.270	0.043	39.622	1	.000	0.763	0.702	0.830
	zSES	-0.050	0.043	1.364	1	.243	0.951	0.875	1.034
	SEXr	0.064	0.071	0.798	1	.372	1.066	0.927	1.226
3 20+ OCCASIONS	Intercept	-3.418	0.072	2,264.927	1	.000			
	zACH	-0.415	0.056	54.760	1	.000	0.661	0.592	0.737
	zSES	-0.045	0.054	0.682	1	.409	0.956	0.859	1.064
	SEXr	0.481	0.093	26.808	1	.000	1.618	1.348	1.941

[a]The reference category is: 0 0 OCCASIONS.

Table 9.8c   Parameter Estimates of Model 2, With Simple Effects and Interaction Terms in the Model

MJ[a]		B	SE	Wald	df	p	Exp(B)	95% CI for Exp(B)	
				Parameter Estimates[a]				Lower Bound	Upper Bound
1 1–2 OCCASIONS	Intercept	−2.273	0.045	2,537.267	1	.000			
	zACH	−0.201	0.052	14.752	1	.000	0.818	0.738	0.906
	zSES	−0.049	0.052	0.901	1	.342	0.952	0.860	1.054
	SEXr	0.174	0.061	8.297	1	.004	1.191	1.057	1.341
	SESACH	−0.157	0.036	19.396	1	.000	0.854	0.796	0.916
	SEXACH	−0.171	0.072	5.667	1	.017	0.842	0.732	0.970
	SEXSES	−0.051	0.072	0.496	1	.481	0.951	0.826	1.094
2 3–19 OCCASIONS	Intercept	−2.566	0.052	2,434.778	1	.000			
	zACH	−0.220	0.060	13.247	1	.000	0.803	0.713	0.904
	zSES	−0.114	0.060	3.659	1	.056	0.892	0.794	1.003
	SEXr	0.064	0.072	0.796	1	.372	1.066	0.926	1.227
	SESACH	−0.254	0.044	33.789	1	.000	0.776	0.712	0.845
	SEXACH	−0.028	0.084	0.109	1	.742	0.973	0.824	1.148
	SEXSES	0.108	0.085	1.621	1	.203	1.114	0.943	1.315
3 20+ OCCASIONS	Intercept	−3.289	0.073	2,013.834	1	.000			
	zACH	−0.341	0.087	15.402	1	.000	0.711	0.600	0.843
	zSES	−0.030	0.085	0.125	1	.723	0.970	0.821	1.147
	SEXr	0.452	0.095	22.630	1	.000	1.571	1.304	1.893
	SESACH	−0.368	0.060	37.612	1	.000	0.692	0.615	0.779
	SEXACH	−0.066	0.112	0.347	1	.556	0.936	0.752	1.166
	SEXSES	−0.125	0.110	1.281	1	.258	0.882	0.711	1.096

[a]The reference category is: 0 0 OCCASIONS.

Figure 9.8b   Interaction of SEX and Achievement Predicting Use of Marijuana 1–2 Times

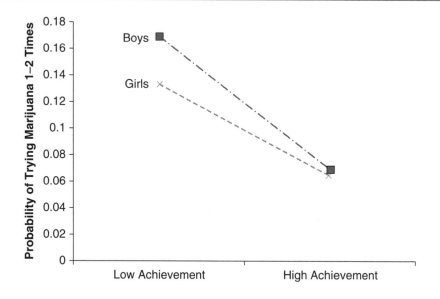

have different strengths of relationship. For example, within the low-achieving students, SES has the strongest effect on the probability of using marijuana 20 or more times, whereas within the high-achieving group of students, SES seems to have the strongest effect on the probability of reporting having used marijuana 3–19 times.

Figure 9.8c   Interaction of Achievement and Socioeconomic Status Predicting Use
of Marijuana

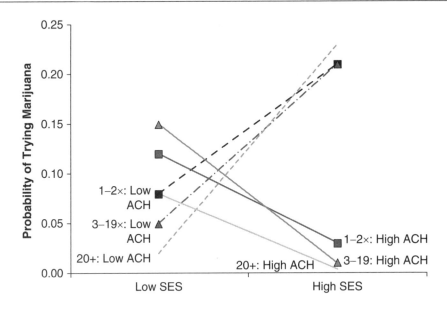

Can These Effects Replicate?

Isn't it fun to see all of these interesting effects emerge from data? I personally find the ability to explore nuances in data exciting, as we all know intuitively that not everyone in the world experiences things in similar ways. Our entire lives are filled with nuance and our data should be too! However, we must be careful to remember that a hallmark of good science is the replication of effects (a caution I wager Fleischman and Pons and many other researchers wish they had paid more attention to before disseminating their assertions about "cold fusion"; Close, 2014; Fleischmann & Pons, 1989). I have already introduced the concept of replication in Chapter 7, and we should continue to be concerned with the issue. Interactions can be interesting and fun, but before we get too excited about them, we should feel confident that the effects will replicate. If you have another data set (or a data set large enough to split in two for independent analysis and replication), it is my opinion that you should replicate the results before publishing.

If you do not have a data set large enough for this, I will introduce you to the concept of bootstrap analysis later in the book that could give you information as to how likely an effect is to replicate. As a rule, the smaller the data set, and the smaller (or more complex) the effect, the more nervous I get that another researcher would not be able to replicate the results. We will explore replication of many different types of effects in Chapter 13, including bootstrap replication.

Post Hoc Probing of Interactions

Those of you familiar with classic regression texts (Aiken & West, 1991; Cohen, Cohen, West, & Aiken, 2003) or tradition within ANOVA analyses may be familiar with the notion of post hoc probing of interaction effects. Let us return to some general labels for variables to facilitate the discussion. We will call our two IVs X and Z, and the DV will be Y. Recall the earlier discussion that a significant effect of XZ indicates that the effect of X is dependent on the level of Z, and the effect of Z is dependent on the level of X. In ANOVA, this is more concrete because the IVs have a limited range of discrete values. In regression, where the variables are both continuous, we have represented the interactions as a set of exemplar lines; in reality, the interaction is a *surface* representing all possible values of X and Z, and the lines are specific points along those continua.

There are two classic questions that are sometimes asked when there is a significant interaction in any type of analysis: (a) are the slopes for X at two different values of Z significantly different, and (b) are slopes of X significantly different from zero at a particular value of Z? Depending on the nature of the analysis and research questions, these might be important for theory or practice. For example, if age and dosage of a drug interact in predicting a health outcome, we might be interested in whether the slope is significantly different for people aged 50 versus 60 years. It might also be important to know whether some ages have a nonsignificant effect, whereas others have significant effects.

Interestingly, Aiken and West (1991, pp. 19–21) demonstrate that the first question is redundant where an interaction effect is significant. Although it is somewhat counterintuitive, they argue (and present mathematical evidence) that by definition a significant interaction effect indicates that any two simple slopes will be significantly different at the same level as the overall effect. They conclude that when both variables are continuous, the significant coefficient for the interaction term indicates that the regression of Y on X varies across the range of Z, and no further test of whether two slopes are different is required.

The question of whether a slope of X for a particular value of Z is significantly different is easily calculated, although given the controversial nature of null hypothesis statistical testing, this might not be a desirable question to ask in the modern era. The answer is going to depend largely on power and effect size. If you have a sample with only 100 participants, it is not going to be easy to detect slopes that are significantly different from zero, and if you have many thousands, as some of our data sets have, it will be relatively easy to reject a null hypothesis for even minuscule effects. Furthermore, if one chooses relatively extreme values of Z, one may be more likely to produce significant effects than if one chooses more common and realistic values. All this is to say that post hoc probing of interactions may not be the most useful way to spend your time.

To explore post hoc probing of interactions, we will turn to a simple data set commonly distributed with SPSS, called CARS, which lists more than 300 makes and models of cars along with interesting statistics such as weight, horsepower, and fuel efficiency (miles per gallon [MPG]). In this example, we will predict MPG from horsepower and weight, both of which were converted to z-scores prior to creating the interaction term. In addition, when performing the analysis, I asked for the covariance matrix to facilitate hand calculation (one method of probing interactions). First, let us review the results of the analysis (note that because this is a classic data set, no data cleaning was performed).

As you might imagine, weight and horsepower were significant predictors of fuel efficiency, as you can see summarized in Table 9.9 (and Figure 9.9). As you can see, both predictors are significant, accounting for a substantial amount of the variance in MPG, and the interaction between the two predictors was also significant. The interaction is graphed in Figure 9.9.

Let us imagine that we want to probe this interaction a little. For example, it looks to me that when cars are very powerful (high horsepower), weight does not seem to have much of an effect on fuel efficiency. However, when cars have very low power, weight seems to matter quite a bit. Thus, it might be interesting to see whether the regression coefficients for the two groups are significantly different from zero (I suspect the first is not, and the second is). There are two ways to perform simple slopes analysis (referring to Aiken & West, 1991; Cohen et al., 2003; Preacher, Curran, & Bauer, 2006).

Table 9.9   Predicting a Car's Fuel Efficiency (Miles per Gallon) From Weight and Horsepower

Model Summary[a]

Model	R	R^2	Adjusted R^2	SE of the Estimate	R^2 Change	F Change	df1	df2	Sig. F Change
					\multicolumn{5}{c}{Change Statistics}				
1	0.822[b]	0.675	0.674	4.459	0.675	404.583	2	389	.000
2	0.849[c]	0.720	0.718	4.143	0.045	62.518	1	388	.000

[a]Dependent variable: MPG.
[b]Predictors: (constant), zWEIGHT, zHORSE.
[c]Predictors: (constant), zWEIGHT, zHORSE, INT.

Coefficients[a]

Model		Unstandardized Coefficients B	SE	Standardized Coefficients Beta	t	p	95% CI for B Lower Bound	Upper Bound
1	(Constant)	23.397	0.225		103.858	.000	22.954	23.840
	zWEIGHT	−4.286	0.437	−0.551	−9.818	.000	−5.144	−3.427
	zHORSE	−2.353	0.441	−0.299	−5.335	.000	−3.221	−1.486
2	(Constant)	21.814	0.290		75.316	.000	21.245	22.384
	zWEIGHT	−3.743	0.411	−0.481	−9.098	.000	−4.552	−2.934
	zHORSE	−4.099	0.466	−0.521	−8.805	.000	−5.015	−3.184
	INT	1.828	0.231	0.270	7.907	.000	1.374	2.283

[a]Dependent variable: MPG.

Coefficient Correlations[a]

Model			zHORSE	zWEIGHT	INT3
2	Correlations	zHORSE	1.000	−0.823	−0.474
		zWEIGHT	−0.823	1.000	0.167
		INT3	−0.474	0.167	1.000
	Covariances	zHORSE	0.217	−0.158	−0.051
		zWEIGHT	−0.158	0.169	0.016
		INT3	−0.051	0.016	0.053

[a]Dependent variable: MPG.

Figure 9.9   Interaction of Horsepower and Weight Predicting Miles per Gallon

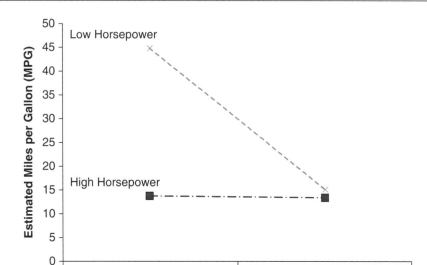

The classic method involves hand calculating a t test for the simple slope, using a hand-calculated standard error of the simple slope. The simple slope calculation is presented in Equation 9.7a. For simplicity, we will consider weight to be X, MPG to be Y, and horsepower to be Z. Thus, the coefficient b_1 is the regression coefficient for X (or weight), b_2 is the coefficient for Z (or horsepower), and b_3 is the coefficient for the interaction of XZ (or weight × horsepower).

$$\text{Simple slope of Y on X at } Z = b_1 + b_3 Z$$

$$= -3.742846 + 1.828424(\text{zHORSE})$$

$$\text{Simple slope}_{Z = -2} = -7.399$$

$$\text{Simple slope}_{Z = +2} = -0.086 \tag{9.7a}$$

If you substitute −2 for "low" and +2 for "high" horsepower, you will get a simple slope estimate of −7.399 for weight when horsepower is "low" and −0.087 for weight when horsepower is "high." Intuitively, one could imagine the slope for "high" horsepower to be not significantly different from zero (unless power is amazingly high), and likely to be significantly different from zero for "low" horsepower. To test the second type of hypothesis described above, you can use the output to calculate the standard error of b_1 at Z via Equation 9.7b to calculate a t statistic that will give us a hypothesis test (Equation 9.7c)[12]:

$$SE_{b \text{ at } Z} = \sqrt{SE_{b_{11}}^2 + 2Z\text{cov}_{b_{13}} + Z^2 SE_{b_{33}}^2} \tag{9.7b}$$

---

12 Note that these equations are only for two-way linear interactions. Aiken and West (1991) have examples for other types of interactions.

where the SE^2 terms are on the diagonals of the covariance matrix (0.169, 0.053 for b_{11} and b_{33}, respectively) and the covariance (0.016) is off diagonal.

Let us calculate the standard error for both Z = −2 and +2:

$$SE_{b\,at\,-2} = \sqrt{0.169 + 2(-2)0.016 + (-2)^2(0.053)}$$
$$= 0.563$$

$$SE_{b\,at\,+2} = \sqrt{0.169 + 2(-2)0.016 + (-2)^2(0.053)}$$
$$= 0.667$$

$$t_{b\,at\,Z} = \frac{b_1 + b_3 Z}{SE_{b\,at\,Z}} \text{ with } df = n - k - 1$$

$$t_{(Z=-2)} = -7.399 / 0.563$$
$$= -13.14$$

$$t_{(Z=+2)} = -0.087 / 0.667$$
$$= -0.13 \qquad\qquad (9.7c)$$

At more than 300 degrees of freedom (df) for this test, the criterion t for significance at $p < .05$ is ±1.968. As anticipated, the slope at Z = −2 is significantly different from zero, and the slope at Z = +2 is not. Somewhere in between these two numbers, there will be a threshold where the slope becomes significant, and researchers wanting to play around (or game the system) could experiment with values to find that point.

In fact, using the above equations and a simple Excel spreadsheet, you can examine a large range of Z easily. I could not resist. My calculations using a simple Excel spreadsheet and the equations above show that the slope is significantly different from zero somewhere below zHORSE = 1.50 and below, is not significant between about 1.50 and 2.90, and then becomes significant again in the opposite direction above 2.90. This leads to the concept of "regions of significance."

Regions of Significance

What might be a more useful and modern way to approach this question is through a relatively recent notion of regions of significance. Rather than engaging in the potentially problematic process of picking possible values for simple slopes, authors such as Preacher et al. (2006) have essentially reversed the question instead, asking what ranges of Z provide a significant relationship between X and Y. They provide formulae for different types of regression procedures, or you can probe as I have done on a more ad hoc basis. The summary of my exploration is presented in Figure 9.10. Note that the region of significance in the lower part of the graph *begins* with a relatively extreme z = 2.90, which is probably not useful or practical. All probing of this nature must be done using the lens of common sense and a touch of skepticism.

Figure 9.10   Visual Representation of Regions of Significance

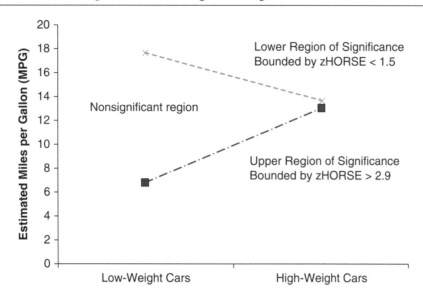

Using Statistical Software to Produce Simple Slopes Analyses

Aiken and West (1991) presented a way to perform the same analysis via modern statistical software. Specifically, this involves computing a new conditional value of Z (Z_{cv}), in which we subtract the value of Z we wish to evaluate. For example, in our example, using zHORSE as "Z," if we wish to evaluate the simple slope of b_1 at $Z = 2.00$, we would compute $Z_{cv} = Z - 2.00$ and also compute a new interaction term (XZ_{cv}) and then enter X, Z_{cv}, and XZ_{cv} into the equation. A brief excerpt from the SPSS output of this analysis shows that the results are within rounding error of what we calculated above. As you can see in Table 9.10, following this modification to zHORSE (now called zHORSE_CV) and the interaction term (now called INT_CV),

Table 9.10   Using Statistical Software to Produce Simple Slopes Estimates

Coefficients[a]

Model		Unstandardized Coefficients		Standardized Coefficients	t	p	95% CI for B	
		B	SE	Beta			Lower Bound	Upper Bound
1	(Constant)	18.690	0.916		20.395	.000	16.889	20.492
	zWEIGHT	−4.286	0.437	−0.551	−9.818	.000	−5.144	−3.427
	zHORSE_CV	−2.353	0.441	−0.299	−5.335	.000	−3.221	−1.486
2	(Constant)	13.616	1.066		12.769	.000	11.519	15.712
	zWEIGHT	−0.086	0.668	−0.011	−0.129	.898	−1.400	1.228
	zHORSE_CV	−4.099	0.466	−0.521	−8.805	.000	−5.015	−3.184
	INT_CV	1.828	0.231	0.424	7.907	.000	1.374	2.283

[a]Dependent variable: MPG.

the simple slope for weight is $b = -0.086$, with a SE of 0.668, and a $t = -0.129$, all of which are within rounding error of our prior calculations. This could be repeated for other values (e.g., -2.0) instead of performing hand calculations. This also presents a general approach that one can use with other possible aspects of the generalized linear model or more complex models.

SUMMARY

Once we have multiple IVs in our analyses, it is a natural extension to begin wondering if they interact in some fashion. In this chapter, we reviewed interactions between continuous variables and between categorical variables in a variety of models. The last model, examining marijuana use, naturally begs the question of whether that analysis should be performed as an ordinal logistic regression, as we did in Chapter 6 when there was only one IV. I leave that for you to experiment with as an enrichment exercise.

As our results get more and more complex, you will have to experiment with the best way to present them. For example, when preparing to report on the multinomial example with the SES × ACH interaction, I debated between three individual graphs rather than a single one for the last set of results, but they were so similar I thought it was silly to take up that much space. Also, it is sometimes interesting to be able to directly compare results. For example, having all three interaction effects on the same graph allows us to see how similar the high-SES groups are, and how the low-SES groups differ in interesting ways. There are no rules to guide you here except some of the principles I presented early in the chapter. Remember that the goal of dissemination is to clearly communicate results. That must be the guiding principle.

It is also easy to forget to test assumptions and clean data when models get complex. However, we must continue to test important assumptions and clean data in order to prevent creating errors of inference or misestimating models. For example, without data cleaning in one of our examples, a very complex three-way interaction could have been interpreted as significant (or nearly so). In our last example, the SEX × ACH interaction might have been missed without data cleaning (the model was also much more powerful after some minimal cleaning). Thus, in the context of one chapter, we have an example of where judicious data cleaning could prevent both Type I and Type II error.

It is also important to remember that failing to meet basic assumptions of OLS regression (e.g., normality of residuals or homoscedasticity) is not ignorable—in other words, if you fail to meet those assumptions, you should take some action. Cleaning the data does not always solve the problem, nor does transformation of the variables. But if neither solves the problem, it might be the case that the model is misspecified or that you need a different model with different assumptions.

ENRICHMENT

1. Download the data from the book's website and replicate the results in the chapter.

2. Explain why you cannot interpret an interaction term in a regression equation without the main effects (simple variables) in the equation.

3. Using NELS88 data, examine whether zSES and zACH predict educational aspirations (BYPSEPLN).

4. Using the NELS88 data, examine whether zSES and race interact in predicting high school graduation.

5. Using the NELS88 MJ data, examine whether it would be appropriate to analyze the MJ data using ordinal logistic models (keeping the same predictors and interactions as in Table 9.8b).

   a. Make sure to test the assumption of parallel lines.
   b. If assumptions are met, interpret analysis in APA format, using graphs where appropriate.

6. Download the NHIS data set for this chapter from the book website and complete the following.

   a. Explore whether body mass index (BMI; *z*-scored) interacts with smoking status in predicting diabetes status. If so, graph the interaction and explain the effect so that a nontechnical audience could understand the interaction.

7. Using the same data set as question 6, see if age and BMI interact in predicting diabetes.

   a. Is there a simple interaction?

8. Using the 2013 CDC Natality data, test whether, for full-term infants, mother weight gain (zMomWT) and race (MOM_RACE) interact in predicting infant birth weight (BIRTHWT, in grams).

   a. Produce simple slopes for two reasonable points, interpreting the results appropriately.

9. Using your own (or an adviser's) data, explore whether your variables interact.

REFERENCES

Aiken, L. S., & West, S. (1991). *Multiple regression: Testing and interpreting interactions.* Thousand Oaks, CA: SAGE.

Close, F. (2014). *Too hot to handle: The race for cold fusion.* Princeton, NJ: Princeton University Press.

Cohen, J., Cohen, P., West, S., & Aiken, L. S. (2003). *Applied multiple regression/correlation analysis for the behavioral sciences.* Mahwah, NJ: Lawrence Erlbaum.

Fleischmann, M., & Pons, S. (1989). Electrochemically induced nuclear fusion of deuterium. *Journal of Electroanalytical Chemistry and Interfacial Electrochemistry, 261*(2, Part 1), 301–308. doi: 10.1016/0022-0728(89)80006-3

Preacher, K. J., Curran, P. J., & Bauer, D. J. (2006). Computational tools for probing interactions in multiple linear regression, multilevel modeling, and latent curve analysis. *Journal of Educational and Behavioral Statistics, 31*(4), 437–448. doi: 10.3102/10769986031004437

CURVILINEAR INTERACTIONS BETWEEN INDEPENDENT VARIABLES

10

Advance Organizer

To this point in the book, we have explored both curvilinear effects in simple models and relatively simple nonadditive effects (interactions, or moderator effects). In Chapter 9, we explored linear interactions, in which the linear effect of one variable is different, or modified, by another variable. However, there is no reason to expect that effects are always linear, and there is no reason to expect that curvilinear effects are identical across all groups or across the entire range of another variable. Thus, we have to accept the possibility that there are curvilinear effects that are moderated by, or contingent on, another independent variable (IV). I will refer to these effects as nonlinear, or curvilinear, interactions. This chapter will explore how you can test for these types of effects within your data.

In their excellent opus on mediation and moderation, Baron and Kenny (1986) briefly mention the possibility of a curvilinear moderator effect, and Aiken and West (1991) have a more expansive and excellent discussion on how complex terms can be incorporated into analyses such as this. However, few texts that I have seen include deep treatment of the topic. Broad searches of the research literature in various disciplines yield a small percentage of published articles that explore curvilinear effects; almost none in the entire corpus of scientific research that I have examined report curvilinear interactions. I included a relatively brief introduction to these complex effects in my previous book on logistic regression, and I decided to include an entire chapter on this topic in this book because I think exploration of these effects is both important and fun. I hope that along the way, you will see the beauty of these complex effects and appreciate that they are not substantially harder to analyze than interactions or curvilinear effects, yet they hold the potential to yield very important, nuanced results from your data.

In this chapter, we will gently bring you through example analyses that incorporate curvilinear interactions in most of the quadrants of the organizing table we have used in previous chapters, reproduced as Table 10.1.

In this chapter, we will cover the following topics:

- Conceptualizing curvilinear interactions
- Procedures for managing complexity

Table 10.1 Examples of the Generalized Linear Model as a Function of Independent Variable and Dependent Variable Type

	Continuous DV	*Binary DV*	*Unordered Multicategory DV*	*Ordered Categorical DV*	*Count DV*
Continuous IV	OLS regression	Binary logistic regression	Multinomial logistic regression	Ordinal logistic regression	OLS, Poisson regression
Mixed continuous and categorical IV					
Binary/ categorical IV only	ANOVA	Log-linear models	Log-linear models		Log-linear models

ANOVA, analysis of variance; DV, dependent variable; IV, independent variable; OLS, ordinary least squares.

- Testing curvilinear interactions in ordinary least squares (OLS) regression
- Testing curvilinear interactions in logistic models
- Interpreting and graphing curvilinear interaction effects
- Examples of American Psychological Association (APA)–compliant summaries of analyses

Guidance on how to perform these types of analyses in various statistical packages will be available online at study.sagepub.com/osbornerlm.

What Is a Curvilinear Interaction?

By this point, you know what a curvilinear effect is—a natural extension of linear modeling without the (often untested) assumption that all effects will be monotonic and linear. You also know what an interaction effect is—a condition where the effect of one variable is dependent on the value of another variable. Combining the two, curvilinear interactions are a natural extension of the processes and concepts from the last chapter: a condition where the nature of a curvilinear effect is dependent on another variable. In other words, we could have a case where the shape or nature of a curve is different for different groups or is different at different levels of another continuous variable. Let us recall that when there is a simple curvilinear effect with only one IV (X, in this case) in the model, a linear regression equation will look something like Equation 10.1a, which presents a general example of a curvilinear effect in OLS regression:

$$\hat{Y} = b_0 + b_1 X_1 + b_2 X_1^2 + b_3 X_1^3 + \cdots + b_K X_1^k \qquad (10.1a)$$

Remember, in this example, X, X², and X³ are different aspects of the same variable (the linear and the curvilinear). Recall also that when we have two or more predictors in the equation (following Aiken and West's lead, and for simplicity in this complex world we are moving into, let us refer to the outcome as Y, the first predictor as X, and the second predictor as Z), a simple regression equation could look like Equation 10.1b, if we include the *linear* interaction between X and Z:

$$\hat{Y} = b_0 + b_1X + b_2Z + b_3XZ \tag{10.1b}$$

The complexity in testing for curvilinear interactions comes in the varieties of combinations of linear and curvilinear components we can have. There are many nonlinear forms we could explore, but to keep our examples simple, we will explore just having quadratic (squared) and cubic (cubed) terms in our curvilinear interactions.

A Quadratic Interaction Between X and Z

Let us assume that there is curvilinearity in X only, and that we want to fully explore whether X and Z interact. Thus, to test for an interaction between X and Z where only X has curvilinear components, we not only must have the basic linear and curvilinear components in the equation (X, X^2, X^3, Z): we also have to cross each aspect of X with Z to fully test whether X and Z interact and whether the nature of the interaction is simply linear (e.g., XZ) or there is a curvilinear component to the interaction. In Equation 10.1b, we began that process by including a linear interaction between X and Z. Let us expand on that model to add X^2, fully integrated into the model to allow us to test for a simple curvilinear interaction. As you can see in Equation 10.1c, we must cross Z with both aspects of X:

$$\hat{Y} = b_0 + b_1X + b_2X^2 + b_3Z + b_4XZ + b_5X^2Z \tag{10.1c}$$

Of course, while we are having fun with our data exploring curvilinear interactions, why would we assume that only one variable has a curvilinear component? If both X and Z are continuous variables, and even if X^2 and Z^2 were not significant, there is still a possibility that there could be a curvilinear interaction involving these terms. So let us add the Z^2 term to Equation 10.1c. In this case, we need both linear and quadratic terms for both X and Z in the equation (X, Z, X^2, Z^2); to fully explore any possible interaction between the two variables, we must cross each aspect of each variable with each other aspect of each other variable (XZ, X^2Z, XZ^2, X^2Z^2), as we see in Equation 10.1d:

$$\hat{Y} = b_0 + b_1X + b_2X^2 + b_3Z + b_4Z^2 + b_5XZ + b_6X^2Z + b_7XZ^2 + b_8X^2Z^2 \tag{10.1d}$$

As you consider what we are doing, remember that there can be a good deal of diversity in the shape and form of interactions, and these are even more complex than typical linear interactions. Thus, we do not require all aspects of these interactions to be significant in order to explore the effect. What we require is that, when entered into the model as a group, the block as a whole will account for a significant improvement in the model.

A Cubic Interaction Between X and Z

Let us not stop with only quadratic effects! If you are still with me at this point, it is a simple generalization from the models in Equations 10.1c and 10.1d to add cubic effects. The equation may look intimidating, but just remember that we are merely testing whether the effect of X is dependent on Z in some way, and we are not restricting that way to being only linear without testing whether that restriction is reasonable. Thus, we can easily cross the various components of X (X, X^2, X^3) with the various components of Z (Z, Z^2, Z^3). As you might have noticed from previous equations, as we add terms like this we are dramatically expanding the number of terms in the model. To make sure that I move through this type of analysis methodically and completely, I usually use a table such as Table 10.2, to manage complexity and ensure that I have created and entered all appropriate terms into the model.

This final regression model, in Equation 10.1e, includes the quadratic and cubic components of both X and Z, as well as the interaction terms crossing each term with each other term.

Table 10.2 Composing All Cross-Products for a Curvilinear Interaction Analysis Between X and Z Using Quadratic and Cubic Components

	X	X^2	X^3
Z	XZ	X^2Z	X^3Z
Z^2	XZ^2	X^2Z^2	X^3Z^2
Z^3	XZ^3	X^2Z^3	X^3Z^3

$$\hat{Y} = b_0 + b_1X + b_2X^2 + b_3X^3 + b_4Z + b_5Z^2 + b_6Z^3 + b_7XZ + b_8X^2Z + b_9XZ^2 + b_{10}X^2Z^2 + b_{11}X^3Z + b_{12}X^3Z^2 + b_{13}XZ^3 + b_{14}X^2Z^3 + b_{15}X^3Z^3 \tag{10.1e}$$

Exploring interactions of any type, but especially curvilinear interactions, is an exercise in managing complexity, and explaining the results is an exercise in reducing complexity for the reader. But nothing we are doing in this chapter is substantially different than anything we have done in Chapters 7 or 9. It just requires a little more care, some methodical attention to detail, and the ability to enter variables in a blockwise fashion and interpret the results graphically. You can see how complexity grows rapidly as one moves from linear interactions (3 terms needed, not counting the intercept) to quadratic interactions (8 terms needed) to cubic interactions (15 terms needed). Of course, if you had more than two IVs, you could do this for different combination of IVs you wished to test. This might seem overwhelming, but after a few practice trials, it will become routine for you. If you consider how much time and effort (by you or someone else) went into gathering the data you want to analyze, spending a few extra minutes fully exploring these interesting effects is, in my opinion, worthwhile.

A Real-Data Example and Exploration of Procedural Details

Let us return to our example from Chapter 9 regarding student achievement and socioeconomic status (SES) predicting high school graduation. If you remember from the last part of Chapter 9, we compared the observed probabilities of graduation with achievement and family SES, reproduced as Figure 10.1a.

The observed data suggest to me that there is a curvilinear effect present in the data, and indeed, there might be a reasonable expectation that we might have different curves in the relationship between SES and graduation as a function of student achievement test scores. For example, it looks as though most students with high achievement test scores tend to graduate regardless of family SES, but SES might have a strong effect on the probability of graduation for students with low levels of achievement. There might be an asymptotic curve where, for these students with low achievement, increasing SES has strong effects on graduation rates as we move from very low to average SES, and then the effect might flatten out. Our analysis of the data via linear interaction in the last chapter produced a result that captures a difference in effect, but in a linear fashion, as reproduced in Figure 10.1b.

Although this interaction effect indicates that the probability of graduation for students with high achievement changes little as a function of SES and a great deal for students with low achievement scores, the linear-only approach most researchers take to modeling of interactions can leave something to be desired. Let us see if we gain any further understanding of the relationship between these variables if we add curvilinear components to the model.

Following my personal recommendation to take each step one at a time to tackle this sort of complex analysis, let us start carefully and methodically to move from conceptualization of the analysis to summarization.

Figure 10.1a   Observed Probabilities of Graduation as a Function of Family Socioeconomic Status and Student Achievement Levels

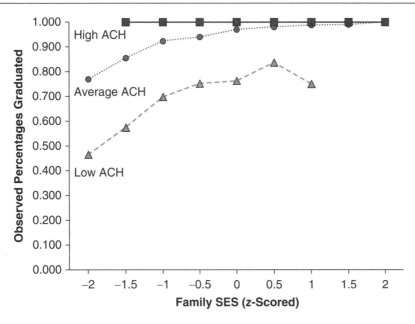

Figure 10.1b   Linear Modeling of the Interaction Between Socioeconomic Status and Student Achievement Predicting Student Graduation

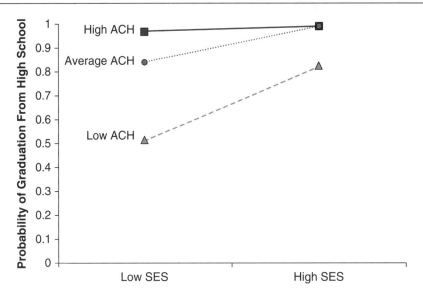

Step 1. Create the Terms Prior to Analysis

Statistical packages vary in how they implement graphical user interfaces, and there can be variations even within packages. For example, in SPSS, you can create interaction terms while creating a logistic regression model; however, when you are creating an OLS regression model, you have to create the cross-product terms prior to analysis. I have talked previously about some of the procedures of creating interaction terms and curvilinear component terms, and I will not reproduce these discussions here. Remember that throughout the rest of this chapter, I will assume you have centered (or converted to z-scores) all continuous variables prior to creating these terms. If we were to fully explore the possible curvilinear interaction between student achievement (zACH) and family SES (zSES), we would need the terms shown in Table 10.3.

Step 2. Build Your Equation Slowly

One of the first mentors I had in regression (Harry Reis, from the University of Rochester) always encouraged us to seek parsimony where possible. He argued that

Table 10.3   Composing All Cross-Products for a Curvilinear Interaction Analysis Between zSES and zACH Using Quadratic and Cubic Components

	$zSES$	$zSES^2$	$zSES^3$
zACH	zSES*zACH	zSES^2*zACH	zSES^3*zACH
zACH^2	zSES*zACH^2	zSES^2*zACH^2	zSES^3*zACH^2
zACH^3	zSES*zACH^3	zSES^2*zACH^3	zSES^3*zACH^3

we want to value the simplest model that best explains the data. Using this principle, we want to evaluate simpler models first, and only accept more complex models when they add significantly to the goodness of fit of the model. Of course, we also have mathematical considerations, such as the fact that we have to enter simpler terms (linear effects of each variable) before we can evaluate the more complex (curvilinear or interaction) effects.

Let us start this analysis with some relatively simple models: linear effects first, then curvilinear effects, then linear interaction terms, and then curvilinear interaction terms. Thus, I would enter zSES and zACH in the first block of variables; $zACH^2$, $zSES^2$, $zACH^3$, and $zSES^3$ on the second block; the linear interaction term on a third step; quadratic interaction terms on the fourth step; and then cubic interaction terms on the final step. My thinking is laid out in Table 10.4a, along with a summary of how these steps are evaluated as contributing or not contributing to the model fit.

Isn't it discouraging that in a chapter focused on curvilinear interactions, the last two steps of our first example are not significant? Never fear, intrepid reader. Your favorite author has the situation well in hand. Continue on to the next section.

Step 3. Clean the Data Thoughtfully to Ensure You Are Not Missing an Interesting Effect

After performing an initial analysis, standardized residuals and Cook's Distance (Cook's D; converted to z-scores) were examined. A small percentage of cases had a Cook's D greater than 4.00, as you can see in Figure 10.1c. Yet even after removing those cases, there were a significant number of cases with extreme residuals, as you can see in Figure 10.1d.

Following data cleaning of cases with Cook's D more than 4 standard deviations (*SD*) from the mean and then cases with standardized residuals greater than 5 *SD*

Table 10.4a Summary of Curvilinear Interaction Model Before Data Cleaning ($N = 16,608$)

Model	*−2LL*	*Δ−2LL*	*p for Δ−2LL*
Intercept only	9,962.22	—	—
Step 1: zSES, zACH	8,059.12	1,903.115	< .0001
Step 2: $zSES^2$, $zACH^2$, $zSES^3$, $zACH^3$	8,043.62	15.482	< .004
Step 3: zSES × zACH	8,023.16	20.46	< .0001
Step 4: $zSES^2$ × zACH, zSES × $zACH^2$, $zSES^2$ × $zACH^2$	8,020.11	3.05	Not significant
Step 5: all cubic interaction terms	8,016.27	3.85	Not significant

SOURCE: National Education Longitudinal Study of 1988 (NELS88) from the National Center for Education Statistics (http://nces.ed.gov/surveys/nels88/).

Figure 10.1c   Cook's Distance, Converted to z-Scores

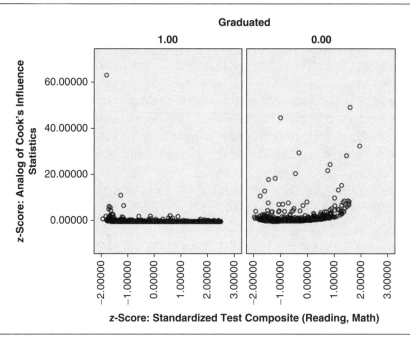

SOURCE: National Education Longitudinal Study of 1988 (NELS88) from the National Center for Education Statistics (http://nces.ed.gov/surveys/nels88/).

Figure 10.1d   Standardized Residuals After Removal of Cases With Cook's Distance (z-Scored) > 4

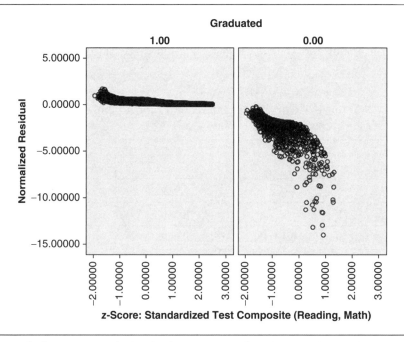

SOURCE: National Education Longitudinal Study of 1988 (NELS88) from the National Center for Education Statistics (http://nces.ed.gov/surveys/nels88/).

from the mean,[1] we are left with 16,463 cases (from an original 16,608, after removing 145; 0.87% of the original cases).

Step 4. After Influential Cases
Are Removed, Perform the Analysis Again

Even following data cleaning, the cubic terms failed to contribute significantly to the model; as such, they and the interaction terms they were associated with were removed from the model. The final summary of the model is shown in Table 10.4b.

As you can see in Table 10.4b, the removal of less than 1% of the cases resulted in some changes to the model—leaving us with generally stronger model fit after the entry of each block of variables, even after removing the cubic terms, and most important for our purposes here in Chapter 10, a significant block of quadratic interaction terms on Step 4. The variable statistics are presented in Table 10.4c.

This leaves us with a final logistic regression equation shown in Equation 10.2:

$$\text{Logit}(\hat{Y}) = 3.953 + 2.638(\text{zSES}) + 2.848(\text{zACH}) + 0.841(\text{SES}^2) +$$
$$0.742(\text{ACH}^2) + 2.637(\text{ACH*SES}) + 0.891(\text{SES}^{2*}\text{ACH}) +$$
$$0.857(\text{SES*ACH}^2) + 0.211 (\text{SES}^{2*}\text{ACH}^2) \qquad (10.2)$$

Table 10.4b   Summary of Curvilinear Interaction Model After Data Cleaning
($N = 16,643$)

Model	After Cleaning Data			Before Cleaning Data		
	−2LL	Δ−2LL	p	−2LL	Δ−2LL	p
Intercept only	9,280.79	—	—	9,962.22	—	—
Step 1: zSES, zACH	6,996.61	2,284.18	< .0001	8,059.12	1,903.115	< .0001
Step 2: SES², ACH²	6,944.85	51.76	< .0001	8,043.62	15.482	< .004
Step 3: zSES × zACH	6,907.15	37.70	< .0001	8,023.16	20.46	< .0001
Step 4: quadratic interaction terms	6,860.31	46.84	< .0001	8,020.11	3.05	Not significant

SOURCE: National Education Longitudinal Study of 1988 (NELS88) from the National Center for Education Statistics (http://nces.ed.gov/surveys/nels88/).

---

1   It may seem arbitrary that I use 4 *SD* for Cook's D and 5 *SD* for residuals, but if you look at the data and graphs, you will see that my principle of removing the fewest cases for the most benefit is what is driving these decisions. If one removes cases at 4 *SD* for standardized residuals, we lose an undesirably large number of cases with little extra benefit.

Table 10.4c   Final Parameter Estimates of Socioeconomic Status and Achievement Predicting High School Graduation Following Data Cleaning and Removal of Cubic Terms

		B	SE	Wald	df	p	Exp(B)	95% CI for Exp(B)	
								Lower	Upper
Step 4	zSES	2.638	0.266	98.680	1	.000	13.990	8.313	23.545
	zACH	2.848	0.288	97.499	1	.000	17.246	9.800	30.350
	SES^2	0.841	0.136	38.139	1	.000	2.320	1.776	3.030
	ACH^2	0.742	0.170	19.158	1	.000	2.101	1.507	2.929
	SESACH	2.637	0.542	23.678	1	.000	13.966	4.829	40.391
	SES^{2}ACH	0.891	0.265	11.316	1	.001	2.437	1.450	4.096
	SESACH^2	0.857	0.302	8.065	1	.005	2.356	1.304	4.257
	SES^{2}ACH^2	0.211	0.149	1.991	1	.158	1.235	0.921	1.655
	Constant	3.953	0.134	867.972	1	.000	52.111		

Variables in the Equation

SOURCE: National Education Longitudinal Study of 1988 (NELS88) from the National Center for Education Statistics (http://nces.ed.gov/surveys/nels88/).

Step 5. Provide Your Audience With a Graphical Representation of These Complex Results

When representing curvilinear effects, more than two points per line are required. We also want to be careful not to extend our predicted values far outside the majority of the data, so we will use values ranging from −2 to 2 (using 0.5 increments[2]) for SES and −2 (low) and 2 (high) and 0 (average) for ACH to attempt to replicate the observed probabilities in Figure 10.1a and represent the nature of the interaction across a reasonable range of the variables without complicating the graph unnecessarily.

As you can see in Figure 10.1e, the quadratic interaction gets much closer to the observed probabilities than the linear interaction, although it is not a perfect representation. This should not be surprising because I am sure there are many factors that influence graduation from high school beyond family affluence and achievement test scores.

Step 6. Summarize the Results Coherently Using the Graphs as Guides

This is how I would summarize this analysis:

> To examine the curvilinear interaction between student achievement scores and family SES, both were first converted to z-scores and the various quadratic and cubic and interaction terms were computed. Variables were entered into the logistic regression equation in blocks as follows: simple effects, curvilinear terms, linear interaction terms, quadratic interaction terms, and cubic interaction terms (as described in Table 10.3). Following this initial analysis, standardized residuals and Cook's D were examined for inappropriately influential cases.

---

2   When graphing lines, two points will completely describe the effect. When graphing curves, more data points allow for more clarity in the nature of the curve. It really does not take much effort to predict and graph more than the minimum number of data points to create a well-elaborated graph.

Cases with residuals greater than |4| were removed first, and then cases with Cook's D greater than 5 SD from the mean were removed. This left 16,463 cases from an original 16,608 (removing 145, or 0.87% of the original cases).

As you can see in Table 10.4b, following data cleaning, all blocks of variables significantly and impressively improved the model except the last block containing the cubic interaction terms. Thus, all cubic terms were removed from the final model, which you can see at the bottom of Table 10.4c. This regression equation was used to compute predicted values for students with low (–2), average (0), and high (2) achievement across a range of family SES values. These predicted logits were then converted to predicted probabilities and are presented graphically in Figure 10.1e.

Examining this curvilinear interaction, we can see that although family SES generally is considered to have a moderately strong influence on student outcomes, our findings indicate that the dynamics might be more complex. For the most part, family SES does not appear to have a strong impact on graduation rates for students with average or high achievement test scores, where graduation rates are relatively consistent and high. The only influence we see from family SES is in the lower range of student achievement. For average students, graduation rates seem to increase from about 90% to almost 100% as family SES increases from 2 SD below the mean to average. For students with high achievement test scores, graduation rates are virtually 100% across the entire range of family SES. It is, however, the students with low achievement test scores that show the strongest effect of SES on graduation rates.

For these students, as SES increases from very low to average, graduation rates more than double, increasing from about 32% to about 77%. As SES increases above the mean, graduation rates increase at a slower rate. Thus, we can conclude that although achievement scores are predictive of graduation rates, and family affluence is also predictive of graduation rates, it is in the combination of the two variables, and particularly when students are below average in both variables, that we see the lowest graduation rates. Having either average or higher achievement or SES seems to be associated with relatively high probabilities of graduation.

Summary

A "simple" linear interaction indicates that the slopes of a regression line are different for different values of another variable. A curvilinear interaction indicates that the shape of the curves is different for different values of another variable. In this case, we see a clear quadratic curve for relatively low-achieving students indicating that the probability of graduation increases as SES increases, flattening out as SES tends toward relatively high values. For students with average or high achievement, the effect of SES is less pronounced.

Using these simple steps in much the same way that we did in simpler analyses in previous chapters, we can uncover interesting and compelling effects in our data.

Curvilinear Interactions Between Continuous and Categorical Variables

Let us now examine an interaction between an unordered categorical variable and a continuous variable. The mechanics of this type of analysis are similar to when we

Figure 10.1e   Curvilinear Interaction Between Achievement and Socioeconomic Status After Data Cleaning

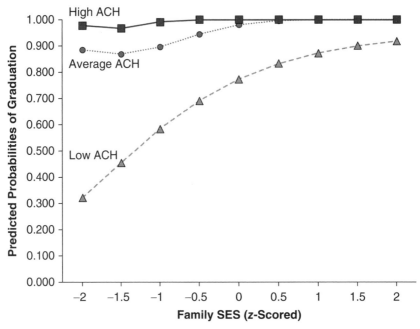

SOURCE: National Education Longitudinal Study of 1988 (NELS88) from the National Center for Education Statistics (http://nces.ed.gov/surveys/nels88/).

explored the simple interaction between the categorical variable RACE and the continuous variable zSES in the previous chapter. Of course, this model of interaction between categorical IV and continuous IV generalizes to the curvilinear interaction model with one caveat: we can only explore curvilinear effects with continuous variables.

Let us turn to our Natality data set from the US Centers for Disease Control and Prevention for a change of pace. In this example, we will predict birth weight (BIRTHWT) for infants at full term (38–40 weeks of gestation) as a function of two variables: the amount of weight the mother gained during pregnancy (zMOMWT, converted to z-scores) and RACE (coded 1 = White, 2 = African American/Black, 3 = Native American/Alaskan Native, 4 = Asian/Pacific Islander).[3] RACE was dummy coded with White as the comparison group, African American/Black as Dum1, Native American/Alaskan Native as Dum2, and Asian/Pacific Islander as Dum3. Thus, our table of cross-products looks like what is presented in Table 10.5a.

---

3   Hispanic ethnicity was coded as a separate variable in this data set. For simplicity, and because this is just a procedural demonstration, that variable was not included in the analysis. This is also a good time to remind all readers yet again that in performing these analyses, I did not take into account any sampling and weighting, so the results should not be interpreted substantively.

Table 10.5a   Composing All Cross-Products for a Curvilinear Interaction Equation With a Categorical and Continuous Variable

	X	X^2	X^3
DUM1	X × DUM1	X^2 × DUM1	X^3 × DUM1
DUM2	X × DUM2	X^2 × DUM2	X^3 × DUM2
DUM3	X × DUM3	X^2 × DUM3	X^3 × DUM3

Once these terms are created, we will use blockwise entry as before, entering the groups of variables in some sort of sensible order of increasing complexity. For this analysis, I entered the main (simple) effects of the two variables (zMOMWT, DUM1–DUM3) on the first block, because they are not of substantial interest given interactions. I then entered the curvilinear effects for zMOMWT (zMOMWT^2, zMOMWT^3). I then entered the three terms that represent the interaction of zMOMWT × RACE, then the three terms that represent zMOMWT^2 × RACE, and then the three terms that represent zMOMWT^3 × RACE. Because the last blocks of variables (the curvilinear interaction effects) are the ones of most interest for this chapter, I do not consider the order of entry of the earlier terms to be a matter of great importance. It does not matter how terms are entered, so long as they are in the model prior to entry of the last two blocks. The results of this analysis (after routine data cleaning and testing of assumptions) are presented in Table 10.5c. As you can see in the top part of the table, Step 5 is significant even though Step 4 was not. Thus, we will retain all five blocks of variables in the model, and we will graph the final model.

Another point that should not be overlooked is that this is *not* a strong model, and these are *not* large effects. The overall variance accounted for is only about 5%,

Table 10.5b   Composing All Cross-Products for a Curvilinear Interaction Equation Between zMOMWT and RACE (Dummy Coded)

	zMOMWT	*zMOMWT^2*	*zMOMWT^3*
DUM1 (African American/Black)	zMOMWT × DUM1	zMOMWT^2 × DUM1	zMOMWT^3 × DUM1
DUM2 (Native American/Alaskan)	zMOMWT × DUM2	zMOMWT^2 × DUM2	zMOMWT^3 × DUM2
DUM3 (Asian/ Pacific Islander)	zMOMWT × DUM3	zMOMWT^2 × DUM3	zMOMWT^3 × DUM3

Table 10.5c   Summary of Curvilinear Interaction Analysis of RACE and zMOMWT

Model Summary

Model[f]	R	R^2	Adjusted R^2	SE of the Estimate	Change Statistics				
					R^2 Change	F Change	df1	df2	Sig. F Change
1	0.217[a]	0.047	0.047	439.908	0.047	302.440	4	24,451	.000
2	0.221[b]	0.049	0.049	439.486	0.002	24.498	2	24,449	.000
3	0.224[c]	0.050	0.050	439.291	0.001	8.222	3	24,446	.000
4	0.224[d]	0.050	0.050	439.286	0.000	1.196	3	24,443	.310
5	0.225[e]	0.051	0.050	439.179	0.001	4.987	3	24,440	.002

[a]Predictors: (constant), DUM3, zMOMWT, DUM2, DUM1.

[b]Predictors: (constant), DUM3, zMOMWT, DUM2, DUM1, $MOMWT^2$, $MOMWT^3$.

[c]Predictors: (constant), DUM3, zMOMWT, DUM2, DUM1, $MOMWT^2$, $MOMWT^3$, INT1b, INT1c, INT1a.

[d]Predictors: (constant), DUM3, zMOMWT, DUM2, DUM1, $MOMWT^2$, $MOMWT^3$, INT1b, INT1c, INT1a, INT2c, INT2b, INT2a.

[e]Predictors: (constant), DUM3, zMOMWT, DUM2, DUM1, $MOMWT^2$, $MOMWT^3$, INT1b, INT1c, INT1a, INT2c, INT2b, INT2a, INT3c, INT3b, INT3a.

[f]Dependent variable: BIRTHWT.

Coefficients[a]

Model	Unstandardized Coefficients		Standardized Coefficients	t	p
	B	SE	Beta		
(Constant)	3,401.799	3.926		866.512	0.000
zMOMWT	102.506	4.521	0.223	22.671	0.000
DUM1	−149.353	9.970	−0.118	−14.980	0.000
DUM2	12.302	35.331	0.003	0.348	0.728
DUM3	−110.069	13.599	−0.061	−8.094	0.000
$zMOMWT^2$	15.126	2.796	0.059	5.409	0.000
$zMOMWT^3$	−7.244	.990	−0.100	−7.317	0.000
$zMOMWT \times DUM1$	−57.720	10.837	−0.054	−5.326	0.000
$zMOMWT \times DUM2$	−104.920	51.415	−0.026	−2.041	0.041
$zMOMWT \times DUM3$	36.316	18.094	0.017	2.007	0.045
$zMOMWT^2 \times DUM1$	−9.431	6.119	−0.019	−1.541	0.123
$zMOMWT^2 \times DUM2$	2.969	20.163	0.001	0.147	0.883
$zMOMWT^2 \times DUM3$	−4.650	12.178	−0.004	−0.382	0.703
$zMOMWT^3 \times DUM1$	7.865	2.155	0.053	3.649	0.000
$zMOMWT^3 \times DUM2$	20.366	14.729	0.018	1.383	0.167
$zMOMWT^3 \times DUM3$	0.949	4.468	0.002	0.212	0.832

[a]Dependent variable: BIRTHWT.

and the model is only significant because the sample is so large. Thus, we can have interesting small effects, as is the case here, but we need to make sure we acknowledge to the reader that the effect is indeed small. Put into concrete terms, the largest effect (for Native American/Alaskan) is about 700 grams, or about 1.4 pounds. However, because birth weight can be a significant predictor of future health, it is not up to me to determine whether that is an important effect. I leave that to those of you who are expert in that field.

$$\begin{aligned}
\hat{Y} = {} & 3401.799 + 102.506(zMOMWT) - 149.353(DUM1) + \\
& 12.302(DUM2) - 110.069(DUM3) + 15.126(zMOMWT^2) - \\
& 7.244(zMOMWT^3) - 57.720(zMOMWT \times DUM1) - \\
& 104.92(zMOMWT \times DUM2) + 36.316(zMOMWT \times DUM3) - \\
& 9.431(zMOMWT^2 \times DUM1) + 2.969(zMOMWT^2 \times DUM2) - \\
& 4.650(zMOMWT^2 \times DUM3) + 7.865(zMOMWT^3 \times DUM1) + \\
& 20.366(zMOMWT^3 \times DUM2) + 0.969(zMOMWT^3 \times DUM3)
\end{aligned} \quad (10.3)$$

Equation 10.3, derived from the analysis summarized in Table 10.5c, might look like an imposing equation, but remember the nature of dummy variables—for the comparison group, all dummy variables and any interaction involving them will be 0 and will drop out of the analysis. Then, for each other group, only one dummy variable will have a 1, and the rest will be zeros, so they and all of the interactions they are associated with will drop out. A simple Excel spreadsheet, once set up, easily creates calculations for each of the four racial groups that can then be transferred to make the graph. I give an example of this in Table 10.5d.

Procedurally, once a section like this is set up appropriately, it is easy to change all of the ones in the "DUM2" column to zeros and to change another column (e.g., DUM3) to all ones to create predicted scores for the next group. I have worked with some analysts who eschew the interaction columns on the right of the table and merely multiply the variables while creating the equation in the last column. I prefer this deliberate, methodical approach I model here because I feel it maximizes the opportunity to identify potential errors. It is easy for me to see, for example, when I change all of the zeros to ones in a DUM1–DUM3 column, I should see corresponding change in the interaction columns and in the predicted weight column.

As you can see in Figure 10.2, the significant curvilinear interaction is due to the variety of different curvilinear effects across the four racial groups. Absent this interaction, each group would have similar curvilinear effects. For example, the African American/Black group has a mostly linear effect, the White and Asian/Pacific Islander curves are similar and cubic, and the Native American group has a cubic effect that is inverted from the previous two.

Table 10.5d Example Spreadsheet for Computing Predicted Values for RACE = 3

zMOMWT	DUM1	DUM2	DUM3	zMOMWT²	zMOMWT³	INT1a	INT1b	INT1c	INT2a	INT2b	INT2c	INT3a	INT3b	INT3c	PRED WT
-3	0	1	0	9	-27	0	-3	0	0	9	0	0	-27	0	3,229.904
-2.5	0	1	0	6.25	-15.625	0	-2.5	0	0	6.25	0	0	-15.625	0	3,328.1985
-2	0	1	0	4	-8	0	-2	0	0	4	0	0	-8	0	3,386.333
-1.5	0	1	0	2.25	-3.375	0	-1.5	0	0	2.25	0	0	-3.375	0	3,414.149
-1	0	1	0	1	-1	0	-1	0	0	1	0	0	-1	0	3,421.488
-0.5	0	1	0	0.25	-0.125	0	-0.5	0	0	0.25	0	0	-0.125	0	3,418.1915
0	0	1	0	0	0	0	0	0	0	0	0	0	0	0	3,414.101
0.5	0	1	0	0.25	0.125	0	0.5	0	0	0.25	0	0	0.125	0	3,419.058
1	0	1	0	1	1	0	1	0	0	1	0	0	1	0	3,442.904
1.5	0	1	0	2.25	3.375	0	1.5	0	0	2.25	0	0	3.375	0	3,495.4805
2	0	1	0	4	8	0	2	0	0	4	0	0	8	0	3,586.629
2.5	0	1	0	6.25	15.625	0	2.5	0	0	6.25	0	0	15.625	0	3,726.191
3	0	1	0	9	27	0	3	0	0	9	0	0	27	0	3,924.008

Figure 10.2   Curvilinear Interaction Between RACE and Weight Gain in Mother (z-Scored)

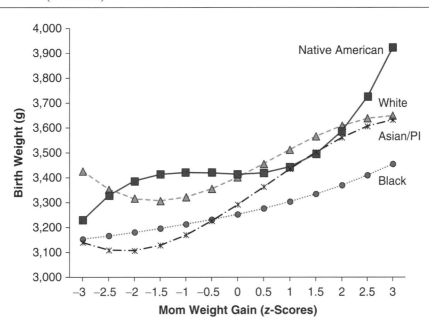

Example Summary of Analysis

To examine the curvilinear interaction between weight gain in the mother (converted to z-scores; zMOMWT) and race (dummy coded with White as the comparison group, DUM1 representing African American/Black, DUM2 representing Native American/Alaskan Native, DUM3 representing Asian or Pacific Islander), zMOMWT was squared and cubed, and interactions were computed between zMOMWT, $zMOMWT^2$, and $zMOMWT^3$ and each dummy variable to fully represent both the linear and curvilinear components of the main effects and interactions. These terms are fully displayed in Table 10.5b. The groups of variables were entered in a blockwise fashion: simple effects of zMOMWT and race; the curvilinear terms of zMOMWT; the three terms that represent the interaction between zMOMWT and race; the three terms that represent the interaction between $zMOMWT^2$ and race; and the three terms that represent the interaction between $zMOMWT^3$ and race.

Following the initial analysis, standardized residuals were examined. Cases with standardized residuals greater than |4| were removed ($N = 34$ of 24,490, or 0.13% of the sample). In addition, one case with an unusual predicted value (more than 8 SD from the mean and far distant from all other cases) was removed. The model was again computed, and the normality of the residuals improved (original skew = 0.12, kurtosis = 1.01; final skew = 0.11, kurtosis = 0.42), meeting the assumption of normality. The plot of predicted values versus residuals was also examined, supporting the assumption of homoscedasticity.

Results indicated a significant but relatively weak effect overall. As you can see in Table 10.5c, the final step with the cubic interaction terms was significant, although they accounted

(Continued)

(Continued)

for only a small percentage of the variance as a group. Furthermore, the overall model accounted for only just over 5% of the variance in infant birth weight. However, because birth weight is a significant predictor of future health outcomes, this may be an important finding. Individual variable parameters are presented in Table 10.5d. The final model was used to create predicted values for each racial group across a range of zMOMWT (–3 to +3 in increments of 0.5) and is presented graphically in Figure 10.2.

Generally, as you can see in Figure 10.2, there was a trend for mothers who gained more weight during the pregnancy to give birth to children who were heavier. However, there were differences across the racial groups in the shape of that relationship. For White and Asian/ Pacific Islander mothers, there was a general increasing cubic trend with a slightly steeper slope for the latter group. Native American/Alaskan mothers also had a clear cubic curve, but of a more dramatic and different nature than the prior two groups. Finally, the relationship for African American/Black mothers was mostly linear in nature.

Summary

As you can see, this type of analysis is merely an extension of the type of interaction analyses we explored in the previous chapter. It is important to remember that not all effects are strong, but depending on the context, even small effects can be important. You, as the expert, are the only one who can determine the importance of an effect given the mastery of the content within a field. I am certainly not an expert in this area and will therefore pass on that type of discussion. Although a curvilinear interaction between a continuous and categorical IV is tricky, it is no more complex than an interaction between two continuous variables, as long as you take each step in turn and manage complexity adequately.

Curvilinear Interactions
With Categorical DVs (Multinomial Logistic)

We can apply the same methodological fun to all other linear models (such as multinomial and ordered logit models). In Chapter 7, we noted a curvilinear effect in our multinomial example predicting marijuana use (0 = never used, 1 = used 1–2 times, 2 = used 3–19 times, 3 = used 20 or more times) from student achievement test scores (zACH). We have also seen family SES and student achievement interacting in interesting ways. Let us see if there is a curvilinear interaction effect to be found in these data by testing whether we can detect a curvilinear interaction between zSES and zACH. To perform this analysis, I computed the necessary terms shown in Table 10.3 to examine possible curvilinear interactions including the cubic effects. I performed an initial multinomial logistic regression analysis, gradually building the model in stages, then performed some data cleaning analyses (via three binary logistic regression analyses), and performed the final analysis by examining a z-scored Cook's D and removing any cases more than 7.5 SD from the mean (initial $N = 14,861$, final $N = 14,786$, 75 or 0.5% of cases removed). Neither analysis supported the conclusion that there were significant cubic effects, and thus all cubic effects were removed from the final model. A summary of these analyses is presented in Table 10.6a. Recall that in SPSS, you have to perform blockwise multinomial analyses individually and calculate model change statistics yourself because the multinomial routine does not provide for blockwise entry. As you can see in Table 10.6a, some judicious data cleaning revealed significant curvilinear effects not detectable prior to data cleaning. The final parameter estimates for that step are presented in Table 10.6b.

For consistency with previous graphs, I will keep zSES on the X axis and graph different lines as a function of low/average/high student achievement. These are presented in Figures 10.3a, 10.3b, and 10.3c. As you can see, the graphs are relatively similar in general pattern across all three analyses (0 versus 1, 0 versus 2, and 0 versus 3). The significant curvilinear interaction is obvious in that different groups (low-, average-, and high-achieving students) have significantly different curves. Because the three marijuana use groups seem to have such a similar interaction pattern, this begs the question as to whether we could have performed an ordinal regression. Many of the regression coefficients are similar in direction and magnitude, suggesting that this type of model might be viable even with a curvilinear interaction effect as complex as observed below.

Table 10.6a   Model Fit Summary for Multinomial Logistic Regression Predicting MJ From zACH and zSES

Model	After Cleaning Data			Before Cleaning Data		
	$-2LL$	$\Delta-2LL$, df	p	$-2LL$	$\Delta-2LL$, df	p
Intercept only	20,506.534	—	—	20,927.044	—	—
Step 1: zSES, zACH	20,362.369	147.44, 6	$< .0001$	20,795.286	131.758, 6	$< .0001$
Step 2: zSES × zACH	20,326.045	36.324, 3	$< .0001$	20,773.470	21.82, 3	$< .0001$
Step 3: SES², ACH²	20,273.673	52.372, 6	$< .0001$	20,761.836	29.44, 6	$< .07$
Step 4: quadratic interaction terms	20,223.379	50.294, 9	$< .0001$	20,737.095	13.99, 9	$< .003$

Table 10.6b   Parameter Estimates From Multinomial Regression

Parameter Estimates

MJ[a]		B	SE	Wald	df	p	Exp(B)	95% CI for Exp(B)	
								Lower Bound	Upper Bound
1 1–2 times	Intercept	−2.080	0.051	1,633.417	1	.000			
	zACH	−0.311	0.044	50.730	1	.000	0.733	0.673	0.798
	zSES	−0.121	0.046	6.901	1	.009	0.886	0.810	0.970
	SESACH	0.135	0.056	5.744	1	.017	1.145	1.025	1.278
	ACH^2	−0.040	0.044	0.841	1	.359	0.961	0.882	1.047
	SES^2	−0.027	0.041	0.428	1	.513	0.974	0.898	1.055
	SES^{2}ACH	0.130	0.041	9.983	1	.002	1.139	1.051	1.235
	SESACH^2	0.040	0.046	0.759	1	.384	1.041	0.951	1.138
	SES^{2}ACH^2	−0.109	0.033	11.145	1	.001	0.896	0.841	0.956
2 3–19 times	Intercept	−2.416	0.060	1,641.252	1	.000			
	zACH	−0.239	0.052	21.524	1	.000	0.787	0.711	0.871
	zSES	−0.110	0.053	4.387	1	.036	0.896	0.808	0.993
	SESACH	0.039	0.063	0.371	1	.543	1.039	0.918	1.177
	ACH^2	−0.045	0.052	0.763	1	.382	0.956	0.864	1.058
	SES^2	0.024	0.047	0.266	1	.606	1.024	0.935	1.122
	SES^{2}ACH	0.145	0.048	8.959	1	.003	1.156	1.051	1.271
	SESACH^2	0.013	0.053	0.060	1	.807	1.013	0.913	1.124
	SES^{2}ACH^2	−0.129	0.041	10.066	1	.002	0.879	0.812	0.952
3 20+ times	Intercept	−2.717	0.073	1,388.573	1	.000			
	zACH	−0.372	0.066	32.194	1	.000	0.689	0.606	0.784
	zSES	−0.140	0.071	3.883	1	.049	0.869	0.757	0.999
	SESACH	0.046	0.088	0.278	1	.598	1.047	0.882	1.244
	ACH^2	−0.116	0.068	2.897	1	.089	0.890	0.779	1.018
	SES^2	−0.116	0.066	3.084	1	.079	0.891	0.782	1.014
	SES^{2}ACH	0.221	0.073	9.142	1	.002	1.247	1.081	1.439
	SESACH^2	0.022	0.078	0.082	1	.775	1.023	0.877	1.192
	SES^{2}ACH^2	−0.172	0.065	6.907	1	.009	0.842	0.741	0.957

[a]The reference category is: 0 0 OCCASIONS.

Curvilinear Interaction Effects in Ordinal Regression

Recall that in order to perform an ordinal regression, or at least the cumulative probability model we have been utilizing, we need to meet the assumption of parallel lines, indicating that the effects are not significantly different across combinations of groups (0 versus 1–3, 0, 1 versus 2, 3, and 0–2 versus 3). In this section, we will replicate the previous multinomial analysis predicting marijuana use from zACH

Figure 10.3a   Multinomial Logistic Regression Predicting Probability of Using
Marijuana (Level 1, Used 1–2 Times) as a Function of Family
Socioeconomic Status and Student Achievement

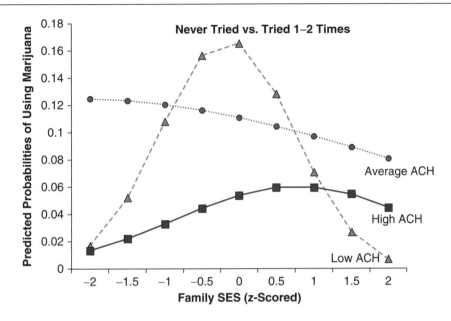

Figure 10.3b   Multinomial Logistic Regression Predicting Probability of Using
Marijuana (Level 2, Used 3–19 Times) as a Function of Family
Socioeconomic Status and Student Achievement

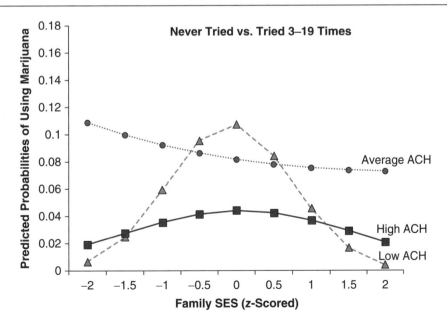

Figure 10.3c   Multinomial Logistic Regression Predicting Probability of Using
Marijuana (Level 3, Used 20 or More Times) as a Function of Family
Socioeconomic Status and Student Achievement

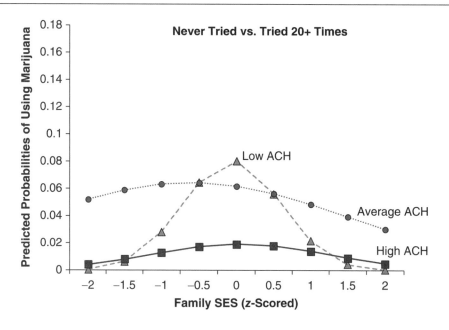

Example Summary

To determine whether family SES and student achievement (both converted to z-scores) predicted the probability that a student would try marijuana (self-reported lifetime use; 0 = never
tried, 1 = tried 1–2 times, 2 = tried 3–19 times, 3 = tried 20 or more times), a multinomial
model was created with quadratic and cubic terms and all linear, quadratic, and cubic interaction terms were entered into the model in a blockwise fashion. At each stage, likelihood
ratio tests evaluated whether the added block of variables significantly improved the model.
As you can see in Table 10.6a, prior to data cleaning, the model was significantly improved
with the addition of each of the first three blocks, but neither block of interaction terms significantly improved the model. Cook's D was examined for inappropriately influential cases,
and cases more than 7.5 SD from the mean were removed, resulting in a removal of 0.5%
of the cases.

Following data cleaning, as you see in Table 10.6a, the quadratic interaction terms significantly improved the model, but no cubic term was significant, so all cubic terms for both
variables were removed from the model. The parameter estimates for the final model are
presented in Table 10.6b. Also, the effects are modeled in Figures 10.3a–10.3c. As you
can see in the three figures, the general pattern of marijuana use is relatively consistent
across all three models. Specifically, family SES seems to have the strongest effect on students who have relatively low achievement scores, with increasing SES generally leading
to increased probability of trying and using marijuana until family SES reaches average,
and then the probabilities decline rather drastically. In contrast, students with average

achievement tend to show a very modest decrease in the probability of trying or using marijuana as SES increases. Finally, for students with relatively high achievement scores, family SES appears to be associated with a modest increase in the probability of trying or using marijuana.

This is not a tremendously strong effect, but it may lend some insight into causes and patterns of marijuana use.

and zSES. Given the similarity of effects across groups in that previous analysis, it is appropriate to wonder if an ordinal regression is appropriate. An initial analysis without data cleaning showed that the assumption was met, and data cleaning (removing cases at $z > 7.5$ or greater) led to a nonsignficant test of parallel lines (as you can see in Table 10.7a). Recall from previous chapters that because ordinal regression examines multiple thresholds and probabilities of being at or below those thresholds, we need to do some simple postprediction processing to arrive at the final conditional probabilities that will represent the effect we graph. The results of this analysis are presented in Table 10.7b.

Table 10.7a   Test of Parallel Lines From Ordinal Regression, With Data Cleaning Using $z = 7.5$ as the Cut Point

Test of Parallel Lines[a]

Model	−2LL	Chi-Square	df	p
Null hypothesis	20,247.903			
General	20,221.852	26.051	16	.053

The null hypothesis states that the location parameters (slope coefficients) are the same across response categories.
[a]Link function: Logit.

Table 10.7b   Ordinal Regression Predicting Marijuana Use From Socioeconomic Status and Achievement

Model Fitting Information

Model	−2LL	Chi-Square	df	p
Intercept only	20,506.534			
Final	20,247.903	258.631	8	.000

Link function: Logit.

Parameter Estimates

		Estimate	SE	Wald	df	p	95% CI	
							Lower Bound	Upper Bound
Threshold	[MJ = 0]	1.267	0.037	1,193.782	1	.000	1.195	1.339
	[MJ = 1]	2.012	0.040	2,477.442	1	.000	1.933	2.091
	[MJ = 2]	3.042	0.052	3,462.046	1	.000	2.941	3.143
Location	zSES	−0.121	0.033	13.385	1	.000	−0.185	−0.056
	zACH	−0.295	0.031	87.514	1	.000	−0.356	−0.233
	SESACH	0.076	0.040	3.689	1	.055	−0.002	0.154
	SES^2	−0.032	0.030	1.151	1	.283	−0.089	0.026
	ACH^2	−0.062	0.032	3.821	1	.051	−0.123	0.000
	SES^{2}ACH	0.150	0.030	24.733	1	.000	0.091	0.209
	SES^{2}ACH^2	−0.118	0.025	22.846	1	.000	−0.166	−0.070
	SESACH^2	0.028	0.033	0.718	1	.397	−0.037	0.093

Link function: Logit.

I will not reproduce each equation for the three thresholds, because they are all the same except for the intercept. The general regression equation will be as presented in Equation 10.4. This process of calculating predicted probabilities for each group is somewhat more complicated than for previous graphing tasks because we have to produce a table with the cumulative probabilities of group (0, 0–1, 0–2), and then perform subtraction to create conditional probabilities for each group at each estimate (zACH at −2, 0, and 2 for low, average, and high achievement, and zSES ranging from −2 to 2 in 0.5 increments, as we did with Figures 10.3a–10.3c). It is not difficult; it just takes a little patience. The spreadsheet I used to create these estimates is extensive, so I will produce only a small part of the table here in Table 10.7c and I will provide a link to the entire spreadsheet on the book's web page.

$$\text{Logit}(\hat{Y}) = b_0 - (-0.121(\text{zSES}) - 0.295(\text{zACH}) + 0.076(\text{zSES}^*\text{zACH}) - 0.032(\text{zSES}^2) - 0.062(\text{zACH}^2) + 0.150(\text{zSES}^2{}^*\text{zACH}) - 0.118(\text{zSES}^2{}^*\text{zACH}^2) + 0.028\,(\text{zSES}^*\text{zACH}^2)) \tag{10.4}$$

As you can see in Table 10.7c, there are some small differences between how I calculated conditional probabilities for this ordinal regression analysis and any other typical logistic regression. First, in the left-hand column, the groups are defined as all cases with scores *less than or equal to* a particular level, rather than simply being in a particular group. As discussed previously, this is due to the nature of this type of cumulative probability model. This necessitates the right-most column. In this final column, we convert cumulative probabilities into conditional probabilities as discussed in prior chapters by subtracting each successive cumulative probability from the prior cumulative probability to isolate the change in probability, which equates to the conditional probability of the last group added to the cumulative prediction.

As you can see in Figures 10.4a–10.4c, the results from this analysis are similar to those from the multinomial logistic regression analysis in the previous section of this chapter. Just as in Chapter 6, when the assumption of parallel lines is met, you will get similar results from either ordinal or multinomial regression analyses. In fact, the three figures (10.4a–10.4c) represent the same effect applied to three different groups with different thresholds. If you examine them closely, you will see that although the magnitudes are different, the general pattern of results is the same across the three figures (of course, because they are all using the same regression equation with different thresholds/ intercepts).

Summary

Some authors recommend presenting an ordinal regression as a single graph without using an intercept as the equation is the same across all groups. Although this is fine if you do not care about the actual magnitude of the probabilities, it can efficiently present the relative patterns of the curvilinear interaction (as you see in Figure 10.4d, which is graphed with no intercept, which functionally makes the intercept 0). Although Figure 10.4d clearly does reflect the general pattern of the

Table 10.7c Partial Spreadsheet for Ordinal Regression Calculations—Low Achievement Only

zSES = −2 Group	Threshold	zSES	zACH	SES*ACH	SES²	ACH²	SES²*ACH	SES²*ACH²	SES*ACH²	Predicted Logit	Cumulative Probability	Conditional Probability
p ≤ 0	1.267	−2	−2	4	4	4	−8	16	−8	3.819	0.978	0.978
p ≤ 1	2.012	−2	−2	4	4	4	−8	16	−8	4.564	0.989	0.011
p ≤ 2	3.042	−2	−2	4	4	4	−8	16	−8	5.594	0.996	0.007
p ≤ 3		≤2	−2	4	4	4	−8	16	−8		1	0.004

zSES = −1.5 Group	Threshold	zSES	zACH	SES*ACH	SES²	ACH²	SES²*ACH	SES²*ACH²	SES*ACH²	Predicted Logit	Cumulative Probability	Conditional Probability
p ≤ 0	1.267	−1.5	−2	3	2.25	4	−4.5	9	−6	2.492	0.923	0.923
p ≤ 1	2.012	−1.5	−2	3	2.25	4	−4.5	9	−6	3.237	0.962	0.0386
p ≤ 2	3.042	−1.5	−2	3	2.25	4	−4.5	9	−6	4.267	0.986	0.0239
p ≤ 3		−1.5	−2	3	2.25	4	−4.5	9	−6		1	0.0138

Figure 10.4b Ordinal Logistic Regression Predicting Probability of Using Marijuana (Level 2, Used 3–19 Times) as a Function of Family Socioeconomic Status and Student Achievement

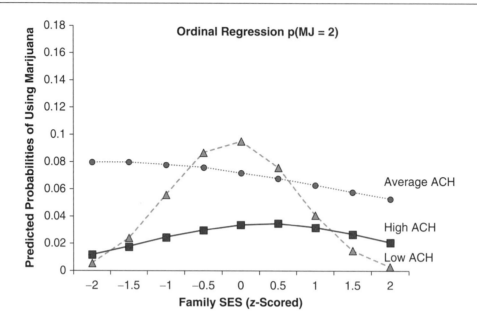

Figure 10.4c Ordinal Logistic Regression Predicting Probability of Using Marijuana (Level 3, Used 20 or More Times) as a Function of Family Socioeconomic Status and Student Achievement

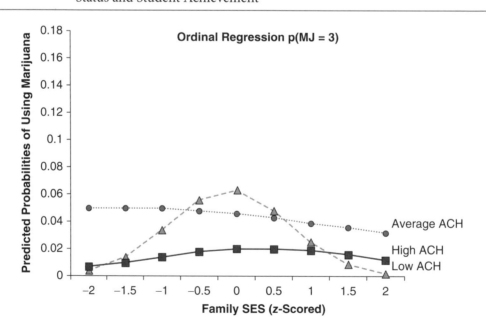

Figure 10.4d   The Ordinal Regression Results Graphed With No Intercept/Threshold (Intercept = 0)

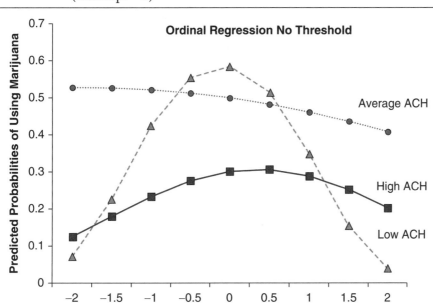

effect, it dramatically distorts the actual probabilities involved, which could be misleading to your audience. For example, the peak of the "low achievement" curve is approximately 60%, whereas in Figure 10.4a, you can see that the highest that curve reaches in the real data is under 14%, because self-reported marijuana use has a low prevalence in these data.

CHAPTER SUMMARY

This was probably the most technical chapter in the book, and if you made it to this point with eagerness in your heart, congratulations! If you feel overwhelmed or intimidated, stay with me and try a few exercises. Once you work through a few problems, it gets easier and easier to perform these tasks. And the outcome can be rewarding—a more interesting and nuanced view of your data.

At this point in the book, we have finished our exploration of the basic skills necessary to take full advantage of the generalized linear model. In the next few chapters, we will explore applications of these techniques, some more advanced aspects of the generalized linear model, and some more esoteric topics.

ENRICHMENT

1. Download data from the book website and replicate the analyses presented in the chapter.

2. Using the NELS88 data, predict graduation from SES and race, examining possible curvilinear interactions.

3. Using AAUP data, determine if there is a curvilinear interaction between number of faculty (NUM_TOT) and Carnegie code (1 = more research extensive, 2 = more balanced in teaching and research, 3 = more teaching focused) in predicting the average salary of associate professors (SAL_ASSOC).

4. Using NHIS data, examine whether there is a curvilinear interaction between smoking status (SMOKE_CAT; 0 = nonsmoker, 1 = former smoker, 2 = occasional smoker, 3 = daily smoker) and body mass index (z-scored into zBMI) predicting diabetes.

REFERENCES

Aiken, L. S., & West, S. (1991). *Multiple regression: Testing and interpreting interactions.* Thousand Oaks, CA: SAGE.

Baron, R. M., & Kenny, D. A. (1986). The moderator–mediator variable distinction in social psychological research: Conceptual, strategic, and statistical considerations. *Journal of Personality and Social Psychology, 51*(6), 1173–1182. doi: 10.1037/0022-3514.51.6.1173

POISSON MODELS

Low-Frequency Count Data
as Dependent Variables

Advance Organizer

Count variables, particularly counts of relatively infrequent events, represent a different category of data that we have not fully explored to this point. Count data are often gathered in the social and behavioral sciences (e.g., how many cigarettes were smoked by adolescents during a day, how many times a student was held back a grade, or how many depressive symptoms were experienced in a given period). In general, count data can be analyzed via ordinary least squares (OLS) regression as long as the assumptions of OLS regression are met. When counts of things represent qualitatively different situations, multinomial regression is an inefficient but appropriate method of analysis. With a particular type of count data, in which we are counting relatively infrequent events, OLS regression may not be appropriate.

Poisson models (named after the French mathematician and physicist Siméon-Denis Poisson) are part of the generalized linear model (GLM) intended to appropriately and efficiently evaluate dependent variables (DVs) composed of these types of counts. Poisson distributions represent an efficient method of estimating probabilities of events, particularly where the population size is large and the probability of an event is relatively low. This technique is often used with highly positively skewed distributions that are *zero inflated* (Cohen, Cohen, West, & Aiken, 2003), although the events do not have to be rare (semi-rare events can also qualify), and the distribution does not have to be highly skewed (Nussbaum, Elsadat, & Khago, 2008). As events become less rare, Poisson distributions can converge with Gaussian (normal) distributions, leaving OLS regression as an appropriate alternative when assumptions of homoscedasticity and normality of residuals are met.

Because Poisson models are not common within many disciplines (even rarer than logistic models!), I have decided to devote a complete chapter to this model in the hope that researchers will begin appropriately using these valuable models and that this chapter can serve as a reference but perhaps not a required part of a course syllabus. Now that you have strong mastery over the art of OLS and logistic regression modeling, Poisson regression should not be a difficult challenge.

As you can see in Table 11.1, we add Poisson regression to the right-hand column for DVs that are count variables where the data fit the Poisson distribution.

Table 11.1 Examples of the Generalized Linear Model as a Function of Independent
Variable and Dependent Variable Type

	Continuous DV	Binary DV	Unordered Multicategory DV	Ordered Categorical DV	Count DV
Continuous IV	OLS regression	Binary logistic regression	Multinomial logistic regression	Ordinal logistic regression	OLS, Poisson regression
Mixed continuous and categorical IV					
Binary/ categorical IV only	ANOVA	Log-linear models	Log-linear models		Log-linear models

ANOVA, analysis of variance; DV, dependent variable; IV, independent variable; OLS, ordinary least squares.

In this chapter, we will cover

- Why Poisson models are necessary
- Basic concepts in Poisson regression
- Examples of Poisson regression with categorical and continuous independent variables (IVs)
- Assumptions and data cleaning for Poisson regression
- Confidence intervals and effect sizes in Poisson regression
- Examples of American Psychological Association (APA)–compliant summaries of analyses
- Dealing with zero-inflated (or deflated) variables
- Dealing with overdispersion

Guidance on how to perform these analyses in various statistical packages will be available online at study.sagepub.com/osbornerlm.

The Basics and Assumptions of Poisson Regression

Poisson regression predicts a count of the number of events that occur in a given time period. The assumption is that this count is due to an underlying rate parameter that can be modeled and the mechanism is consistent with a Poisson process.[1] One special feature of the Poisson distribution is that (in an ideal example of this process) the mean (μ) is equal to the variance, whereas a normal distribution has separate and independent parameters of mean (μ) and variance (σ^2). This is the only assumption in Poisson regression that differs from prior models.

---

1 A Poisson process is essentially a series of time intervals where an event may or may not occur (a Bernoulli yes/no random variable).

The general function for a Poisson distribution is given in Equation 11.1 (adapted from sources such as Cohen et al., 2003; Coxe, West, & Aiken, 2009; Nussbaum et al., 2008), where y is a given value of Y (the DV) and μ is the average (arithmetic mean) of the number of incidents per given time interval.[2] Some examples of a Poisson distribution are presented in Figure 11.1.

$$P(Y = y) = \frac{\mu^y}{y!} e^{-\mu} \tag{11.1}$$

In Poisson regression, we use the natural log link function (ln), modeling the *predicted average count* ($\hat{\mu}$) as a function of specific predictors (as contrasted with logistic regression, in which we are predicting the probability of an outcome). An example Poisson regression equation is presented in Equations 11.2a–11.2c (showing multiple regression, curvilinear regression, and interaction examples, respectively):

$$\ln(\hat{\mu}) = b_0 + b_1 X_1 + b_2 X_2 + b_3 X_3 + \ldots + b_k X_k \tag{11.2a}$$

$$\ln(\hat{\mu}) = b_0 + b_1 X_1 + b_2 X_1^2 + b_3 X_1^3 + \ldots + b_k X_1^k \tag{11.2b}$$

$$\ln(\hat{\mu}) = b_0 + b_1 X_1 + b_2 X_2 + b_3 X_1 X_2 \tag{11.2c}$$

Figure 11.1   Poisson Distributions With Different Average Event Rates

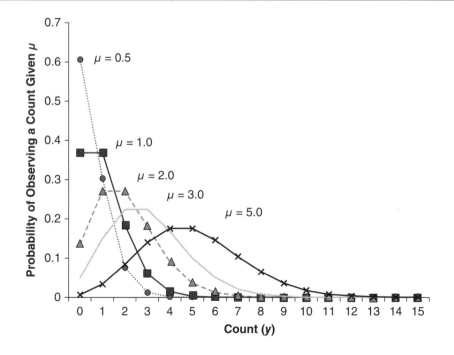

---

2 $y!$ is the factorial of y, which is equal to the value of y multiplied by each lower number between y and 1.

Curvilinearity in Poisson Models

Note that like all other regression models, we assume that the relationship between the logarithm transformed count and the predictor(s) is linear. As in previous models (and Equation 11.2b), we can test whether that is the case. Converting predicted log counts back to counts is slightly simpler than in logistic regression—we can merely exponentiate the predicted log values to arrive at predicted counts (Coxe et al., 2009), making visual representation and interpretation of effects easier on the reader.

Similar to the logic of logistic regression (where a significant predictor means that there are different probabilities of an outcome as a function of the value of the predictor), a significant predictor in Poisson regression indicates a potentially different underlying Poisson process and a different average count as a function of the value of the predictor.

The Nature of the Variables

In Chapter 6, we explored multinomial and ordinal logit models, showing how interesting things can get when you model variables more granularly rather than the yes/no EVER_MJ variable we had used in the binary logistic examples. The "better" marijuana use variable, shown in Table 11.2a, was what we used and appears to be measuring the number of times someone used the substance. This type of variable would have been an excellent candidate for exploration through Poisson modeling *if* the actual reported counts were available, rather than as presented in Table 11.2a. Because this variable does not include actual counts, but rather uneven intervals of counts, it would not be appropriate for this type of analysis. Count data measurement should usually be ratio-level measurement if you are truly counting things or events (because zero would represent zero things or events, making it a meaningful zero), and this variable in Table 11.2a is clearly ordinal measurement because of the unequal intervals due to clustering of responses.

Table 11.2a   In Lifetime, Number of Times R Used Marijuana

Original Category	Frequency	Label
0	13,487	Tried marijuana 0 occasions
1	1,578	Tried marijuana 1–2 occasions
2	1,094	Tried marijuana 3–19 occasions
3	751	Tried marijuana 20 or more occasions

SOURCE: National Education Longitudinal Study of 1988 (NELS88) from the National Center for Education Statistics (http://nces.ed.gov/surveys/nels88/).

Table 11.2b   Number of Female Sexual Partners in Last 3 Months (SEXFNUM)

Original Category	Frequency	Number of Female Partners	Valid Percent	Percent in Perfect Poisson Distribution
0	2,812	None	70.5	67.5
1	954	1	23.9	26.5
2	141	2	3.5	5.2
3	43	3	1.1	0.7
4	16	4	0.3	0.005
5	4	5	0.1	0.00002
6	20	6 or more	0.5	0.000003

SOURCE: National College Health Risk Behavior Survey of 1995 from the US Centers for Disease Control and Prevention (http://www.cdc.gov/healthyyouth/yrbs/data/index.htm). Only students who indicated they had ever had intercourse in their life were included in this analysis.

Figure 11.2   Poisson Distribution of SEXFNUM, the Self-Reported Number of Female Sexual Partners in the Last 3 Months

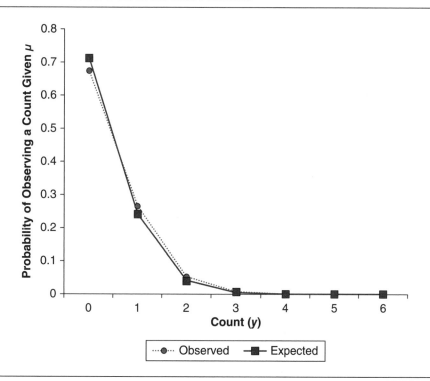

SOURCE: National College Health Risk Behavior Survey of 1995 from the US Centers for Disease Control and Prevention (http://www.cdc.gov/healthyyouth/yrbs/data/index.htm).

In Table 11.2b and Figure 11.2, we have a variable more suited to Poisson distribution. These data are taken from the US Centers for Disease Control and Prevention 1995 National College Health Risk Behavior Survey (NCHRBS1995),[3] which asked college students in the United States about a variety of health- and risk-related habits and behaviors. Some of the questions regarded sexual behavior, and one question in particular ("During the past 3 months, with how many females have you had sexual intercourse?") seemed well suited for this chapter because the distribution is very close to Poisson. In fact, if you compare the actual distribution of the variable with the ideal Poisson distribution calculated using the mean of this distribution ($\mu = 0.3932$; displayed in Table 11.2b and Figure 11.2), the correspondence is so strong that it is difficult to distinguish the observed frequencies from the expected ones.[4] We will use these data for our example below.

Issues With Zeros

Because of the nature of these curves, many Poisson distributions assume a certain number of zeros. Having an unexpectedly large or small number of zeros can lead to misestimation of the model. Note that in other analyses such as OLS regression, zero-inflated variables are a problem that needs to be addressed. In Poisson regression, zero-inflated variables are not a problem unless they violate the reasonable expectation of the curve or represent a distinct population not covered by the proposed Poisson process (e.g., nonsmokers asked how many cigarettes they smoked during the last month). You can also encounter issues such as having too few zeros in the distribution (e.g., in a distribution of $\mu = 1.0$, we expect about 38% of the cases to have zeros). In the case of SEXFNUM (Table 11.2b, Figure 11.2), the college students who have had zero female sexual partners in the past 30 days are composed of two groups. One group has been abstinent during the specified time but has had sex with a female before. The other group has never had sex with a female and may never (e.g., heterosexual females, who are currently included in these data, or other individuals who have been abstinent for whatever reason). Thus, the ability to identify and remove individuals who are not likely to be part of the Poisson process being observed could help refine the model.

Issues With Variance

One characteristic of Poisson models is that the variance is equal to the mean (the assumption of *equidispersion*). When variables being analyzed diverge from this expectation, we have either *overdispersion* (variance significantly larger than the mean) or *underdispersion* (variance significantly lower than the mean), either of which can be dealt with through modification of the model parameters or use

3   Information on NCHRBS1995 is available from the US Centers for Disease Control and Prevention (http://www.cdc.gov/healthyyouth/yrbs/data/index.htm).

4   One very stringent test of correspondence to a distribution shape is the one-sample Kolmogorov-Smirnov (K-S) test, which compares the observed distribution with the theoretical one and provides a z test of the null hypothesis that the observed distribution equals the theoretical one. This test is notoriously sensitive in large samples, but the K-S z was 1.89 with this large a sample, which is impressively close to 0 given the large sample size (N).

of an alternative negative binomial regression model (e.g., Cohen et al., 2003; Coxe et al., 2009). In the case of SEXFNUM, there is mild overdispersion because the mean is 0.3387 but the variance is 0.527. We will address this issue below in more detail.

Why Can't We Just Analyze Count Data via OLS, Multinomial, or Ordinal Regression?

To be sure, many researchers (especially those who have never been introduced to Poisson regression) probably use OLS regression to analyze count data. There may be times when one can legitimately use OLS regression for this type of data. However, one must be cautious in doing so and ensure that assumptions are met and sensible predicted values are obtained. Even if residuals are normally distributed and homoscedastic, we have the thorny problem of the real possibility of predicted values that are negative. Of course, it is not usually possible to have negative counts of something. OLS regression may also produce biased estimates with standard errors that are underestimated, thus leading to overestimated significance tests (Cohen et al., 2003, p. 525).

As you can see in Figure 11.3a, it is easy to get highly non-normally distributed residuals when attempting to perform an OLS regression analysis with a variable

Figure 11.3a   Residuals From an Ordinary Least Squares Regression Analysis of SEXFNUM

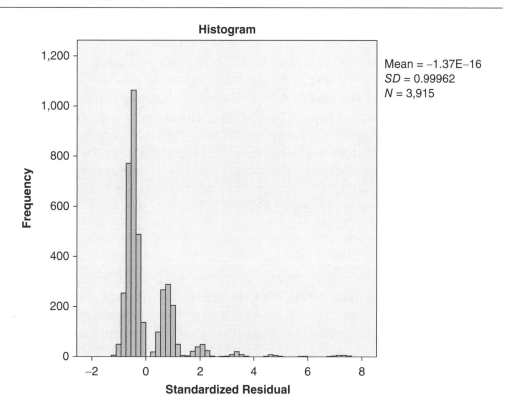

Figure 11.3b   Residuals From an Ordinary Least Squares Regression Analysis of
SEXFNUM Following Box-Cox Transformation

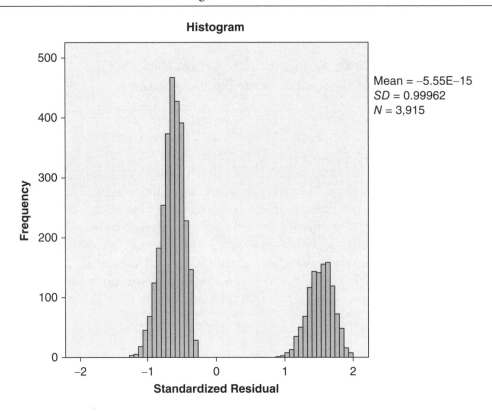

like SEXFNUM as the DV. You may be asking whether a transformation of the count data to improve normality of the residuals might help avoid the use of Poisson regression. The answer is a qualified "yes." This was the strategy (and probably still is) for individuals who have count data that do not meet OLS assumptions and who do not use Poisson regression. However, with the simple access to Poisson regression in modern times, this workaround is no longer desirable. Transformation also does not eliminate the issue of predicted negative counts. In this case, even some rather extreme Box-Cox transformations of the DV did not allow us to meet assumptions, as you can see in Figure 11.3b.

Multinomial or Ordinal Regression

One could use multinomial logistic regression to analyze this type of data, which would result in each group being treated as independent and unordered, creating a defensible but inefficient analysis that would compare each count with another comparison base rate. This would be manageable if there were only a few counts; however, it becomes highly inefficient very quickly if you are analyzing the number of days someone was unemployed during a year (which could range from 0 to 365). Ordinal regression would be a much better choice, but it is still inefficient in that ordinal regression is not taking advantage of the interval/ratio nature of the count variable. Thus, when equidispersion assumptions are met, it seems to me that Poisson regression is one of the most efficient methods of analysis (aside from OLS regression if those assumptions are met).

Hypotheses Tested in Poisson Regression

Model Fit

As with other models such as logistic regression, our initial hypothesis is that the overall model is significantly improved by the entry of the IV (in this case, change in −2 log likelihood [−2LL], which is evaluated as a chi-square) compared with a model with only the intercept (which I sometimes refer to inaccurately as an "empty model"):

$$H_0: \text{Omnibus likelihood ratio } \chi^2 \, (\Delta-2LL) = 0$$

$$H_a: \text{Omnibus likelihood ratio } \chi^2 \, (\Delta-2LL) \neq 0$$

If this initial hypothesis is rejected and we conclude that the IV (or set of IVs) has significantly improved the model fit, then we can examine our subsequent hypotheses for each term. One example is the often-overlooked hypothesis test concerning whether the intercept is zero:

$$H_0: b_0 = 0$$

$$H_a: b_0 \neq 0$$

Another example is the test of whether the slope of each variable in the equation is zero:

$$H_0: b_1 = 0$$

$$H_a: b_1 \neq 0$$

If these look familiar, they should, because they are the same hypotheses from prior chapters. Note that if we fail to reject the null hypothesis for the overall model, then it is not legitimate to examine individual effects.

Poisson Regression With Real Data

For this example, we will analyze the SEXFNUM variable presented above (number of female sexual partners during the last 3 months) as a function of two variables: biological sex (SEX, coded 0 = female, 1 = male) and age of first sexual intercourse (AGESEX, coded 0 = never had sex, 1 = 12 years or younger, 2 = 13/14, 3 = 15/16, 4 = 17/18, 5 = 19/20, 6 = 21–24, 7 = 25 or older. For this analysis, those who had never had sex were excluded and the variable AGESEX was centered at 4 (age 17 or

18 years), resulting in a distribution of −3 to 3). This analysis included 3,990 cases from the NCHRBS1995 database.[5] Some results from these analyses are presented in Table 11.3.

Table 11.3   Poisson Regression Analysis Predicting Number of Female Sex Partners From Biological Sex and Age of First Intercourse

Omnibus Test[a]

Likelihood Ratio Chi-Square	df	p
2,602.915	2	.000

Dependent variable: SEXFNUM.
Model: (intercept), AGESEX_c, SEX.
[a]Compares the fitted model against the intercept-only model.

Parameter Estimates

Parameter	B	SE	95% Wald CI		Hypothesis Test		
			Lower	Upper	Wald Chi-Square	df	p
(Intercept)	−3.898	0.1376	−4.168	−3.628	801.981	1	.000
AGESEX_c	−0.138	0.0201	−0.178	−0.099	47.434	1	.000
SEX	3.830	0.1398	3.555	4.104	750.078	1	.000
(Scale)	1[a]						

Dependent variable: SEXFNUM.
Model: (intercept), AGESEX_c, SEX.
[a]Fixed at the displayed value.

As you can see in Table 11.3, the omnibus test of the model is significant, meaning that the model fit was significantly (and substantially) improved by the entry of the two IVs ($\chi^2_{(2)} = 2,602.92$, $p < .001$). Following this, we can examine the individual effects. The intercept (−3.898) is significantly different from zero, has a relatively narrow confidence interval, and represents the natural log of the predicted mean number of events when both SEX and AGESEX_c = 0 (a female who first had intercourse at age 17 or 18 years). Exponentiation of this number reveals a predicted average count of 0.020 per 3-month period. There were a small but significant number of female college students who responded that they had engaged in sexual intercourse with another female partner at any point in their lives ($N = 110$), and a smaller number ($N = 46$) who had engaged in this activity within the last 3 months. Leaving aside the more prurient aspects of this line of research for a moment, the basic idea behind Poisson regression (and logistic and OLS regression too) is that there are variables (like biological sex) that represent modifications to the Poisson process being observed. This is an excellent example. Female college students engaging in sexual intercourse with other female partners is arguably a different underlying Poisson process than male college students engaging in sexual intercourse with female partners. Thus, we have not only the base rate (intercept) but also a variable (SEX) that is significant ($b_1 = 3.830$, $p < .001$). AGESEX was also a significant predictor ($b_2 = −0.138$, $p < .001$),

---

5   Of course, we are discussing human sexual behavior, which is an important outcome in many fields but is also a sensitive topic for some. I apologize if anyone is uncomfortable with this example, but it was too perfect a pedagogical tool to pass up, as you (hopefully) will see in pages to come.

meaning that the underlying Poisson process was consistent regardless of age of first intercourse. Using the regression equation from the information above (Equation 11.3), we can predict the mean rate for males who initiated sexual activity at age 17–18 years (AGESEX_c = 0):

$$\ln(\hat{\mu}) = -3.898 + 3.830(\text{SEX}) - 0.138(\text{SEXAGE\_c}) \quad (11.3)$$

$$\ln(\hat{\mu}) = -3.898 + 3.830\,(1)$$

$$\ln(\hat{\mu}) = -0.068$$

Exponentiated, this leads us to a predicted average count of 0.934 every 3 months for the males in the sample. Interestingly, as a "reality check" on this prediction, the average for all males in the sample who had ever had sex with a female and initiated sexual activity at age 17–18 years was 0.926, very close to the predicted value of 0.934.

Interactions in Poisson Regression

Let us press further ahead with this analysis and explore how modeling interactions in Poisson regression is very similar to that of other linear models. Using the same sample and data, we can create an interaction term between SEX and AGESEX, entering it into a nested model and comparing the two to see whether there is a significant improvement in model fit as a result. As you can see in Table 11.4, the improvement in model fit is $\chi^2_{(3)} = 2{,}603.093$, whereas the prior model without the interaction term was $\chi^2_{(2)} = 2{,}602.915$, a difference evaluated as $\chi^2_{(1)} < 1$, not significant. This indicates, absent further evaluation of the model, that the effects of neither of the two predictors are significantly impacted by the level of the other.

Table 11.4 Poisson Regression Analysis Predicting Number of Female Sex Partners From Biological Sex and Age of First Intercourse, Including Interaction

Omnibus Test[a]

Likelihood Ratio Chi-Square	df	p
2,603.093	3	.000

Dependent variable: SEXFNUM.
Model: (intercept), AGESEX_c, SEX, AGESEX_c*SEX.
[a]Compares the fitted model against the intercept-only model.

Parameter Estimates

Parameter	B	SE	95% Wald CI		Hypothesis Test		
			Lower	Upper	Wald Chi-Square	df	p
(Intercept)	−3.921	0.1500	−4.216	−3.627	683.023	1	.000
AGESEX_c	−0.187	0.1185	−0.420	0.045	2.499	1	.114
SEX	3.854	0.1529	3.554	4.154	635.561	1	.000
AGESEX_c*SEX	0.051	0.1203	−0.185	0.286	0.177	1	.674
(Scale)	1[a]						

Dependent variable: SEXFNUM.
Model: (intercept), AGESEX_c, SEX, AGESEX_c*SEX.
[a]Fixed at the displayed value.

Data Cleaning in Poisson Regression

Conceptually, there is little that changes between OLS, logistic, and Poisson regression models in terms of data cleaning. However, the details and technicalities vary by software package. SPSS, which I used for these analyses, had some interesting options available through the GLM procedure. I chose to examine Cook's Distance (or Cook's D, a measure of how much all residuals would change if a particular case were removed from the analysis) and the Pearson residual (square root of the contribution to the Pearson chi-square statistic, with the sign of the raw residual). Conceptually, the Pearson residual is more concretely interpretable as an overall influence on model fit, but both should help identify cases that are aberrant.

As you can see in Figure 11.4a, there are some cases within female respondents that seem to significantly separate from the group. For Cook's D, the 99th percentile is 0.005, so there are a number of cases that could be removed.

Conceptually, because Pearson chi-square is significant at about 4.00 with 1 degree of freedom, any case with a Pearson residual greater than |2| could be a candidate for exclusion. As you can see in Figure 11.4b, there are a number of cases that have Pearson residuals greater than |2| (almost 3%), some over 10, which are substantial. However, the cases are primarily in the female group. Thus, we need to be cautious, as we have discussed in the logistic models, that we not eliminate a small but important group of cases. First, let us be sure we are analyzing the right cases.

Figure 11.4a   Cook's Distance Separated for Males and Females

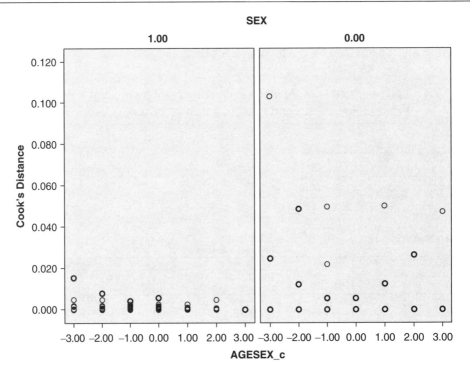

Figure 11.4b   Pearson Residuals Separated for Males and Females

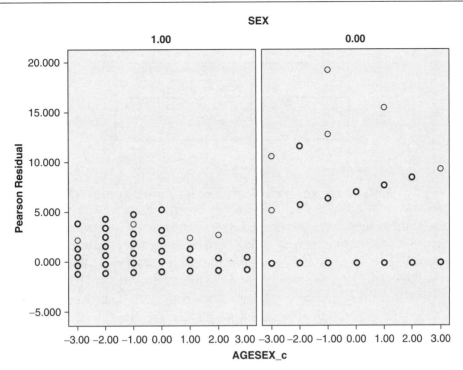

Refining the Model by Eliminating Excess (Inappropriate) Zeros

As I mentioned above, this model probably has excess zeros because we are asking how many times something might have happened (sexual intercourse with a female) of people who have no reasonable expectation of ever engaging in this behavior (heterosexual females). This is similar to asking someone who has never been married how many years they have been married, or asking a nonsmoker how many cigarettes he or she smoked in the past 24 hours. This variable is *zero inflated* because there are many cases included in the analysis where there is no reasonable chance that these individuals will have anything but a zero. Thus, we essentially have two different questions comingled into the same data: (a) is someone ever likely to have intercourse with a female, and (b) for those who answer yes, how many times in the past 3 months did they engage in this behavior.

As you can see in Table 11.5, after we eliminate all cases that indicate they never have engaged in sexual intercourse, we have a large number of female respondents ($N = 2,381$) who are sexually active but not with female partners. Likewise, we have some male respondents ($N = 22$) who indicate they are sexually active but not with female partners. It does not make sense to include these folks in this analysis, because they likely represent different populations. Refining the sample may help remove problematic cases and may create a more informative result.

Table 11.5   Number of Males and Females Who Have Ever Had Intercourse With a Female

SEX*EVERSEXF Cross-Tabulation

Count

		EVERSEXF		Total
		0 (No)	1 (Yes)	
SEX	0 (Female)	2,381	110	2,491
	1 (Male)	22	1,477	1,499
Total		2,403	1,587	3,990

A Refined Analysis With Excess Zeros Removed

Cases indicating they have never had intercourse with a female partner were removed from the analysis, leaving 1,587 (110 females, 1,477 males). Note that this is not considered a "nested" model within the previous set of analyses because the sample has changed. Thus, we cannot compare model fit across analyses. You can see in Table 11.6 that the addition of the variables significantly improves model fit, and that SEX and the interaction of AGESEX_c and SEX are significant.

As you can see in Table 11.6, with a more specific sample, we now have a significant interaction between AGESEX_c and SEX, but the refining of the sample did not completely eliminate the inappropriately influential cases. As you can see in Figures 11.6a and 11.6b, there are still some cases that seem highly influential (the 99th

Table 11.6   Poisson Regression Analysis Predicting Number of Female Sex Partners From Biological Sex and Age of First Intercourse, Including Interaction, Reduced Sample

Omnibus Test[a]

Likelihood Ratio Chi-Square	df	p
84.923	3	.000

Dependent variable: SexFnum.
Model: (intercept), agesex_c, sex, agesex_c*sex.
[a]Compares the fitted model against the intercept-only model.

Parameter Estimates

Parameter	B	SE	95% Wald CI		Hypothesis Test		
			Lower	Upper	Wald Chi-Square	df	p
(Intercept)	−0.636	0.1464	−0.923	−0.350	18.905	1	.000
AGESEX_c	0.146	0.0992	−0.049	0.340	2.157	1	.142
SEX	0.583	0.1493	0.291	0.876	15.269	1	.000
AGESEX_c*SEX	−0.283	0.1013	−0.482	−0.084	7.803	1	.005
(Scale)	1[a]						

Dependent variable: SEXFNUM.
Model: (intercept), AGESEX_c, SEX, AGESEX_c*SEX.
[a]Fixed at the displayed value.

percentile for Cook's D was 0.0094 and for Pearson residuals was 3.82, so cases with Cook's D greater than or equal to 0.01 or Pearson residual equal to or greater than 4.0 will be removed from the analysis, see Figures 10.5a and 10.5b). This data cleaning left us with a sample of 1,558 (29 cases removed). Again, these analyses are not nested because the sample is different, but you can see that the results of the cleaning have clarified the effects in Table 11.7.

Using the parameter estimates from the final analysis presented in Table 11.7, we are left with the following regression line equation, presented in Equation 11.4, from which we can generate a graphical depiction of the nature of the interaction, presented in Figure 11.6.

$$\ln(\hat{\mu}) = -0.709 + 0.616(\text{SEX}) + 0.390(\text{SEXAGE\_c}) - 0.486(\text{SEX*AGESEX\_c}) \tag{11.4}$$

These results tell a different and more nuanced story once we remove inappropriate zeros, refine the sample to include only those cases that are likely to have engaged in the behavior we are interested in studying, and engage in some judicious data cleaning.

Figure 11.5a   Cook's Distance Separated for Males and Females

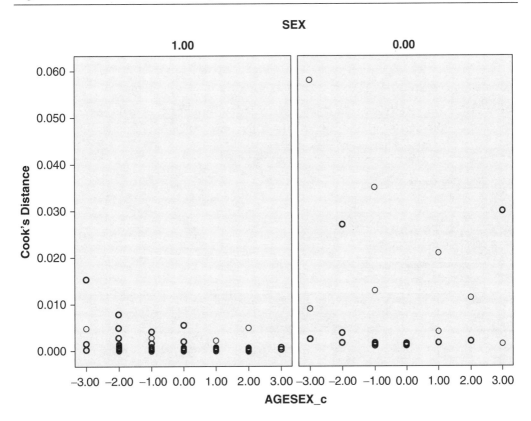

Figure 11.5b   Pearson Residuals Separated for Males and Females

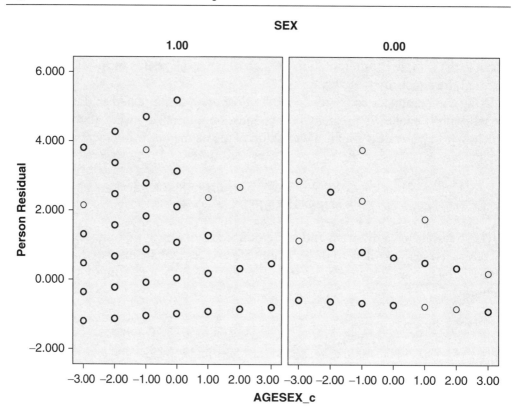

Table 11.7   Poisson Regression Analysis Predicting Number of Female Sex Partners From Biological Sex and Age of First Intercourse, Including Interaction, Reduced Sample After Data Cleaning

Omnibus Test[a]

Likelihood Ratio Chi-Square	df	p
70.625	3	.000

Dependent variable: SEXFNUM.
Model: (intercept), AGESEX_c, SEX, AGESEX_c*SEX.
[a]Compares the fitted model against the intercept-only model.

Parameter Estimates

Parameter	B	SE	95% Wald CI		Hypothesis Test		
			Lower	Upper	Wald Chi-Square	df	p
(Intercept)	−0.709	0.1610	−1.025	−0.394	19.407	1	.000
AGESEX_c	0.390	0.1222	0.151	0.630	10.205	1	.001
SEX	0.616	0.1638	0.295	0.937	14.154	1	.000
AGESEX_c*SEX	−0.486	0.1241	−0.730	−0.243	15.371	1	.000
(Scale)	1[a]						

Dependent variable: SEXFNUM.
Model: (intercept), AGESEX_c, SEX, AGESEX_c*SEX.
[a]Fixed at the displayed value.

Figure 11.6   Interaction Between SEX and Age of First Sexual Experience Predicting Number of Female Sex Partners in Prior 3 Months

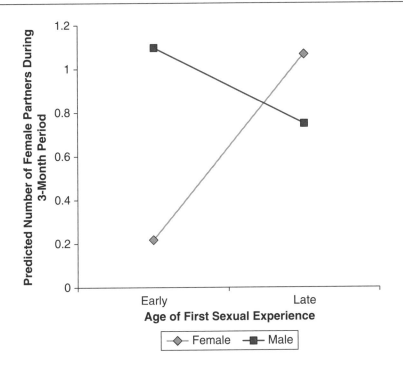

Example Summary of Poisson Regression Analysis

We used Poisson regression to predict the number of female sexual partners during the last 3 months (SEXFNUM) from SEX of the respondent (0 = female, 1 = male) and age of first sexual intercourse (AGESEX, coded 0 = never had sex, 1 = 12 years or younger, 2 = 13/14, 3 = 15/16, 4 = 17/18, 5 = 19/20, 6 = 21–24, 7 = 25 or older, centered at 4 and renamed AGESEX_c). Only cases that indicated they had ever had sex with a female partner were included in the analysis (N = 1,587). Conservative data cleaning based on relatively extreme scores on Cook's D or Pearson residuals reduced the sample size to N = 1,558 (a reduction of 29 cases).

An initial model with only the main effects of the variables showed a significant improvement in model fit over a model with only the intercept ($\chi^2_{(2)} = 56.41$, $p < .001$). Addition of the interaction term improved the likelihood ratio test ($\chi^2_{(3)} = 70.63$, $p < .001$, a difference of $\chi^2_{(1)} = 14.22$, $p < .001$). Based on the parameter estimates (Table 11.7), the regression line equation was used to predict estimated averages for males and females with both relatively early (AGESEX_c = −2) and relatively late (AGESEX_c = 2) initiation of sexual activity. Predicted estimates in Poisson regression represent the natural log of the predicted mean number of events, so each prediction was exponentiated to convert it to a mean rate. These are presented in Figure 11.6. As you can see, because the age of first sexual experience is later for female respondents, the average number of female partners during the preceding 3 months increases. The opposite is true for male respondents: those who had earlier sexual experiences tend to report having more female partners during the preceding 3 months than those who initiated sexual activity relatively later.

Curvilinear Effects in Poisson Regression

We have covered curvilinear effects and curvilinear interactions in depth in the preceding chapters, so I will briefly highlight the fact that these analyses also work in Poisson regression. I will spare you the details, but we will use the same data as the prior analysis (only those 1,587 cases who have ever had a female partner); some conservative data cleaning (Cook's D < 0.02, Pearson residuals < 4.2)[6] results in a sample of 1,567. The results are presented in Table 11.8 and Figure 11.7. If you compare Figures 11.6 and 11.7, you can see that the same general result is present but the data are much more nuanced in the case of the curvilinear interaction.

Table 11.8 Curvilinear Interactions Predicting Number of Female Sexual Partners During Prior 3 Months From Sex of Respondent and Age of First Sexual Experience

Omnibus Test[a]

Likelihood Ratio Chi-Square	df	p
94.312	7	.000

Dependent variable: SEXFNUM.
Model: (intercept), AGESEX_c, SEX, AGESEX2, AGESEX3, AGESEX_c*SEX, SEX*AGESEX2, SEX*AGESEX3.
[a]Compares the fitted model against the intercept-only model.

Parameter Estimates

Parameter	B	SE	95% Wald CI		Hypothesis Test		
			Lower	Upper	Wald Chi-Square	df	p
(Intercept)	−0.492	0.1926	−0.870	−0.115	6.533	1	.011
AGESEX_c	0.665	0.2271	0.220	1.110	8.578	1	.003
SEX	0.372	0.1960	−0.012	0.756	3.598	1	.058
AGESEX2	−0.151	0.0994	−0.346	0.044	2.309	1	.129
AGESEX3	−0.095	0.0480	−0.189	−0.001	3.919	1	.048
AGESEX_c*SEX	−0.895	0.2306	−1.347	−0.443	15.052	1	.000
SEX*AGESEX2	0.163	0.1003	−0.033	0.360	2.650	1	.104
SEX*AGESEX3	0.119	0.0485	0.024	0.214	6.048	1	.014
(Scale)	1[a]						

Dependent variable: SEXFNUM.
Model: (intercept), AGESEX_c, SEX, AGESEX2, AGESEX3, AGESEX_c*SEX, SEX*AGESEX2, SEX*AGESEX3.
[a]Fixed at the displayed value.

AGESEX_c*SEX Cross-Tabulation

Count

		SEX		Total
		0 (Female)	1 (Male)	
AGESEX_c	−3.00	8	79	87
	−2.00	19	202	221
	−1.00	39	438	477
	.00	25	456	481
	1.00	9	182	191
	2.00	3	84	87
	3.00	2	21	23
Total		105	1,462	1,567

---

6 I wanted to be particularly conservative with this data cleaning because the sample of female respondents (N = 110) was already small, and with such complex effects, losing small numbers could cause harm to the results. These results are similar to the pre–data cleaning results, just slightly stronger and clearer. As you can see from the cross-tabulations below, there are small numbers, particularly in the female respondent group, in some of the extreme age categories.

Figure 11.7   Curvilinear Interaction Between SEX and AGESEX_c After Data Cleaning

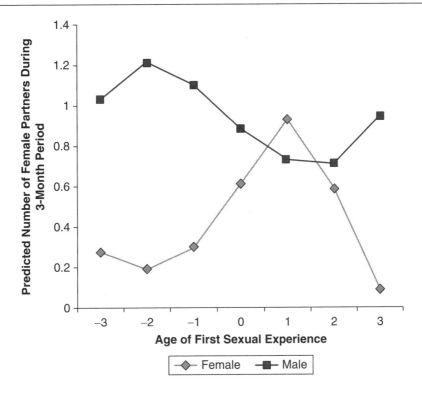

Dealing With Overdispersion or Underdispersion

If your data fail to meet the assumption of *equidisperson* (i.e., the mean and the variance are identical), there are several alternative models you can use if your data fail to meet the restrictive assumptions of the Poisson distribution (one, the negative binomial model, we will briefly discuss below). If this assumption is violated, the estimated standard errors can be too small, leading to overestimation of significance tests. The converse is true for underdispersion.

You can also adjust the Poisson model to account for some overdispersion or underdispersion. Coxe et al. (2009) and Cohen et al. (2003) are excellent and accessible resources for beginning to explore this issue. In general, the scaling parameter (φ) is assumed to be 1.0; in fact, in the parameter estimate tables thus far in the chapter, the scale parameter is always 1.0. However, in SPSS and other implementations of Poisson regression (via the GLM), we can specify a different scaling parameter to account for overdispersion or underdispersion. The simple calculation to test for dispersion is presented in Equation 11.5.

$$\varphi = \chi^2_{\text{(Pearson)}} / df \qquad (11.5)$$

In fact, during the last analysis, SPSS provided the calculation for me (although in the interests of brevity, I have excluded some of the output). As you can see in

Table 11.9 Test of the Assumption of Equidispersion

Goodness of Fit[a]	Value	df	Value/df
Deviance	1,104.223	1,559	.708
Scaled deviance	1,104.223	1,559	
Pearson chi-square	991.990	1,559	.636
Scaled Pearson chi-square	991.990	1,559	
Log likelihood[b]	−1,790.517		
Akaike's information criterion (AIC)	3,597.034		
Finite sample corrected AIC (AICC)	3,597.126		
Bayesian information criterion (BIC)	3,639.889		
Consistent AIC (CAIC)	3,647.889		

Dependent variable: SEXFNUM.
Model: (intercept), AGESEX_c, SEX, AGESEX2, AGESEX3,
AGESEX_c*SEX, SEX*AGESEX2, SEX*AGESEX3.
[a]Information criteria are in small-is-better form.
[b]The full log likelihood function is displayed and used in computing
information criteria.

Table 11.10 Curvilinear Interaction Analysis With Scale Adjusted for Underdispersion

Omnibus Test[a]		
Likelihood Ratio Chi-Square	df	p
147.363	7	.000

Dependent variable: SexFnum.
Model: (Intercept), agesex_c, sex,
agesex2, agesex3, agesex_c*sex,
sex*agesex2, sex*agesex3.
[a]Compares the fitted model against the
intercept-only model.

Parameter Estimates			95% Wald CI		Hypothesis Test		
Parameter	B	SE	Lower	Upper	Wald Chi-Square	df	p
(Intercept)	−0.492	0.1540	−0.794	−0.190	10.207	1	.001
AGESEX_c	0.665	0.1817	0.309	1.021	13.404	1	.000
SEX	0.372	0.1568	0.064	0.679	5.622	1	.018
AGESEX2	−0.151	0.0795	−0.307	0.005	3.609	1	.057
AGESEX3	−0.095	0.0384	−0.170	−0.020	6.123	1	.013
AGESEX_c*SEX	−0.895	0.1845	−1.256	−0.533	23.518	1	.000
SEX*AGESEX2	0.163	0.0802	0.006	0.320	4.141	1	.042
SEX*AGESEX3	0.119	0.0388	0.043	0.195	9.450	1	.002
(Scale)	.64[a]						

Dependent variable: SEXFNUM.
Model: (intercept), AGESEX_c, SEX, AGESEX2, AGESEX3, AGESEX_c*SEX, SEX*AGESEX2, SEX*AGESEX3.
[a]Fixed at the displayed value.

Table 11.9, the Pearson chi-square divided by the degrees of freedom is less than 1.0, meaning that we can have underdispersion.

Effects of Adjusting the Scale Parameter

Adjusting the scale parameter to 0.64 leads to a slightly different model outcome. Taking the model and exact same sample ($N = 1,567$) from the curvilinear interaction analysis presented in the preceding section (Table 11.8, Figure 11.7), these results are generally stronger, as presented in Table 11.10. Note that the actual parameter estimates are unaffected by this adjustment. Rather, the standard errors are reduced compared with the unadjusted model, and thus the Wald chi-squares and the p values are higher.

Negative Binomial Model

The negative binomial model is a more flexible model that can deal with the same data and types of questions as Poisson regression, but it allows for models that do not meet the assumption that the variance of the distribution is equal to the mean. In this model, we retain the assumption that a Poisson model applies to each individual, but allow rates for individuals (μ) to vary across individuals. This allows for additional variance in rate parameters across individuals. There are also semiparametric mixed Poisson models that can deal with distributions that do not meet the assumptions of traditional Poisson models. Most of these options are easily accessible in any statistical software package that includes Poisson and GLM.

SUMMARY AND CONCLUSIONS

In many fields, we look at counts of things as important outcomes. Poisson regression fits a specific niche in the GLM where we are interested in modeling an outcome that is a highly positively skewed count variable. Although many researchers would be tempted to model this within an OLS regression model, the odds are that assumptions of OLS regression will not be met with this type of variable. If you have a count variable that produces normally distributed residuals and meets all of the other assumptions of OLS regression, then feel free to use it.

Multinomial logistic regression would be another option, as would ordinal logistic regression, although these are a bit more kludgy than necessary given how simple Poisson regression is to use and interpret. Multinomial and ordinal logistic regression both ignore some value in the data when a variable is measured at the interval level, although interval measurement is not required for Poisson regression.

Although Poisson is rare in many fields, my hope is that this book will help bring this tool to mainstream research as an appropriate and valuable aspect of the GLM.

ENRICHMENT

1. Download the MJ data used in the examples from the chapter and replicate the results.

2. Replicating and extending the analysis predicting SEXFNUM, add SUICIDE ("Have you ever seriously contemplated committing suicide in previous 12

months?," coded 0 = no, 1 = yes) and FRAT ("Do you belong to a fraternity or sorority?," coded 0 = no, 1 = yes) and see if those variables (and associated inter-actions) predict the number of female sex partners.

3. Download data from my colleague Dr. Philip Gabel, who studies physical ther-apy and health. In one study I was collaborating with him on (he gave permis-sion to share the deidentified data for this book), we were modeling the number of exercises a patient could do (EX_SUM, ranging from 0 to 5; exercises included squats, leg extensions, sit-ups, etc.) as a function of patient sex (SEX; 0 = female, 1 = male) and amount of lower back pain (LBP; 0 = none, 1 = some, 2 = mostly).

   a. Is it appropriate to model EX_SUM as a function of SEX, LBP, and SEX*LBP in a Poisson model or do you need a different model?
   b. Are any of the terms significant predictors of EX_SUM? Summarize the findings in APA-compliant (or similar) format.
   c. If the interaction is significant, graph the results.

4. Download data from the NCES Education Longitudinal Study of 2002. There are several potential Poisson outcomes to model:

   • GRDRPT: number of grades a student repeated from kindergarten to 10th grade
   • RISKFC: number of academic risk factors in 10th grade
   • NSPORT: number of interscholastic sports participated in (varsity or junior varsity only)
   • Potential IVs include SEX (0 = female, 1 = male), RACE, zSES, and ACH (achievement test score)

   a. Choose a DV that interests you, and evaluate whether it meets the assumption of Poisson regression.
   b. Choose some IVs and model that DV as a function of the main effects, interactions, curvilinear effects, and so forth.
   c. Some example analyses will be provided in the Chapter 11 answer key on the website.

REFERENCES

Cohen, J., Cohen, P., West, S., & Aiken, L. S. (2003). *Applied multiple regression/correlation analysis for the behavioral sciences.* Mahwah, NJ: Lawrence Erlbaum.

Coxe, S., West, S. G., & Aiken, L. S. (2009). The analysis of count data: A gentle introduction to Poisson regression and its alternatives. *Journal of Personality Assessment, 91*(2), 121–136. doi: 10.1080/00223890802634175

Nussbaum, E. M., Elsadat, S., & Khago, A. H. (2008). 21 best practices in analyzing count data. In J. Osborne (Ed.), *Best practices in quantitative methods* (pp. 306–323). Thousand Oaks, CA: SAGE.

LOG-LINEAR MODELS

General Linear Models When All of Your Variables Are Unordered Categorical

12

Advance Organizer

How would you analyze your data if you wanted to explore whether students were retained or dropped out as a function of *only categorical variables* (like race, sex, or marijuana use)? Logistic regression would certainly be an option; however, with variables like race, which can often lead to multiple dummy variables, interactions can become cumbersome to test. The same issue is present when you have a dependent variable (DV) that is continuous and only categorical independent variables (IVs). Analysis of variance (ANOVA) is a great solution for that situation, because it prevents a good many headaches when you are looking at interactions. Because you already know that it is a special case of the generalized linear model (GLM), as long as assumptions are met, it will produce the same results as modeling the same data in ordinary least squares (OLS) regression and is thus a valuable tool for making life easier in certain situations.

In this chapter, we look at an ANOVA analogue for the situation in which all variables are binary or categorical—log-linear models. This represents the last new model we will explore in this book and completes a toolbox that can cover almost any situation where you have data with independent observations (i.e., no repeated measures and no clustering of data).[1] Technically, if you are good at logistic or multinomial regression (and particularly with dummy coding and creating interactions between dummy-coded variables), you can model these types of analyses within a general logistic regression framework. In fact, we will do that as a comparison for the log-linear model analysis. However, this special tool is useful and valid (again, when assumptions are met), and there is no reason to avoid this model if it provides valid results and makes life easier.

Log-linear analysis is a generalization of a simple two-way cross-tabulation of the type often seen in introductory textbooks and conforms to the final piece of the GLM framework: both IVs and DV all being categorical. As you can see in Table 12.1, the most common log-linear models are really most appropriate when there are unordered variables.

---

1 If you have clustered data, the next chapter on hierarchical linear modeling will help with that, and analysis of repeated data is a rich topic that is beyond the scope of the book . . . at least for the current edition!

Table 12.1 Examples of the Generalized Linear Model as a Function of Independent Variable and Dependent Variable Type

	Continuous DV	*Binary DV*	*Unordered Multicategory DV*	*Ordered Categorical DV*	*Count DV*
Continuous IV	OLS regression	Binary logistic regression	Multinomial logistic regression	Ordinal logistic regression	OLS, Poisson regression
Mixed continuous and categorical IV					
Binary/ categorical IV only	ANOVA	Log-linear models	Log-linear models		Log-linear models

ANOVA, analysis of variance; DV, dependent variable; IV, independent variable; OLS, ordinary least squares.

Although log-linear analyses are not common in many fields, I hope that having all of the tools available to you as a result of the GLM is useful. You will find many similarities between what we have already covered and what is in this chapter, such as log link functions and evaluation of nested (or hierarchical) models.

In this chapter, we will cover

- Basic concepts in log-linear models
- Assumptions and data cleaning for log-linear models
- A three-variable log-linear model
- Confidence intervals and effect sizes in log-linear models
- Examples of American Psychological Association (APA)–compliant summaries of analyses
- A demonstration that logistic regression can replicate the results of a log-linear analysis

Guidance on how to perform these analyses in various statistical packages will be available online at study.sagepub.com/osbornerlm.

The Basics of Log-Linear Analysis

There are some unusual aspects to log-linear models but nothing that will be difficult for you to grasp if you have survived the first 11 chapters of this book. There are two types of log-linear models often implemented in statistical software packages: hierarchical log-linear models, and general log-linear models. In this chapter, we will present general log-linear models, which most closely mirror typical implementations of ANOVA for pedagogical ease and which also (at least within SPSS) most closely mirror the spirit of the GLM.

Consider a simple cross-tabulation of counts of two variables. For the sake of clarity, we will use data from the 2013 Youth Risk Behavior Survey from the US Centers for Disease

Control and Prevention,[2] which surveys high school students nationally on factors and behaviors that relate to wellness and health. As a simple example, Table 12.2 contains the simple frequency counts cross-tabulating E_BULLIED (Q25, "During the past 12 months, have you ever been electronically bullied? [include being bullied through e-mail, chat rooms, instant messaging, Web sites, or texting]," 0 = no, 1 = yes) and HOPELESS12 (Q26, "During the past 12 months, did you ever feel so sad or hopeless almost every day for 2 weeks or more in a row that you stopped doing some usual activities?," 0 = no, 1 = yes).

The popular media and much research indicate that bullying through social media and other electronic venues can be a significant stressor leading to negative outcomes for those receiving the bullying, even including suicide ideation. Thus, it is not surprising that in Table 12.2a, we see a higher percentage of those reporting bullying also feeling hopeless (58.1%) than those not reporting bullying (25.8%). If you remember our binary logistic regression chapter, I hope you were beginning to think in terms of odds, odds ratios (ORs), or relative risk (RR). You can certainly calculate odds, and the appropriate OR for these data (3.986) or the RR (2.25). Even eyeballing the data,[3] common sense dictates that there is an effect here, and that the effect is as expected: students subjected to bullying are also more likely to report symptoms of depression. When performing a cross-tabulation, you can request a chi-square test to examine whether the expected percentages or probabilities are independent or not influenced by the other variable (Pearson $\chi^2_{(1)} = 796.54$, $p < .001$; the likelihood ratio $\chi^2_{(1)} = 732.489$, $p < .001$). Of course, if we were to perform the same analysis in logistic regression, we would be testing whether the conditional probabilities are equal across all levels of the IV. The results of this analysis are presented in Table 12.2b.

As you can see in Table 12.2b, the OR calculated here matches that calculated from the cross-tabulation, and the chi-square test for the improvement of fit for the model matched that of the likelihood ratio ($\chi^2_{(1)} = 732.489$, $p < .001$) from the cross-tabulation. In other words, the lessons learned in Chapter 5 are still valid.

Table 12.2a   Cross-Tabulation of E_BULLIED and HOPELESS12

E_BULLIED*HOPELESS12 Cross-Tabulation

			HOPELESS12		Total	Odds
			0 No	1 Yes		
E_BULLIED	0 No	Count	8,602	2,987	11,589	0.347
		% within E_BULLIED	74.2	25.8	100.0	
	1 Yes	Count	786	1,088	1,874	1.384
		% within E_BULLIED	41.9	58.1	100.0	
Total		Count	9,388	4,075	13,463	
		% within E_BULLIED	69.7	30.3	100.0	

2   Data from the 2013 Youth Risk Behavior Survey are available from the US Centers for Disease Control and Prevention (http://www.cdc.gov/healthyyouth/data/yrbs/index.htm).

3   My friend and colleague Brett Jones calls this the "ocular test."

Table 12.2b Logistic Regression Predicting HOPELESS12 From E_BULLIED

Variables in the Equation

		B	SE	Wald	df	p	Exp(B)
Step 1[a]	E_BULLIED	1.383	0.051	723.699	1	.000	3.986
	Constant	−1.058	0.021	2,480.472	1	.000	0.347

[a]Variable(s) entered on step 1: E_BULLIED.

What Is Different About Log-Linear Analysis?

There are aspects of log-linear analysis that require a slightly different way of thinking. For example, in log-linear analysis, we do not distinguish between IVs and DVs; we merely look at associations between all variables. They are all merely factors (categorical variables) entered into the analysis.[4] As you can see in Table 12.2c, this does not change the basic information, but it does change how we might think about the information. Note also in Table 12.2c that SPSS adds 0.50 to all counts in the saturated model (model with all variables in the model, in which the expected and observed counts are identical, leaving no residuals). This is to prevent counts of zero in cells, which can wreak havoc on this type of model. These two nuances are mostly surface features of the analysis. What happens "under the hood" is more akin to Poisson or logistic regression.

The name "log-linear" analysis refers to the fact that we are modeling a linear model by converting frequency counts within a contingency table (like in Table 12.2c) to natural logarithms. Thus, we have a linear model using the natural logarithm as a link function, and we can create a linear model predicting the natural log of the expected frequency counts (similar to Poisson regression, in which we predicted the natural log of the mean count).

As you can see in Equations 12.1a (routine notation in log-linear tradition) and 12.1b (converted to typical GLM notation we have used in this book), we model the natural log of the expected frequencies of each cell in the table as a function of an intercept (typically noted as μ, although b_0 would also be appropriate) and the effects of each of the two variables in the equation (λ_{A_i} and λ_{B_j} or b_1X_1 and b_2X_2) and their interaction or nonadditive effect ($\lambda_{AB_{ij}}$ or $b_3X_1X_2$).

$$\text{LN}(F_{ij}) = \mu + \lambda_{A_i} + \lambda_{B_j} + \lambda_{AB_{ij}} \tag{12.1a}$$

$$\text{LN}(F_{ij}) = b_0 + b_1X_1 + b_2X_2 + b_3X_1X_2 \tag{12.1b}$$

---

4   Some authors argue that logistic regression should be used if DVs and IVs are used in a more explicit fashion. However, as you will see (and as I have argued throughout this book), the statistical analysis is neutral on the topic of "causality" or theoretical independence or dependence. The same arguments have been historically made with correlation and regression, although the outcomes are identical. Thus, I disagree with that stance and encourage you to use this analysis *if* appropriate to the data.

Table 12.2c   Log-linear Analysis of HOPELESS12 and E_BULLIED

Cell Counts and Residuals

E_BULLIED	HOPELESS12	Observed		Expected		Residuals	Std. Residuals
		Count[a]	%	Count	%		
0 No	0 No	8,602.500	63.9	8,602.500	63.9	.000	.000
	1 Yes	2,987.500	22.2	2,987.500	22.2	.000	.000
1 Yes	0 No	786.500	5.8	786.500	5.8	.000	.000
	1 Yes	1,088.500	8.1	1,088.500	8.1	.000	.000

[a]For saturated models, .500 has been added to all observed cells.

Parameter Estimates[a,b]

Parameter	Estimate	SE	Z	p	95% CI	
					Lower Bound	Upper Bound
Constant	6.993[c]					
[E_BULLIED = .00]	1.010	0.035	28.518	.000	0.940	1.079
[E_BULLIED = 1.00]	0[d]	.	.	.	.	.
[HOPELESS12 = .00]	−0.325	0.047	−6.944	.000	−0.417	−0.233
[HOPELESS12 = 1.00]	0[d]	.	.	.	.	.
[E_BULLIED = .00]*[HOPELESS12 = .00]	1.383	0.051	26.903	.000	1.282	1.483
[E_BULLIED = .00]*[HOPELESS12 = 1.00]	0[d]	.	.	.	.	.
[E_BULLIED = 1.00]*[HOPELESS12 = .00]	0[d]	.	.	.	.	.
[E_BULLIED = 1.00]*[HOPELESS12 = 1.00]	0[d]	.	.	.	.	.

[a]Model: multinomial.

[b]Design: constant + E_BULLIED + HOPELESS12 + E_BULLIED*HOPELESS12.

[c]Constants are not parameters under the multinomial assumption. Therefore, their standard errors are not calculated.

[d]This parameter is set to zero because it is redundant.

In ANOVA, the intercept is the grand mean, or the mean of all cell means. In log-linear analysis, the intercept is the mean of the natural log of the frequencies for each cell. In each cell, the intercept baseline mean is then modified by the particular combination of variables. Referring back to Table 12.2c, the parameter estimates can be a bit confusing to interpret at first; once we understand that each categorical variable has an assigned reference group, and that each other group is compared with that, we can use our knowledge of dummy-coded variables to understand what we are seeing. Interestingly, you can see that the SPSS default assigns the last (1 = yes) group to be the comparison group, which may not be ideal and may not be true for all implementations within different statistical computing packages. To summarize, the linear model is presented in Equation 12.1c:[5]

$$LN(F_{ij}) = 6.992556 + 1.009636 \text{ (E\_BULLIED)} - 0.324963(\text{HOPELESS12}) +$$
$$1.382579(\text{E\_BULLIED}* \text{HOPELESS12}) \qquad (12.1c)$$

---

5   As in previous chapters, when creating prediction equations in log units, I used more decimals of precision than standard SPSS output provides. Using the figures rounded to three decimal places works well to approximate the actual counts, but using more precision allows you to reproduce the numbers more exactly.

Table 12.3   Predictions for Each Group Based on E_BULLIED and HOPELESS12

Group	Constant	E_BULLIED	HOPELESS12	Interaction	Prediction	Frequencies
Not bullied, not hopeless	6.992556	1.009636	−0.324963	1.382579	9.059808	8,602.499
Bullied, not hopeless	6.992556	0	−0.324963	0	6.667593	786.5002
Not bullied, hopeless	6.992556	1.009636	0	0	8.002192	2,987.499
Bullied and hopeless	6.992556	0	0	0	6.992556	1,088.5

Now the trick becomes the counterintuitive action of interpreting the parameter estimates. You can see that the parameter estimate for E_BULLIED, for example, is 1.009636 when the variable's value is 0 (not bullied) and 0 otherwise, which is the opposite of how we have performed this type of analysis thus far. Thus, to create a prediction, we can proceed as in Table 12.3.

You can see that although the way SPSS creates this analysis is a bit confusing, it is relatively simple to create predicted values that match the expected (or observed) frequencies.

Hypotheses Being Tested

As with every other type of analysis we have looked at thus far, we have an overall model test and individual hypotheses related to each variable. Let's start with the overall model. The default model is a *saturated* model, in which every term is in the analysis (in this case, both main effects and the interaction). In log-linear analyses, a saturated model always exactly replicates the observed frequencies so that the expected frequencies match, leaving perfect fit. What researchers often do at this point is attempt to create parsimony by subtracting terms (e.g., the interaction) to see if the model significantly degrades. If it does not, we can conclude that the interaction effect is not important and does not significantly improve the model fit. At that point, we could continue subtracting main effects to test them for contribution to goodness of fit. Of course, this hierarchical model testing follows the same rules (although in reverse) that we used in previous chapters (e.g., interactions): we cannot remove a main effect of a variable if that effect is required for an interaction. If we had multiple main effects and multiple interactions, it might be possible to eliminate an interaction, as well as some constituent main effects, without adversely affecting model fit.

The overall model fit is either summarized by the Pearson chi-square or the likelihood ratio (L^2 or G^2, depending on your reference), with the latter being more commonly used because this is the statistic that maximum likelihood estimation is working to minimize during iterations. The likelihood ratio also has important advantages in that the sum of all of the effects in the model is equal to the total likelihood ratio, whereas the effects are not additive under some common conditions for the chi-square. This simple computation is presented in Equation 12.2:[6]

$$L^2 = 2 \Sigma f_0 \ln \left(\frac{f_0}{F_e} \right) \qquad (12.2)$$

where f_0 represents the observed frequencies of a cell and F_e is the expected frequency. Summed over all cells, this number is used for the model likelihood ratio tests. L^2 also follows a chi-square distribution with degrees of freedom equal to the number of terms set equal to zero in this analysis. Thus, the larger the statistic relative to the available degrees of freedom, the more the model departs from the observed frequencies.

The null hypothesis for the model is that all cells have equal frequencies, or that total or model L^2 will equal zero:

$$H_0: L_T^2 = 0$$

$$H_a: L_T^2 \neq 0$$

Individual Parameter Estimates

In this analysis, we also have significance tests for each effect, as with all other prior types of analyses. In this case, the variable hypothesis is that the likelihood ratio for a particular effect (L_A^2, for example, representing the likelihood ratio for variable A) is zero, or the coefficient for that variable (e.g., b_A) is zero.

$$H_0: L_A^2 = 0$$

$$H_a: L_A^2 \neq 0$$

$$H_0: b_A = 0$$

$$H_a: b_A \neq 0$$

With log-linear analyses, there are often initial evaluations of the model in which entire classes of effects (e.g., interactions) are removed to test for change to model fit. As you know by now, a significant degradation in model fit can mean that all terms removed are significant, that only some terms removed are significant, or, in some

---

6   This formula and some other information throughout this chapter are from Tabachnick and Fidell (2001, Chapter 7).

rare cases, that none of the individual terms are significant. Thus, when a significant likelihood ratio indicates significant degradation of the model as a result of removing multiple terms, those individual terms should be examined. When a block of terms are removed from the model and no significant degradation occurs, it is safe to assume that none of those terms are significant.

As you can see in Table 12.2c, each effect is significant, meaning that each is significantly different from zero. Thus, there is a reasonable expectation that removing any effect from the model will degrade fit. As expected (and shown in Table 12.4), the likelihood ratio went from perfect fit (0.00) to 732.49 (or Pearson chi-square from 0.00 to 796.544), both of which are significant degradations of the model. Thus, it would not be desirable to remove the interaction.

Table 12.4   Model Fit After Interaction Removed

Goodness-of-Fit Tests[a,b]			
	Value	df	p
Likelihood ratio	732.489	1	.000
Pearson chi-square	796.544	1	.000

[a]Model: multinomial.

[b]Design: constant + E_BULLIED + HOPELESS12.

Assumptions of Log-Linear Models

This is a relatively simplistic and nonparametric analysis, and as such has few assumptions. As with most of our analyses, we assume independence of observations (i.e., no repeated measures or nested/clustered data). We also assume that there is an adequate sample to support the analysis. Some authors suggest you have five times the number of cells in the analyses as a minimum, but this guideline would lead to small samples in which a random change of one individual to a different cell could dramatically influence the results. Thus, as with all other chapters, I argue that larger samples will provide a more stable and replicable result.[7]

Finally, you cannot have null cells (with counts of 0), which is why many software programs add 0.5 to all cell counts by default—to avoid this issue of sparse data. Further, no more than 20% of the cells should have an expected frequency of less than 5. Uneven data such as described here will dramatically reduce power.

A Slightly More Complex Log-Linear Model

Let us have some fun with this model and add two other variables: SEX (biological sex; 0 = female, 1 = male) and SUICIDE ("During the past 12 months, did you ever

---

7   As you will see in Chapter 15, even relatively large samples can lead to results of questionable replication probabilities.

seriously consider attempting suicide?," 0 = no, 1 = yes). Even this analysis, with all of the two-, three- and four-way interactions, would be doable within a logistic regression framework, particularly because this analysis is composed entirely of binary variables. Things would quickly get out of hand if each variable were instead a multicategory variable needing to be dummy coded (e.g., race, religion, and education attainment).

Our typical approach to this model will be to start with a fully saturated model and then eliminate effects that do not significantly degrade the model. I will summarize the models in Table 12.5a for easy review.

As you can see in Table 12.5a, removing the four-way interaction does not significantly degrade the model and allows us to assert that it is not a significant effect. Removing all three-way interactions does significantly degrade the model—in particular, two three-way interactions have significant effects (MBH and MHS were both $p < .05$). Thus, it would be advisable to revisit Model 3 and add those two interactions back into the equation, leaving the two without significant effects out of the model for the sake of parsimony. Removal of all two-way interactions led to a devastating loss of model fit and would not be advisable.

In the final analysis (which I labeled Model 3a) presented in Table 12.5b, HOPELESS12*SUICIDE was not significant as a two-way interaction. However, we cannot remove that effect to improve parsimony because that effect is part of the SEX*HOPELESS12*SUICIDE interaction that is significant and is retained in the final model.

In considering how to report the results of this analysis, I would focus on the two three-way interactions that are most interesting and complex, because all other effects are qualified by those interactions. Tables 12.5c and 12.5d present the frequency counts for each of those interactions.

Table 12.5a Overview of Log-Linear Analysis of SEX, E_BULLYING, HOPELESS12, and SUICIDE

Model	Variables in Model	Likelihood Ratio (L^2)	df	p
1	Saturated	0.00	0	—
2	MBH, MBS, MHS, BHS MB, MH, MS, BH, BS, HS M, B, H, S	2.64	1	.104
3	MB, MH, MS, BH, BS, HS M, B, H, S	20.03	5	.001
4	M, B, H, S	4,313.80	11	.001
3a	MBH, MHS, MB, MH, MS, BH, BS, HS M, B, H, S	3.90	3	.272

NOTE: Variables in the model are abbreviated as follows: M (SEX), B (E_BULLYING), H (HOPELESS12, and S (SUICIDE).

Table 12.5b   Final Parameter Estimates for SEX, E_BULLYING, HOPELESS12, and SUICIDE Log-Linear Analysis

Parameter	Estimate	SE	Z	p	95% CI Lower Bound	95% CI Upper Bound
					Lower Bound	Upper Bound
Constant	5.070[d]					
[E_BULLIED = .00]	.919	.074	12.488	.000	.775	1.063
[HOPELESS12 = .00]	−1.798	.110	−16.278	.000	−2.015	−1.582
[SEX = .00]	1.117	.075	14.924	.000	.970	1.264
[SUICIDE = .00]	−.219	.074	−2.974	.003	−.363	−.075
[E_BULLIED = .00]*[HOPELESS12 = .00]	1.101	.094	11.705	.000	.917	1.285
[SEX = .00]*[E_BULLIED = .00]	−.577	.080	−7.228	.000	−.733	−.420
[E_BULLIED = .00]*[SUICIDE = .00]	.909	.063	14.522	.000	.786	1.031
[SEX = .00]*[HOPELESS12 = .00]	−.123	.133	−.927	.354	−.384	.137
[HOPELESS12 = .00]*[SUICIDE = .00]	2.514	.088	28.716	.000	2.342	2.686
[SEX = .00]*[SUICIDE = .00]	−.223	.068	−3.269	.001	−.356	−.089
[SEX = .00]*[E_BULLIED = .00]*[HOPELESS12 = .00]	−.307	.111	−2.758	.006	−.525	−.089
[SEX = .00]*[HOPELESS12 = .00]*[SUICIDE = .00]	−.291	.114	−2.555	.011	−.514	−.068

Parameter Estimates[a,b,c]

[a]Model: multinomial.
[b]Design: constant + E_BULLIED + HOPELESS12 + SEX + SUICIDE + E_BULLIED*HOPELESS12 + SEX*E_BULLIED + E_BULLIED*SUICIDE + SEX*HOPELESS12 + HOPELESS12*SUICIDE + SEX*SUICIDE + SEX*E_BULLIED*HOPELESS12 + SEX*HOPELESS12*SUICIDE.
[c]Redundant parameters were removed for ease of presentation.
[d]Constants are not parameters under the multinomial assumption. Therefore, their standard errors are not calculated.

Table 12.5c   Frequency Counts for SEX*E_BULLIED*HOPELESS12 Interaction

SEX	E_BULLIED	HOPELESS12	Observed Count	Observed %
			Count	%
0 Female	0 No	0 No	3,494.500	26.0
		1 Yes	1,785.500	13.3
	1 Yes	0 No	496.500	3.7
		1 Yes	801.500	6.0
1 Male	0 No	0 No	5,099.500	37.9
		1 Yes	1,200.500	8.9
	1 Yes	0 No	290.500	2.2
		1 Yes	287.500	2.1

Table 12.5d   Frequency Counts for SEX*HOPELESS12*SUICIDE Interaction

SEX	HOPELESS12	SUICIDE	Observed	
			Count	%
0 Female	0 No	0 No	3,696.500	27.5
		1 Yes	293.500	2.2
	1 Yes	0 No	1,408.500	10.5
		1 Yes	1,174.500	8.7
1 Male	0 No	0 No	5,166.500	38.4
		1 Yes	226.500	1.7
	1 Yes	0 No	927.500	6.9
		1 Yes	559.500	4.2

There are some interesting patterns in Table 12.5c. For example, more females report being the victim of bullying than males (1,297 versus 577). Of those who report being victims of bullying, females are much more likely to feel hopeless (801 or 61.8% of females reporting bullying versus 49.7% of males). Converted to RRs, of those students reporting being the victims of electronic bullying, females are about 24% more likely to report feeling hopeless. Thus, females are either disproportionately impacted by e-bullying or are more likely to report negative effects of it.

Table 12.5d is equally concerning. Females seem much more likely to report feeling hopeless than males (2,582 of 6,571 females [39.3%] versus 1,486 of 6,878 males [21.6%]). This is not new to the psychological literature; however, what would be concerning to me (if I did substantive research in this area) is that of those who report feeling hopeless, females seem more likely to have suicidal thoughts than boys (1,174 [45.5%] versus 559 [37.6%]). In other words, for those who feel hopeless, the RR for females reporting suicidal thoughts is 1.21 compared with males (i.e., females are about 21% more likely to have suicidal thoughts if they reported feeling hopeless).[8]

Can We Replicate These Results in Logistic Regression?

This question is not quite as simple as it might first appear; we must remember that there is no DV in log-linear analysis, whereas we have to identify one in logistic regression. Let us start with a few analyses that replicate the spirit of what we were trying to do in the prior analysis. First, let us replicate the four-way analysis that was not significant (which, in logistic regression, would be a three-way interaction predicting the outcome DV).

---

8 Of course, as with every other chapter, none of these results should be interpreted substantively, because I did not do the appropriate weighting and accounting for cluster sampling in these data. These results are merely to introduce the concepts.

Summary of Log-Linear Analysis

Using data from 13,419 participants with complete data from the 2013 Youth Risk Behavior Surveillance System, we examined four variables of interest: SEX (biological sex; 0 = female, 1 = male), E_BULLIED (Q25, "During the past 12 months, have you ever been electronically bullied? [include being bullied through e-mail, chat rooms, instant messaging, Web sites, or texting]," 0 = no, 1 = yes), HOPELESS12 (Q26, "During the past 12 months, did you ever feel so sad or hopeless almost every day for 2 weeks or more in a row that you stopped doing some usual activities?," 0 = no, 1 = yes), and SUICIDE ("During the past 12 months, did you ever **seriously** consider attempting suicide?," 0 = no, 1 = yes). There were 164 cases with incomplete data and these cases were eliminated from the analysis. Because all variables were binary in nature, log-linear analysis was appropriate for examining relationships among these variables.

An initial saturated model (all main effects and all interactions) was estimated and then, following typical hierarchical modeling procedures, model fit was examined when more complex effects were removed from the model. As you can see in Table 12.5a, removing the four-way interaction does not significantly degrade the model ($\chi^2_{(1)} = 2.64$, $p < .104$). Removing all three-way interactions does ($\chi^2_{(5)} = 20.03$, $p < .001$); in particular, two three-way interactions have significant effects (SEX*E_BULLIED*HOPELESS12 and SEX*HOPELESS12*SUICIDE were both $p < .05$). Thus, it would be advisable to revisit Model 3 and add those two interactions back into the equation. Removal of all two-way interactions led to a devastating loss of model fit and would not be advisable. In the final analysis, as you can see in Table 12.5b, HOPELESS12*SUICIDE was not significant as a two-way interaction; however, because that effect is part of the SEX*HOPELESS12*SUICIDE interaction that is significant, we cannot remove it.

If you examine Table 12.5c, you can see some interesting patterns. For example, more females report being the victim of bullying than males (1,297 versus 577). Of those who report being victims of bullying, females are much more likely to feel hopeless (801 females [61.8%] reporting bullying versus 287 males [49.7%]). Converted to RRs, of those students reporting being the victims of electronic bullying, females are about 24% more likely to report feeling hopeless. Thus, females are either disproportionately impacted by e-bullying or are more likely to report negative effects of it.

Table 12.5d is equally concerning. Females seem much more likely to report feeling hopeless than males (2,582 of 6,571 [39.3%] versus 1,486 of 6,878 [21.6%]). This is not new to the psychological literature, by the way. What is concerning to me is that of those who report feeling hopeless, females seem more likely to have suicidal thoughts than boys (1,174 [45.5%] versus 559 [37.6%]). In other words, for those who feel hopeless, the RR for females reporting suicidal thoughts is 1.21 compared with males (i.e., females are about 21% more likely to have suicidal thoughts if they reported feeling hopeless).

If we construct a logistic regression analysis predicting SUICIDE from the other three variables, all two-way interactions, and the three-way interaction, we should closely replicate the four-way interaction. Recall from Table 12.5a that removal of that interaction did not significantly degrade the model; in fact, the likelihood ratio was 2.64. In the binary logistic regression, the three-way interaction was not significant when entered alone on a final step ($\chi^2_{(1)} = 2.64$, $p < .104$). These numbers match the results from the

Table 12.5e   Logistic Regression Results Predicting SUICIDE From SEX, E_BULLIED, and HOPELESS12

		Variables in the Equation					
		B	SE	Wald	df	p	Exp(B)
Step 1[a]	SEX	−0.529	0.095	31.160	1	.000	0.589
	E_BULLIED	0.914	0.127	51.651	1	.000	2.495
	HOPELESS12	2.243	0.081	760.704	1	.000	9.418
	SEX by E_BULLIED	0.137	0.135	1.025	1	.311	1.147
	SEX by HOPELESS12	0.272	0.115	5.539	1	.019	1.312
	E_BULLIED by HOPELESS12	−0.065	0.140	0.214	1	.643	0.937
	Constant	−2.691	0.068	1,587.804	1	.000	0.068

[a]Variable(s) entered on step 1: SEX*E_BULLIED , SEX*HOPELESS12, and E_BULLIED*HOPELESS12.

log-linear analysis reported earlier in the chapter. Removing that effect from the analysis, the results from the first two steps in the logistic regression are presented in Table 12.5e.

As you can see in Table 12.5e, SEX*E_BULLIED and E_BULLIED*HOPELESS12 were not significant, with the same p values as the prior analysis (represented by the three-way interactions of SEX*E_BULLIED*SUICIDE and E_BULLIED*HOPELESS12*SUICIDE), and the SEX*HOPELESS12 interaction was identical in significance (to the SEX*HOPELESS12*SUICIDE interaction) as in the prior analysis. In other words, there is very close correspondence between the results of the two analyses, as one would expect from sister analytic models within the GLM.

Data Cleaning in Log-Linear Models

In log-linear models, the only data we have are frequency counts. Thus, residuals correspond to an entire cell, or group, and are the difference between the expected (predicted) value for a cell and the observed value (specifically, the difference between the expected and observed frequencies). This leaves us unable to perform data cleaning on individual cases. Standardized residuals (the difference between the observed and expected frequencies divided by the square root of the expected frequency; see Equation 12.3) can indicate which cells are better accounted for by a particular model. If cells have relatively large standardized residuals, this is an indication that the model is not accounting for that group well. Instead of deleting those cases (which would produce a cell with a count of zero, not a desirable situation in log-linear analysis), you might consider adding another variable to the model in an attempt to account for the differences across groups.

Standardized residual

$$(z) = \frac{(f_0 - f_e)}{\sqrt{F_e}} \tag{12.3}$$

Of course, there are no residuals when the model is saturated and thus has perfect fit. Once one starts removing terms, however, residuals can be informative. For example, returning to the example from before analyzing SEX, E_BULLIED, HOPELESS12,

Table 12.6a Residuals After Removing Four-Way Interaction From Model

Cell Counts and Residuals[a,b]

SEX	E_BULLIED	HOPELESS12	SUICIDE	Observed N	%	Expected N	%	Residual	zResidual
Female	0 No	0 No	0 No	3,262	24.3	3,267.473	24.3	−5.473	−0.110
			1 Yes	227	1.7	221.527	1.7	5.473	0.371
		1 Yes	0 No	1,090	8.1	1,084.527	8.1	5.473	0.173
			1 Yes	687	5.1	692.473	5.2	−5.473	−0.214
	1 Yes	0 No	0 No	428	3.2	422.527	3.1	5.473	0.271
			1 Yes	66	0.5	71.473	0.5	−5.473	−0.649
		1 Yes	0 No	315	2.3	320.473	2.4	−5.473	−0.309
			1 Yes	484	3.6	478.527	3.6	5.473	0.255
Male	0 No	0 No	0 No	4,901	36.5	4,895.527	36.5	5.473	0.098
			1 Yes	190	1.4	195.473	1.5	−5.473	−0.394
		1 Yes	0 No	794	5.9	799.473	6.0	−5.473	−0.200
			1 Yes	400	3.0	394.527	2.9	5.473	0.280
	1 Yes	0 No	0 No	253	1.9	258.473	1.9	−5.473	−0.344
			1 Yes	35	0.3	29.527	0.2	5.473	1.008
		1 Yes	0 No	129	1.0	123.527	0.9	5.473	0.495
			1 Yes	158	1.2	163.473	1.2	−5.473	−0.431

[a]Model: multinomial.
[b]Design: constant + SEX + E_BULLIED + HOPELESS12 + SUICIDE + E_BULLIED*HOPELESS12 + SEX*E_BULLIED + E_BULLIED*SUICIDE + SEX*HOPELESS12 + HOPELESS12*SUICIDE + SEX*SUICIDE + SEX*E_BULLIED*SUICIDE + SEX*E_BULLIED*HOPELESS12 + E_BULLIED*HOPELESS12*SUICIDE + SEX*HOPELESS12*SUICIDE.

Table 12.6b Residuals After Removal of Nonsignificant Three-Way Interactions

Cell Counts and Residuals[a,b]

SEX	E_BULLIED	HOPELESS12	SUICIDE	Observed N	Observed %	Expected N	Expected %	Residual	zResidual
Female	0 No	0 No	0 No	3,262	24.3	3,267.207	24.3	-5.207	-0.105
			1 Yes	227	1.7	221.793	1.7	5.207	0.353
		1 Yes	0 No	1,090	8.1	1,092.302	8.1	-2.302	-0.073
			1 Yes	687	5.1	684.698	5.1	2.302	0.090
	1 Yes	0 No	0 No	428	3.2	422.793	3.2	5.207	0.257
			1 Yes	66	0.5	71.207	0.5	-5.207	-0.619
		1 Yes	0 No	315	2.3	312.698	2.3	2.302	0.132
			1 Yes	484	3.6	486.302	3.6	-2.302	-0.106
Male	0 No	0 No	0 No	4,901	36.5	4,892.357	36.5	8.643	0.155
			1 Yes	190	1.4	198.643	1.5	-8.643	-0.618
		1 Yes	0 No	794	5.9	795.134	5.9	-1.134	-0.041
			1 Yes	400	3.0	398.866	3.0	1.134	0.058
	1 Yes	0 No	0 No	253	1.9	261.643	1.9	-8.643	-0.540
			1 Yes	35	0.3	26.357	0.2	8.643	1.685
		1 Yes	0 No	129	1.0	127.866	1.0	1.134	0.101
			1 Yes	158	1.2	159.134	1.2	-1.134	-0.090

[a]Model: multinomial.
[b]Design: constant + SEX + E_BULLIED + HOPELESS12 + SUICIDE + E_BULLIED*HOPELESS12 + SEX*E_BULLIED + E_BULLIED*SUICIDE + SEX*HOPELESS12 + HOPELESS12*SUICIDE + SEX*SUICIDE + SEX*E_BULLIED*HOPELESS12 + SEX*HOPELESS12*SUICIDE.

Table 12.6c Residuals After Removal of All Interactions

Cell Counts and Residuals[a,b]

SEX	E_BULLIED	HOPELESS12	SUICIDE	Observed N	Observed %	Expected N	Expected %	Residual	zResidual
Female	0 No	0 No	0 No	3,262	24.3	3,279.414	24.4	−17.414	−0.350
			1 Yes	227	1.7	659.581	4.9	−432.581	−17.273
		1 Yes	0 No	1,090	8.1	1,421.126	10.6	−331.126	−9.289
			1 Yes	687	5.1	285.828	2.1	401.172	23.986
	1 Yes	0 No	0 No	428	3.2	530.339	4.0	−102.339	−4.534
			1 Yes	66	0.5	106.666	0.8	−40.666	−3.953
		1 Yes	0 No	315	2.3	229.821	1.7	85.179	5.667
			1 Yes	484	3.6	46.223	0.3	437.777	64.502
Male	0 No	0 No	0 No	4,901	36.5	3,429.911	25.6	1471.089	29.114
			1 Yes	190	1.4	689.850	5.1	−499.850	−19.540
		1 Yes	0 No	794	5.9	1,486.343	11.1	−692.343	−19.044
			1 Yes	400	3.0	298.945	2.2	101.055	5.911
	1 Yes	0 No	0 No	253	1.9	554.677	4.1	−301.677	−13.082
			1 Yes	35	0.3	111.561	0.8	−76.561	−7.279
		1 Yes	0 No	129	1.0	240.368	1.8	−111.368	−7.248
			1 Yes	158	1.2	48.345	0.4	109.655	15.799

[a]Model: multinomial.
[b]Design: constant + SEX + E_BULLIED + HOPELESS12 + SUICIDE.

and SUICIDE, once we remove the four-way interaction (which led to a nonsignificant likelihood ratio of $\chi^2_{(1)} = 2.64$, $p < .104$), the residuals are as presented in Table 12.6a.

As you can see in Table 12.6a, most of the standardized residuals are within a reasonable range (about 0.10 to 1.01 in magnitude). Removing the three-way interactions (SEX*E_BULLIED*SUICIDE and E_BULLIED*HOPELESS12*SUICIDE) that were not significant (which led to a nonsignificant likelihood ratio of $\chi^2_{(1)} = 3.90$, $p < .272$) modified the residuals slightly. As you can see in Table 12.6b, the magnitude of the standardized residuals (absolute value of the residuals) now ranges from about 0.04 to 1.69, a reasonable range. However, when we take the extreme action of removing all interaction terms, the standardized residuals (presented in Table 12.6c) become problematic and substantial in magnitude, ranging from less than 1.00 to over 64.00. This of course is also reflected in terrible model fit.

SUMMARY AND CONCLUSIONS

Log-linear analysis is sometimes overlooked in the social and behavioral sciences, but it is an integral part of the GLM and provides an attractive alternative when all variables are categorical, particularly when some of the variables have multiple categories. Although it is possible to replicate these types of analyses within a logistic regression framework (as it is possible to replicate ANOVA-type analyses within an OLS regression framework), this provides a convenient alternative to extensive dummy coding and hand coding of interaction terms.

There are certain drawbacks to this analysis aside from it being rather rare in some disciplines. Specifically, it does not allow for the designation of a DV; as such, some reviewers and scholars may object, seeing it as more atheoretical and exploratory than other types of analysis, such as logistic regression. In fact, some authors recommend use of logistic regression when modeling DVs and IVs.[9]

This reflects an inaccurate mindset present in other areas of quantitative methods (e.g., arguments concerning use of correlation versus regression). However, the examples in this chapter completely refute this mindset. Although one does not get to designate a DV, the results obtained are identical to those obtained from logistic regression. Furthermore, because this technique is part of the GLM, there is no reason to think that any particular part of the GLM is more or less adept at modeling theoretical dependencies. Model choice should be driven by what is most appropriate to model the types of variables present in the data, not misguided beliefs about which analyses are most appropriate for modeling of correlational or causal data.

Depending on the software you use and whether you are comfortable coding syntax rather than staying with point-and-click methods of analysis, you may have access to more advanced features such as specifying the comparison group.

---

9   I will decline to cite or name them to spare their reputations, but they are easily found if you care to search.

One of the difficulties in reading the output from SPSS presented in this chapter was the fact that the last group was coded as the comparison group. There are syntax-only methods of modifying this, and there are also more direct methods. For example, knowing this, we could have altered the order of the groups within our variables. If you are going to use this technique, you will quickly become adept at deciphering the output or adapting your analyses to create more easily decipherable results.

ENRICHMENT

1. Download the data used in the examples from the chapter and replicate the results.

2. Download the ELS2002 data from Chapter 11 and find a Poisson outcome that you would like to model as a function of RACE and SEX (and the interaction of the two). Be sure to use Poisson rather than multinomial as the model specification.

   a. Various examples of these analyses will be presented on the book website.

REFERENCES

Tabachnick, B. G., & Fidell, L. S. (2001). *Using multivariate statistics* (4th ed.). New York, NY: Harper Collins.

A BRIEF INTRODUCTION TO HIERARCHICAL LINEAR MODELING

13

Advance Organizer

No Child Left Behind (NCLB) was an initiative within the United States to attempt to improve the educational system—particularly to ensure that lower-performing students are brought up to standards that are considered at least minimally acceptable. Although there is great controversy over whether NCLB was a good idea or a poor one, and whether it was helpful or harmful, one policy was to evaluate whether schools were being successful (and by proxy, whether teachers were being successful) by evaluating student achievement. This has long been a central question, and central conflict, within the culture and community of researchers—how best to measure the effects of schools and teachers on students (and vice versa). It makes sense that teachers (and schools) matter. This is one reason why many parents are willing to pay thousands of dollars a year to send their children to private schools when they deem the public schools lacking, and why the quality of the local schools is one of the key factors in housing choice.

The central questions in this debate are multilevel questions—does a teacher in a classroom have an effect on the students within the class (and if so, is it the same effect for every student)? Do schools really influence those within their walls in some way? Although researchers have been attempting to answer these questions for decades, it is only in the past couple of decades that multilevel modeling (MLM; also called hierarchical linear modeling [HLM]) can be used to more appropriately test and answer these questions. In this chapter, we will briefly explore some very basic hierarchical linear models and answer the same types of questions that we have explored throughout the book—often in a more nuanced and appropriate manner.

The analyses presented in this chapter should be possible in almost any HLM-capable software package. The package I have used since the 1990s is called "HLM" and there is a free "student" download available from Scientific Software International (http://www.ssicentral.com/hlm/student.html) that is capable of analyzing this example as well as the other examples at the end of the chapter. More statistical computing packages also include HLM modeling capabilities (e.g., SPSS and SAS) and I will have guides for how to perform analyses via various software packages on the book's website. As with every other chapter, I use specific software for the analyses but attempt to write the chapter in a software-neutral manner so that you should be able to use it with almost any package you like.

There are entire, very good books devoted to the topic of HLM. It cannot be covered adequately in one chapter. My goal here is to expose you to some basic concepts and motivate you to want to learn more.

In this chapter, we will cover

- Why HLM models are necessary
- Basic concepts in HLM
- Examples of HLM with

    o Continuous dependent variables (DVs)
    o Categorical DVs

- Assumptions and data cleaning in HLM
- Examples of American Psychiatric Association (APA)–compliant summaries of analyses

Guidance on how to perform these analyses in various statistical packages will be available online at study.sagepub.com/osbornerlm.

Why HLM Models Are Necessary

What Is a Hierarchical Data Structure?

People (and most living creatures, for that matter) tend to exist within organizational structures, such as families, schools, business organizations, churches, towns, states, and countries. In education, students exist within a hierarchical social structure that can include family, peer group, classroom, grade level, school, school district, state, and country. Workers exist within production or skill units, businesses, and sectors of the economy, as well as geographic regions. Health care workers and patients exist within households and families, medical practices and facilities (e.g., a doctor's practice or hospital), counties, states, and countries. Many other communities exhibit hierarchical data structures as well. Raudenbush and Bryk (2002) also discuss two other types of data hierarchies that are less obvious but equally important and well served by HLM: repeated-measures data and meta-analytic data. In this case, we can think of repeated measures as data that are nested or clustered within individuals, and meta-analytic data similarly involve clusters of data or subjects nested within studies. Once one begins looking for hierarchies in data, it becomes obvious that data repeatedly gathered on an individual are hierarchical, as all of the observations are nested within individuals.

Why Is Hierarchical or Nested Data an Issue?

Hierarchical, or nested, data present several problems for analysis. First, people or creatures that exist within hierarchies tend to be more similar to each other than people randomly sampled from the entire population. For example, students in a particular third-grade classroom are more similar to each other than to students randomly sampled from the school district as a whole, or from the national population of third-graders. This is because (in many countries) students are not randomly assigned to classrooms from the population; rather, they are assigned to schools based on geographic factors (e.g., home location) or other characteristics (e.g., aptitude test scores). When assigned based on geography, students within a particular classroom tend to come from a community or community segment that is more homogeneous in terms of morals and values, family background, socioeconomic status (SES), race or ethnicity, religion, and even educational preparation than a similar-sized sample randomly sampled from the entire population as a whole. When assigned based on similarity in other characteristics, students are obviously more homogeneous than a random sample of the entire population. Furthermore, regardless of similarity or dissimilarity of background, students within a particular classroom share the experience of being in the same environment—the same teacher, physical environment, curriculum, and similar experiences, which may increase homogeneity over time.

The Problem of Independence of Observations

This discussion could be applied to any level of nesting, such as the family, school district, county, state, or even country. Based on this discussion, we can assert that individuals who are drawn from a group, such as a classroom, school, business, town or city, or health care unit, will be more homogeneous than if individuals were randomly sampled from a larger population. This is often referred to as a *design effect*. Because these individuals tend to share certain characteristics (environmental, background, experiential, demographic, or otherwise), observations based on these individuals are not fully independent, yet most statistical techniques require independence of observations as a primary assumption for the analysis. Because this assumption is violated in the presence of hierarchical or nested data, most linear models produce standard errors (*SEs*) that are too small (unless these so-called design effects are incorporated into the analysis). In turn, this leads to an inappropriately higher probability of rejection of a null hypothesis than if (a) an appropriate statistical analysis was performed, or (b) the data included truly independent observations.

The Problem of How to Deal With Multilevel Data

It is often the case in the social and behavioral (and health) sciences that a researcher is interested in understanding how environmental variables affect individual outcomes. For example, in some health sciences, researchers are trying to understand how environmental variables (e.g., availability of healthy food) influence health outcomes (e.g., obesity). In education, we certainly wonder about how classroom, school, and family variables influence student outcomes.

Given that outcomes are gathered at the individual level and other variables are centered at higher levels, the question arises as to what the unit of analysis should be and how to deal with the cross-level nature of the data. One strategy (called disaggregation) would be to assign classroom or teacher (or other group-level) characteristics to all students (i.e., to bring the higher-level variables down to the student level). The problem with this approach is that all students within a particular classroom assume identical scores on a variable, clearly violating assumptions of independence of observation.

Another way to deal with this issue (called aggregation) would be to aggregate up to the level of the classroom, school, district, and so forth. Thus, we could talk about the effect of teacher or classroom characteristics on average classroom achievement. However, there are several problems with this approach: (a) much (up to 80%–90%) of the individual variability on the outcome variable is lost, which can lead to dramatic underestimation or overestimation of observed relationships between variables (Raudenbush & Bryk, 2002); and (b) the outcome variable changes significantly and substantively from individual achievement to average classroom achievement.

Neither of these strategies constitutes a best practice, although both have been commonly found in top journals in a variety of fields. Neither of these strategies allows the researcher to ask truly important questions—such as what is the effect of a particular teacher variable on student learning? A third approach (HLM) becomes necessary in this age of more sophisticated hypotheses.

How Do Hierarchical Models Work? A Brief Primer

The goal of this chapter is to introduce the concept of hierarchical modeling and explicate the need for the procedure. This chapter cannot fully communicate the nuances and procedures needed to actually perform a hierarchical analysis. The reader is encouraged to refer to Raudenbush and Bryk (2002) and the other suggested readings for a full explanation of the conceptual and methodological details of HLM.

The basic concept behind HLM is similar to that of ordinary least squares (OLS) regression. On the base level (usually the individual level, or the level where repeated measures are taken within a particular individual, referred to here as level 1, the lowest level of your data), the analysis is similar to that of OLS regression: an outcome variable is predicted as a function of a linear combination of one to k level 1 variables, plus an intercept, as in Equation 13.1a:

$$Y_{ij} = \beta_{0j} + \beta_{1j}X_1 + \ldots + \beta_{kj}X_k + r_{ij} \tag{13.1a}$$

where β_{0j} represents the intercept of group j, β_{1j} represents the slope of variable X_1 of group j, and r_{ij} represents the residual for individual i within group j. On subsequent levels, the level 1 slope(s) and intercept become DVs being predicted from k level 2 variables, as in Equation 13.1b:

$$\beta_{0j} = \gamma_{00} + \gamma_{01}W_1 + \ldots + \gamma_{0k}W_k + u_{0j}$$

$$\beta_{1j} = \gamma_{10} + \gamma_{11}W_1 + \ldots + \gamma_{1k}W_k + u_{1j}$$

$$\beta_{kj} = \gamma_{k0} + \gamma_{k1}W_1 + \ldots + \gamma_{km}W_m + u_{kj} \tag{13.1b}$$

and so forth, where γ_{00} and γ_{10} are intercepts, and γ_{01} and γ_{11} represent slopes predicting β_{0j} and β_{1j}, respectively, from variable W_1. Through this process, we accurately model the effects of level 1 variables on the outcome and the effects of level 2 variables on the outcome. In addition, because we are predicting slopes as well as intercepts (means), we can model cross-level interactions, whereby we can attempt to understand what explains differences in the relationship between level 1 variables and the outcome. Those of you who are more mathematically inclined will also note that several different error terms (i.e., r and u terms) are computed in this process, rather than just a simple residual present in OLS regression.

The advantages of HLM over aggregation and disaggregation have been explored in many places (e.g., Osborne, 2000, 2008). In brief, failing to appropriately model multilevel data can lead to underestimation of *SEs*, substantial misestimation of effects and variance accounted for, and errors of inference.

Generalizing the Basic HLM Model

As many authors have discovered in the years since HLM became available, there are many applications for these analyses. Generalizations to three- and four-level (or more) models are available, as are logistic regression analogues (e.g., HLM with binary or polytomous outcomes), applications for meta-analysis, and powerful advantages for longitudinal analysis (compared with other methods such as repeated-measures analysis of variance [ANOVA]); many of the fun aspects of OLS regression (e.g., modeling curvilinear effects) are possible in HLM as well. The types of hypotheses we test in this type of analysis are similar to those tested in other linear models, so we will save the space and not repeat them here.

There is little downside to HLM, aside from the learning curve. If one were to use HLM on data where no nesting, dependence, or other issues were present, one would get virtually identical results to simple linear regression models. The advantage comes when there is nesting or dependence, in which case the model will be misspecified and produce erroneous results if MLM is not used.

The rest of this chapter is devoted to two simple examples that represent common questions within educational (and many areas of social science) research: (a) how do individual- and school-level variables affect student achievement, and (b) can we understand growth or change in an individual as a function of individual or environmental traits? I will assume you understand that these models can be generalized to multinomial models, Poisson models, models with interactions, and so forth.

Because of the large variety of models you can create within HLM (essentially, anything we have modeled thus far within the limits of the HLM software you use), the assumptions of any given model—and the data cleaning issues around any given model—will vary depending on the model. Thus, I will skip those topics for this chapter, because you can refer back to previous chapters for the appropriate assumptions and tips on data cleaning.

Example 1. Modeling a Continuous DV in HLM

The first example is analogous to an OLS regression analysis with independent variables (IVs) at two levels. Specifically, this example will have a continuous DV, two variables at the student level (family SES and student locus of control), and two school-level variables (percentage of students who meet a particular definition of economic need in the United States, as measured by the percentage of students receiving free lunch in school, and the percentage of students who belong to traditionally disadvantaged racial minority groups) predicting student standardized achievement test scores. In this section, we illustrate the outcomes achieved by each of the three possible analytic strategies for dealing with hierarchical data. The first two are anachronistic methods of analyzing multilevel data without modeling the data at multiple levels (a poor practice, as we will empirically demonstrate), and the third involves modern modeling through HLM:

1. Disaggregation (bringing school-level data down to the individual level),

2. Aggregation (bringing individual-level data in summarized fashion up to the school level), and

3. MLM (appropriately modeling variables at the level at which they were gathered).

Data for this example were drawn from the National Education Longitudinal Survey of 1988 (NELS88).[1] The analysis we performed predicted composite achievement test scores (math, reading combined) from student SES (family SES), student locus of control (LOCUS), the percentage of students in the school who are members of racial or ethnic minority groups (%MINORITY), and the percentage of students in a school who received free lunch (%LUNCH, an indicator of school poverty). We expect SES and LOCUS to be positively related to achievement, and %MINORITY and %LUNCH are expected to be negatively related to achievement. In these analyses, 995 of a possible 1,004 schools had sufficient data to be included. Cases were not weighted appropriately, so the results should not be interpreted substantively.

Disaggregated Analysis

To perform the disaggregated analysis, the level 2 values were assigned to all individual students within a particular school. A standard OLS multiple regression was performed via SPSS entering all predictor variables simultaneously. The resulting model was significant, with $R = 0.56$, $R^2 = 0.32$, $F_{(4, 22,899)} = 2,648.54$, $p < .0001$. The individual regression weights and significance tests are presented in Table 13.1.

In this analysis, all four variables were significant predictors of student achievement. As expected, SES and LOCUS were positively related to achievement, whereas %MINORITY and %LUNCH were negatively related.

---

1   These data were from the National Education Longitudinal Study of 1988 (NELS88) from the National Center for Education Statistics (http://nces.ed.gov/surveys/nels88/).

Table 13.1   Comparison of Three Analytic Strategies

Variable	Disaggregated			Aggregated			Hierarchical		
	b	*SE*	*t*	*b*	*SE*	*t*	*b*	*SE*	*t*
SES	4.97[a]	0.08	62.11***	7.28[b]	0.26	27.91***	4.07[c]	0.10	41.29***
LOCUS	2.96[a]	0.08	37.71***	4.97[b]	0.49	10.22***	2.82[a]	0.08	35.74***
%MINORITY	−0.45[a]	0.03	−15.53***	−0.40[a]	0.06	−8.76***	−0.59[b]	0.07	−8.73***
%LUNCH	−0.43[a]	0.03	−13.50***	0.03[b]	0.05	0.59	−1.32[c]	0.07	−19.17***

NOTE: *b* refers to an unstandardized regression coefficient and is used for the HLM analysis to represent the unstandardized regression coefficients produced therein, although these are commonly labeled as betas and gammas. *b*s with different superscripts were found to be significantly different from other *b*s within the row at $p < .05$. ***$p < .0001$. This analysis was first presented in Osborne (2000).

Aggregated Analysis

To perform the aggregated analysis, all level 1 variables (achievement, LOCUS, SES) were aggregated up to the school level (level 2) using school-based means. A standard multiple regression was performed via SPSS entering all predictor variables simultaneously. The resulting model was significant, with $R = 0.87$, $R^2 = 0.75$, $F_{(4, 999)} = 746.41$, $p < .0001$. Again, as expected, both average SES and average LOCUS were positively related to achievement, and %MINORITY was negatively related. In this analysis, %LUNCH was not a significant predictor of average achievement. Absent any other analysis, it is difficult to know which of the two analyses is the best summary of these effects. They are strongly divergent in terms of effect size(s) and in whether %LUNCH has any effect.

HLM Analysis

Finally, a multilevel analysis was performed via HLM, in which the respective level 1 and level 2 variables were modeled appropriately. Note also that all level 1 predictors were centered at the group mean, and all level 2 predictors were centered at the grand mean (following recommendations in places such as those of Raudenbush and Bryk, 2002). The resulting model demonstrated goodness of fit (chi-square for change in model fit = 4,231.39, 5 degrees of freedom, $p < .0001$). As seen in the right-hand portion of Table 13.1, this analysis reveals expected relationships—positive relationships between achievement and the level 1 predictors (SES and LOCUS) and strong negative relationships between achievement and the level 2 predictors (%MINORITY and %LUNCH). Furthermore, the analysis revealed significant interactions between SES and both level 2 predictors, indicating that the slope for SES gets weaker as %LUNCH and %MINORITY increase. In addition, there was an interaction between LOCUS and %MINORITY, indicating that as %MINORITY increases, the slope for LOCUS weakens. There is no clearly equivalent analogue to R and R^2 available in HLM, so it is not possible to compare effect size for the overall model.

Comparison of the Three Analytic Strategies and Conclusions

We assume that the third analysis represents the best estimate of what the "true" relationships are between the predictors and the outcome. Unstandardized regression coefficients (b in OLS, β and γ in HLM) were compared statistically via procedures outlined in Cohen, Cohen, West, and Aiken (2003).

Neither of the first two analyses appropriately modeled the relationships of the variables. The disaggregated analysis significantly overestimated the effect of SES and significantly and substantially underestimated the effects of the level 2 effects. The *SE*s in this analysis are generally lower than they should be, particularly for the level 2 variables (a common issue when assumptions of independence are violated).

The second analysis substantially overestimated the multiple correlation, overestimated the regression slope by 79% for SES and by 76% for LOCUS, and underestimated the slopes by 32% for %MINORITY and by 98% for %LUNCH.

These analyses reveal the need for multilevel analysis of multilevel data. Neither OLS analysis accurately modeled the true relationships between the outcome and the predictors. In addition, HLM analyses provide other benefits, such as easy modeling of cross-level interactions, which allows for more interesting questions to be asked of the data. For example, in this final analysis we could examine how family and school poverty interact, something not possible unless the multilevel data are modeled correctly.

Example 2. Modeling Binary Outcomes in HLM[2]

For this example, let us take another set of variables from the NELS88 data. We will explore predictors of whether a student drops out before completing his or her high school diploma (twelfth grade, in the United States). Thus, our IV is DROPOUT (1 = yes, 0 = no). We will also select the following student variables: zBYACH (eighth-grade composite achievement test score, converted to *z*-scores), zBYSES (family SES, converted to *z*-scores), and EVER_MJ (whether the student has ever admitted to trying marijuana, coded 0 = no, 1 = yes). Because this is an HLM example, we will also include one school-level variable: %LUNCH (the percentage of students in the school who are eligible for free or reduced price lunch, an indicator of poverty in the United States). Thus, the question we can ask is whether school environment is a separate predictor of student dropout above that of student-level variables such as family SES, achievement, and so forth. Similar to simple logistic regression, when performing HLM with a binary outcome, the DV is the log of the odds of DROPOUT, denoted η_{ij}.

On the first level (the most granular level of your data, the individual level in this case), the analysis is familiar to anyone familiar with regression analyses: an outcome variable is predicted as a function of a linear combination of one or more level 1 variables, plus an intercept, as in Equation 13.2a:

$$Y_{ij} = \beta_{0j} + \beta_{1j}X_1 + \dots + \beta_{kj}X_k + r_{ij} \tag{13.2a}$$

---

2   This section of the chapter is adapted from a similar chapter in my 2015 book on logistic regression (Osborne, 2015).

where β_{0j} represents the intercept of group j, β_{1j} represents the slope of variable X_1 of group j, and r_{ij} represents the residual for individual i within group j. Adapting our equation to our example, our level 1 equation will look something like Equation 13.2b:

$$\eta_{ij} = \beta_{0j} + \beta_{1j}{}^*(zBYACH_{ij}) + \beta_{2j}{}^*(zBYSES_{ij}) + \beta_{3j}{}^*(EVER\_MJ_{ij}) + r_{ij} \quad (13.2b)$$

As with the previous example, HLM then estimates a level 2 equation where we enter the higher-level variables, which in this case is the school-level variable %LUNCH (also converted to z-scores). In level 2 equations, the level 1 slope(s) and intercept become DVs being predicted from m level 2 variables, as in Equation 13.2c:

$$\beta_{0j} = \gamma_{00} + \gamma_{01} W_1 + \dots + \gamma_{0m} W_m + u_{0j}$$

$$\beta_{1j\,seq} = \gamma_{10} + \gamma_{11} W_1 + \dots + \gamma_{1m} W_m + u_{1j}$$

$$\beta_{kj} = \gamma_{k0} + \gamma_{k1} W_1 + \dots + \gamma_{km} W_m + u_{1j} \quad (13.2c)$$

and so forth, where γ_{00} and γ_{10} to γ_{k0} are intercepts, and γ_{01} and γ_{11} to γ_{1m} represent slopes predicting β_{0j} and β_{1j} to β_{kj}, respectively, from variable W_1 and from all other variables. Note that there will be a level 2 equation, as shown above, for each level 1 parameter estimated (intercept, slope of each predictor at level 1). Going back to our example, our level 2 equations would look like Equation 13.2d:

$$\beta_{0j} = \gamma_{00} + \gamma_{01}{}^*(z\%LUNCH_j) + u_{0j}$$

$$\beta_{1j} = \gamma_{10} + \gamma_{11}{}^*(z\%LUNCH_j) + u_{1j}$$

$$\beta_{2j} = \gamma_{20} + \gamma_{21}{}^*(z\%LUNCH_j) + u_{2j}$$

$$\beta_{3j} = \gamma_{30} + \gamma_{31}{}^*(z\%LUNCH_j) + u_{3j} \quad (13.2d)$$

These two sets of equations can then be combined to achieve the final single equation that will be estimated (the u_{1j}–u_{3j} were left out of the equation for technical reasons beyond the scope of this chapter) presented in Equation 13.2e:

$$\eta_{ij} = \gamma_{00} + \gamma_{01}{}^*z\%LUNCH_j + \gamma_{10}{}^*zBYACH_{ij} + \gamma_{11}{}^*z\%LUNCH_j{}^*zBYACH_{ij}$$
$$+ \gamma_{20}{}^*zBYSES_{ij} + \gamma_{21}{}^*z\%LUNCH_j{}^*zBYSES_{ij} + \gamma_{30}{}^*EVER\_MJ_{ij} +$$
$$\gamma_{31}{}^*z\%LUNCH_j{}^*EVER\_MJ_{ij} + u_{0j} \quad (13.2e)$$

Through this process, we accurately model the effects of level 1 variables on the outcome and the effects of level 2 variables on the outcome. In addition, as we are predicting slopes as well as intercepts (means), we can model cross-level interactions, whereby we can attempt to understand what explains differences in the relationship between level 1 variables and the outcome.

Residuals in HLM

Those of you more mathematically inclined will also note that several different error terms (i.e., r and u terms) are computed in this process, rather than just a simple residual present in the types of analyses we have been familiar with. Each of these residuals is similar to the residuals we have seen in logistic regression. However, in HLM we also get two different data files containing the residuals from each level. Of particular interest, we get unstandardized level 1 residuals and other metrics of influence and leverage in the level 1 data file (depending on which software package you are using). If you are using HLM, you may want to refer to Raudenbush and Bryk (2002) and the HLM user manual for more detailed information about what each data file contains. In the example we are discussing, there were some hugely influential data points, one with a residual 12 standard deviations outside the normal range of residuals.

In HLM, the level 2 residual file presents Mahalanobis distance (MDIST) as a measure of influence for the residuals. MDIST is a common multivariate indicator of leverage or influence that ranges from 0 to ∞. Generally, cases with MDIST over 4 may be considered suspect, with increasingly large MDIST values relating to increasingly suspect cases. I generally clean the level 1 data file of unacceptably influential data points and then rerun the analysis before looking at level 2 residuals. Of course, other indices of leverage/influence, when available, are also appropriate for this purpose.

Results of DROPOUT Analysis in HLM

Results of our HLM analyses are presented in Table 13.2.[3] As you can see in the table, the results are similar to the type of output you would see in a simple logistic regression analysis. The logistic regression coefficients are in the first column. In HLM, with all variables appropriately centered (which allows us to interpret the intercept as the mean, because the mean is 0), we interpret γ_{00} as the overall intercept or grand mean. Because you are all experts by now at interpreting logistic regression output, you will know that this means the probability of someone who is average on all other variables dropping out of school is 0.004. The second line is the effect of z%LUNCH on dropout ($\gamma_{01} = 1.63$, odds ratio [OR] of 5.10 with a corresponding confidence interval [CI] of [3.36, 7.75]). This is interpreted identically to a simple logistic regression OR—that as you increase %LUNCH 1 standard deviation (SD) (increasing school-level poverty), the odds of a student dropping out are 5.10 times that of a student in a school 1 SD lower in %LUNCH. All other things being equal (average), a student at a relatively affluent school (z%LUNCH = −1) will have a dropout probability of 0.0016, whereas a student at a less affluent school (z%LUNCH = 1) will have a dropout probability of 0.042, or a relative risk (RR) of 26.05.[4]

---

3   Note that these data are a random subsample of cases from the full NELS88 data and are not weighted appropriately to reflect the complex sampling. Thus, as with all other examples in this book, you should not interpret these results substantively; rather, use them *solely* as an example of binary logistic regression analysis in HLM.

4   This is very similar to the odds calculation you would make figuring the odds of someone dropping out at a school 1 SD below the mean and 1 SD above the mean ($5.104^2 = 26.01$).

Table 13.2   Results of DROPOUT Analyses in Hierarchical Linear Modeling

Fixed Effect	Coefficient	SE	t-Ratio	df	p	OR	CI
For INTRCPT1, β_0							
INTRCPT2, γ_{00}	−5.483036	0.168747	−32.493	195	< .001	0.004157	(0.003, 0.006)
z%LUNCH, γ_{01}	1.630046	0.211962	7.690	195	< .001	5.104110	(3.360, 7.754)
For zBYACH slope, β_1							
INTRCPT2, γ_{10}	−1.776596	0.175515	−10.122	1,968	< .001	0.169213	(0.120, 0.239)
z%LUNCH, γ_{11}	−0.016114	0.233401	−0.069	1,968	.945	0.984015	(0.622, 1.556)
For zBYSES slope, β_2							
INTRCPT2, γ_{20}	−1.034264	0.129101	−8.011	1,968	< .001	0.355488	(0.276, 0.458)
z%LUNCH, γ_{21}	0.504195	0.152356	3.309	1,968	< .001	1.655652	(1.228, 2.232)
For EVER_MJ slope, β_3							
INTRCPT2, γ_{30}	0.726309	0.271777	2.672	1,968	.008	2.067436	(1.213, 3.524)
z%LUNCH, γ_{31}	−0.135861	0.246234	−0.552	1,968	.581	0.872964	(0.538, 1.415)

SOURCE: National Education Longitudinal Study of 1988 (NELS88) from the National Center for Education Statistics (http://nces.ed.gov/surveys/nels88/).

The next section of the output deals with zBYACH, the eighth-grade student achievement composite. The intercept (γ_{10}) represents the average effect of zBYACH—which is −1.78. In other words, the odds of dropping out decrease as student achievement increases. This is not surprising, because we have seen similar results in prior chapters. The corresponding OR (0.17 [CI of 0.12, 0.24]) is interpreted similarly to that of z%LUNCH. As a student increases 1 *SD* in achievement test scores, the odds that student will drop out are 0.17 that of a student 1 *SD* lower in achievement. Again, calculations show that the probability of a student dropping out when 1 *SD* below the mean are markedly higher than that of a student 1 *SD* above the mean (0.049 versus 0.001, for a RR of 34.92; note that this is still holding all other variables constant at the mean).

The next line is the effect of %LUNCH on ZBYACH—in other words, testing whether there is an interaction between achievement and school poverty. It is not significant, indicating that there is no moderating effect of poverty on the effect of achievement. Put another way, each of the two effects is constant across the range of the other.

We can interpret the next sections relating to SES and marijuana use similarly (I leave the calculations and interpretation of probabilities to you). The effect of zBYSES on DROPOUT (γ_{20}) is significant and negative, also indicating that increasing family affluence is associated with decreasing dropout, as we saw in previous chapters. In the case of this variable, there is a significant interaction between zBYSES and z%LUNCH. Let us come back to this interaction after discussing the other variable, EVER_MJ. The effect of marijuana use (γ_{30}) is significant and positive, indicating that (after controlling for all other variables) students who admit to using marijuana are more likely to drop out than those not admitting to using. The OR for this variable indicates that for those who admit to using marijuana, the odds of dropping out are approximately twice that of those who do not admit to having used marijuana. In addition, there is no significant interaction with %LUNCH for this variable.

Cross-Level Interactions in HLM Logistic Regression

Recall that there is a significant interaction between %LUNCH and zBYSES— between school environment and family environment. I prefer to graph out results such as this, as I have described and demonstrated in previous chapters. Because all continuous variables are converted to z-scores prior to analysis, we can form the regression equation from the coefficients in the output, substituting the mean (0) for variables that are not relevant to this interaction (ZBYSES, Z%LUNCH, and the interaction between the two).

For our analysis, we could constitute the equation to graph the interaction from the output, using the intercept (γ_{00}) and coefficients presented in Table 13.2. For this example, the equation becomes as shown in Equation 13.2f:

$$\hat{\eta}_{ij} = \gamma_{00} + \gamma_{01} {}^{*}z\%LUNCH_j + \gamma_{20} {}^{*}zBYSES_{ij} + \gamma_{21} {}^{*}z\%LUNCH_j {}^{*}zBYSES_{ij} \qquad (13.2f)$$

By substituting values for gammas, we get

$$\hat{\eta}_{ij} = -5.48 + 1.63 {}^{*}z\%LUNCH_j - 1.034 {}^{*}zBYSES_{ij} + 0.504 {}^{*}z\%LUNCH_j {}^{*}zBYSES_{ij}$$

Following the procedure I have suggested in previous chapters for graphing interactions, we would substitute a relatively conservative number for low and high (± 1.5 in this case, to keep the estimates conservative and reasonable) for the percentage of free and reduced lunch (school poverty) and family SES.

As Figure 13.1 shows, when graphed in probabilities, and controlling for all other variables in the equation, those in low-poverty schools tend to have a very low probability of dropping out regardless of family SES, whereas those in high-poverty schools are much more affected by family SES.

So What Would Have Happened If These
Data Had Been Analyzed via Simple Logistic
Regression Without Accounting for the Nested Data Structure?

For a comparison to the prior analysis, I analyzed the same data using the strategy of disaggregation, mentioned above as a common (not recommended) practice when dealing with nested data. As you can see in Table 13.3, these results are not a complete

Figure 13.1   Interaction Between z%LUNCH and zBYSES Graphed in Probability

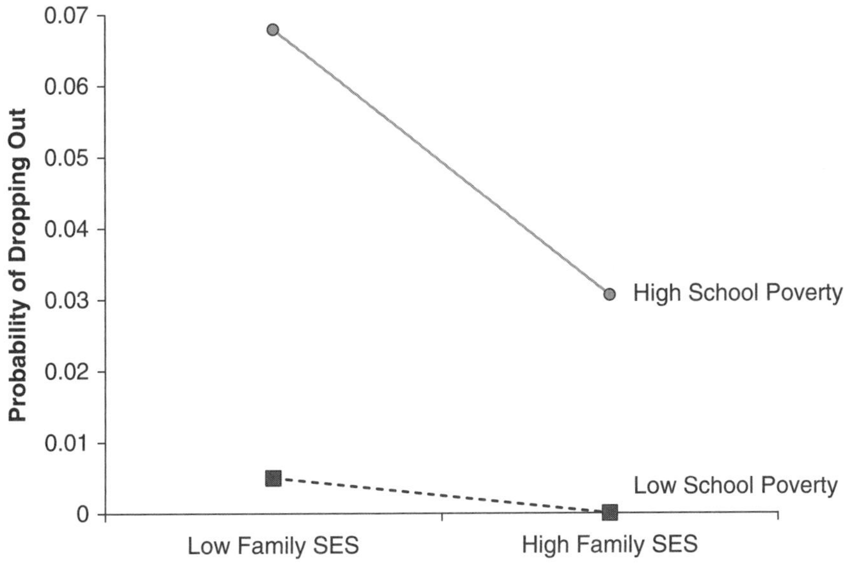

SOURCE: National Education Longitudinal Study of 1988 (NELS88) from the National Center for Education Statistics (http://nces.ed.gov/surveys/nels88/).

Table 13.3   Inappropriately Disaggregated Analysis

	B	SE	Wald	df	p	Exp(B)	95% CI for Exp(B)	
							Lower	Upper
z%LUNCH	2.273	0.457	24.708	1	.000	9.706	3.961	23.780
zBYACH	−2.402	0.417	33.229	1	.000	0.090	0.040	0.205
zBYSES	−1.371	0.316	18.877	1	.000	0.254	0.137	0.471
EVER_MJ	0.511	0.505	1.021	1	.312	1.667	0.619	4.489
zBYACH by z%LUNCH	0.039	0.328	0.014	1	.906	1.039	0.547	1.975
zBYSES by z%LUNCH	0.567	0.242	5.501	1	.019	1.763	1.098	2.832
EVER_MJ by z%LUNCH	0.208	0.406	0.261	1	.609	1.231	0.555	2.729
Constant	−7.291	0.707	106.249	1	.000	0.001		

SOURCE: National Education Longitudinal Study of 1988 (NELS88) from the National Center for Education Statistics (http://nces.ed.gov/surveys/nels88/).

departure from the appropriately modeled analysis; however, there are some important differences. For example, the intercept itself is altered substantially from the appropriate HLM analysis, the effect of the disaggregated z%LUNCH is overestimated significantly (1.630 versus 2.273), as are the effects of zBYACH (−1.777 versus −2.402) and zBYSES (−1.034 versus −1.371). The effect of EVER_MJ is underestimated and also represents a Type II error here, being significant in the HLM analysis and about half again as strong (0.726 versus 0.511). The interaction effects are largely concordant, in that only the zBYSES × z%LUNCH interaction is significant, but it is overestimated in this case.

You may also notice that some of the *SE*s are larger than in the HLM analysis, often twice as large or more. This leads to CIs that are much broader than in the HLM analysis. Overall, this again shows how appropriate modeling of nested data via HLM/MLM is important to the quality of your conclusions.

SUMMARY AND CONCLUSIONS

HLM is an important tool in every researcher's toolbox. HLM software is increasingly user friendly and is often incorporated into the statistical computing software already being used (SPSS, SAS, STATA, R, etc.). HLM can deal with data and analyses of the type we have already explored in this book: OLS-type analyses, logistic analyses, and so forth. You can model curvilinear effects, interactions, and curvilinear interactions easily. Newer software applications can combine latent variable modeling (SEM) with HLM to produce high-quality, modern analyses, assuming that the data are of good quality and the sample is large enough to support this type of analysis. I hope this chapter motivates you to begin exploring HLM as a next step in your development as a researcher.

HLM-type analyses can also provide analyses not easily produced in other linear models, such as appropriately modeled repeated-measures designs. Although repeated-measures ANOVA is fine for many applications, it has restrictive assumptions and cannot appropriately handle missing data well. HLM has much more flexibility in these matters, because it is able to model data where some data are missing at different time points and can model complex effects like curvilinear growth models with individual- or higher-level variables. Think about being able to model different growth curves for different individuals or groups, and then predict what factors influence the shape of the curve. If you were excited by Chapter 10, where we explored curvilinear interactions, you would really enjoy HMLM (repeated-measures HLM) with growth curves.

ENRICHMENT

1. Replicate the analyses in the chapter using the data provided on the book's website. If you do not have access to HLM or similar software, you can download a free version of HLM from Scientific Software (http://www.ssicentral.com/hlm/student.html), which will allow you to explore the topic prior to committing to an expensive, full version of that or similar software.

2. After replicating the two HLM analyses, perform routine data cleaning and see how the results change or do not change as a result.

3. Your job is to help us understand student dropout in high school. Use data from NELS88 to create a data file for the following variables of interest:

Level 1 (L1)

- Sch_ID: school ID
- DROPOUT: 0 = not dropped out, 1 = dropped out
- zBY2XCOMP: standardized eighth-grade achievement test score
- zBYSES: standardized eighth-grade family SES
- RACEBW: race: 0 = white, 1 = African American
- RACEHW: race: 0 = white, 1 = Latino/latina

Level 2 (L2)

- zG8ENROL: standardized school size
- zG8LUNCH: standardized percentage of students on free/reduced lunch
- zG8MINOR: standardized percentage of racial minority students in the school

After creating the HLM2 data file (you should tell HLM there are missing data, and tell it to delete the missing data when RUNNING the analysis), you should see something similar to this when creating MDM:

```
LEVEL-1 DESCRIPTIVE STATISTICS

VARIABLE NAME N MEAN SD MINIMUM MAXIMUM

DROPOUT 3511 0.10 0.30 0.00 1.00

ZBY2XCOM 3412 0.00 1.00 -1.95 2.50

ZBYSES 3511 -0.00 1.00 -3.80 2.53

RACEBW 3511 0.12 0.32 0.00 1.00

RACEHW 3511 0.15 0.35 0.00 1.00

LEVEL-2 DESCRIPTIVE STATISTICS

VARIABLE NAME N MEAN SD MINIMUM MAXIMUM

ZG8ENROL 200 -0.00 1.00 -2.19 1.63

ZG8LUNCH 200 0.00 1.00 -1.42 2.09

ZG8MINOR 200 0.00 1.00 -1.15 2.13
```

Under Basic Settings, be sure to tell the computer you have a 0,1 DV. Also be sure to set up the two residual files with all relevant variables so you can use them for a second round of analyses if need be. Add ZBY2XCOM and ZBYSES group centered, and two race variables uncentered, as routine.

At level 2, add zg8enrol to all equations, grand centered. This is what your screen should look like before you run the analysis (make sure to save it in a place you can find so you can access the residual files prior to running!):

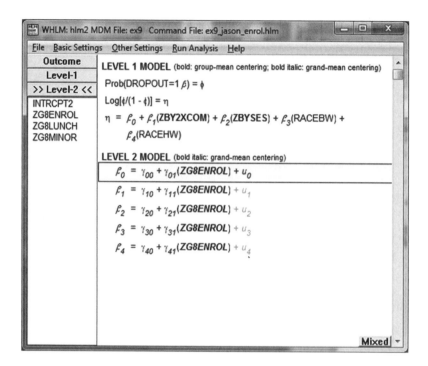

Reporting

- Focus on the population with robust *SEs*. Make a table with the unstandardized coefficients, significance test information, and ORs at a minimum. If you have any cross-level interactions (you should have one), graph it out using unstandardized coefficients.

Hints

- There are some extreme L1 residuals. Remove them. Once gone, rerun the model using the newly cleaned data. You should see the cross-level interaction become more significant. You will also see that only two schools have MDIST >4, and not by much. As my gift to you, you may leave them as is and interpret the second (L1 cleaned) run for your write-up.
- Be sure to interpret the nonsignificant race effects—they are somewhat unexpected and important conceptually. Note that for these effects, the 95% CIs include an OR of 1.00—which means they are not significant.

- Graph the cross-level interaction using the unstandardized coefficients. Be sure to convert the predicted values to probabilities.
- When graphing the interaction, you can drop the race effects because they are not significant.

REFERENCES

Cohen, J., Cohen, P., West, S., & Aiken, L. S. (2003). *Applied multiple regression/correlation analysis for the behavioral sciences*. Mahwah, NJ: Lawrence Erlbaum.

Osborne, J. W. (2000). Advantages of hierarchical linear modeling. *Practical Assessment, Research & Evaluation, 7*(1), 1–3.

Osborne, J. W. (2008). A brief introduction to hierarchical linear modeling. In J. W. Osborne (Ed.), *Best practices in quantitative methods*. Thousand Oaks, CA: SAGE.

Osborne, J. W. (2015). *Best practices in logistic regression*. Thousand Oaks, CA: SAGE.

Raudenbush, S. W., & Bryk, A. S. (2002). *Hierarchical linear models: Applications and data analysis methods* (Vol. 1). Thousand Oaks, CA: SAGE.

MISSING DATA IN LINEAR MODELING

14

Advance Organizer

Every study has the potential for incomplete or missing data. Missing data—the absence of an answer or response where one was expected—can occur for many reasons. For example, participants can fail to respond to questions (legitimately or illegitimately—more on that later), equipment and data collecting or recording mechanisms can malfunction, subjects can withdraw from studies before they are completed, and data entry errors can occur. Data cleaning can also create missingness, because data points that are deemed outside the bounds of reasonable range (or inappropriately influential) can be deleted.

The issue with missingness is that nearly all classic and modern statistical techniques (including the various types of linear modeling we are discussing in this book) require complete data, and most common statistical computing packages default to some of the least desirable options for dealing with missing data: deletion of the case from the analysis. This approach is also called "complete case analysis" and has been the standard response to missing data for most researchers across most fields. In my book on data cleaning (Osborne, 2013) I made the argument that complete case analysis can be a suboptimal choice in the modern age of statistical computing in which we have ready access to more desirable methods for dealing with missing data, such as single and multiple imputation. The poorest option is what was at one time considered progressive—mean substitution.[1] The evidence regarding mean substitution and complete case analysis is compelling.

Vach and Blettner (1995) present compelling examples of complete case analysis introducing significant bias into the analysis when individuals with complete cases are a biased subsample of all possible or present participants. As Greenland and Finkle (1995) point out, even when complete cases are a representative subsample of the overall sample, the loss of participants can cause increases in variance and standard errors (SEs), leading to misestimation or Type II errors. No modern author writing

---

[1] Sadly, this appears to still be the case. In a recent survey of top-tier journals, authors seemed to ignore missingness (defaulting to complete case analysis) most of the time, but when they deviated from this, it was almost always to use mean substitution.

on missing data is a proponent of mean substitution, which can cause significant misestimation and errors of inference because it leads to substantially reduced *SE*s and underestimated variance. In my book on data cleaning, I used simulations from existing data sets empirically demonstrating that mean substitution is just not acceptable as a best practice in the 21st century.

Modern scholars point to advanced techniques such as regression imputation (using present data and a multiple regression framework to estimate what the missing values are likely to be) and multiple imputation (in which multiple copies of the data set are generated with varying values inserted for the missing data, and then all data sets are analyzed and likely estimates of the parameters are produced), which are valuable tools in the analyst's toolbox. Although imputation is relatively simple and can easily be performed without special software, multiple imputation is not implemented in all statistical software packages (or is only available via costly add-ons; for example, SAS implements multiple imputation nicely, but researchers are required to purchase an expensive missing data module in SPSS).

The ideal situation is to minimize exposure to the problem by attempting to reduce the opportunity for missing data. Where that fails, my argument here is that all researchers should examine their data for missingness, and researchers wanting the best (i.e., the most replicable and generalizable) results from their research need to be prepared to deal with missing data in the most appropriate and productive way possible. Through the course of this chapter, we will utilize real data to explore the benefits of appropriately handling missing data in two common linear models: ordinary least squares (OLS) regression and logistic regression. The lessons learned will also generalize to other linear models.

In this chapter, we will cover

- A basic overview of missing data concepts
- Legitimate missingness versus illegitimate missingness
- Determining whether data are missing completely at random (MCAR), missing at random (MAR), or missing not at random (MNAR)
- Comparison of traditional and modern techniques for dealing with missing data

  o Complete case analysis (listwise deletion)
  o Mean substitution
  o Regression imputation
  o Multiple imputation

- Why you might want to examine missingness as an interesting variable itself

Guidance on how to perform these analyses in various statistical packages will be available online at study.sagepub.com/osbornerlm.

Not All Missing Data Are the Same

The issue before us is whether we have complete data from all research participants on all variables (at all possible time points, if it is a repeated-measures design). If any data on any variable from any participant (at any time point) are not present, the researcher is dealing with missing or incomplete data. In many types of research, it is the case that there can be *legitimate missing data,* or an absence of data when it is appropriate for there to be an absence. This can come in many forms, for many reasons. Imagine you are filling out a survey that asks you whether you are in a committed romantic relationship, and if so, how long you have been in this particular committed relationship. If you answer "no" to the first question (indicating you are not currently in a committed romantic relationship), it is legitimate for you to skip the follow-up question on how long you have been in that relationship. If a survey asks you whether you voted in the last election, and if so, what party the candidate was from, it is legitimate to skip the second part if you did not vote in the last election. If a survey asks if you are a smoker, and you answer no, it is legitimate for you to skip the next questions on whether you smoke cigarettes, pipes, cigars, or hookahs and how often. In a long-term study of people receiving a particular type of treatment, if you are no longer receiving treatment because you are cured, that might be a legitimate form of missing data. Or perhaps you are following employee satisfaction at a company. If an employee leaves the company (and thus is no longer an employee), to me it seems legitimate that this person should no longer be responding to employee satisfaction questionnaires.

Utility of Legitimately Missing Data for Data Checking

Large data sets, especially government data sets or large survey studies with complex skip patterns, are full of legitimately missing data, and researchers need to be thoughtful about handling this issue appropriately. Also note that even in the case of legitimate missingness, missingness is meaningful. Missingness in this context informs and reinforces the status of a particular individual and can even provide an opportunity for checking the validity of an individual's responses. In cleaning the data from a survey on adolescent health risk behaviors many years ago, I came across some individuals who indicated on one question that they had never used illegal drugs, but later in the questionnaire, when asked how many times they had used marijuana, they answered that question indicating a number greater than 0. Thus, what should have been a question that was legitimately skipped was answered with an unexpected number. What could this mean? One possibility is that the respondent was not paying attention to the questions and answered carelessly or in error. Another possibility is that the initial answer ("Have you ever used illegal drugs?") was answered inaccurately. Finally, it is possible that some subset of the population did not include marijuana in the category of "illegal drugs"—an interesting finding in itself and one way in which researchers can use data cleaning to improve their subsequent research or at least raise interesting questions.

There are different ways to deal with legitimate missing data. Survival analysis is particularly good at dealing with some types of missing data.[2] Another (perhaps more common) method of dealing with this sort of legitimate missing data is adjusting the

---

2 Which can deal with issues like participants leaving the study (right-censored or truncated data) or entering the study at a particular point (left-censored or truncated data).

denominator. Again taking the example of the relationship survey, we could eliminate those not in relationships from the particular analysis looking at length of relationship (in this case, the question is appropriate only to those with complete data on that item), but we would leave all respondents in the analysis when looking at issues relating to being in a relationship or not. Thus, instead of asking a slightly silly question of the data ("How long, on average, do all people, even people not in a relationship, stay in a relationship?"), we can ask two more refined questions: "What are the predictors of whether someone is currently in a committed relationship?" and "Of those who are currently in a committed relationship, how long on average have they been in that relationship?" In this case, it makes no sense to include those not in a relationship when asking questions about how long someone has been in a relationship.

Illegitimately missing data are relatively common in all types of research. Sensors fail or become miscalibrated, leaving researchers without data until that sensor is replaced or recalibrated. Research participants choose to skip questions on surveys that the researchers expect everyone to answer. Participants drop out of studies before they are complete. At times, sampling frames purposefully administer select items to subsamples to help control length of interviews or for other purposes. Few authors seem to appropriately deal with the issue of illegitimately missing data, despite its obvious potential to substantially skew the results (Cole, 2008). For example, in a recent survey my students and I performed in highly regarded journals from a variety of fields (Osborne, Kocher, & Tillman, 2012), we found that between 30% and 50% of articles discussed the issue of missing data with surprisingly little variation across fields. Of those researchers who did report something to do with missing data, most reported having used the classic but anachronistic methods of listwise deletion (complete case analysis) or mean substitution, neither of which is a particularly effective practice (Schafer & Graham, 2002). In only a few cases did researchers report doing anything constructive with the missing data, such as imputation. And in no case did we find that researchers analyzed the missingness to determine whether it was MCAR, MAR, or MNAR. As I suggested in my book on data cleaning (Osborne, 2013), this suggests to me that there is a mythology in quantitative research that (a) individuals with incomplete data cannot contribute to the analyses, and (b) removing them from the analyses is an innocuous action. I hope to convince you that neither of these points is supported in the literature or by empirical evidence, and that it is relatively simple to make your logistic regression analysis more methodologically sound by appropriately dealing with missing data.

Categories of Missingness:
Why Do We Care If Data Are MCAR or Not?

When exploring missing data, it is important to come to a conclusion about the *mechanism of missingness*—in other words, the hypothesized reason for why data are missing. This can range from arbitrary or random influences to purposeful patterns of nonresponse (e.g., most women in a study refuse to answer a question that is offensive or sensitive to women but that does not affect men in the same way).

If we can infer the data are MCAR/MAR, then the nonresponse is deemed *ignorable*. In other words, random missingness can be problematic from a power perspective (in that it often reduces sample size or degrees of freedom for an analysis), but it does not potentially bias the results. However, data that are MNAR could potentially be a strong biasing influence (Osborne, 2013; Rubin, 1976).

Let us take an example of an employee satisfaction survey given to schoolteachers in a local district as an example of MCAR, MAR, and MNAR. Imagine that all teachers are surveyed in September (X) (near the beginning of the traditional school year in the United States), and then teachers are surveyed again in January (Y). MCAR would mean that missingness in January is completely unrelated to any variable, including September satisfaction level, age, years of teaching, satisfaction level at follow-up, and so forth. At times, missingness of this type is a function of the study design. For example, because of a lack of resources, the survey designers could have decided that only 50% of all respondents from September would be randomly sampled to respond to the survey again in January, with all potential respondents completing surveys at both time points. In this case, having data for Y present or absent is completely explained by random selection (if all selected respond to the survey). In other words, missingness has no systematic relation to any variable present or unmeasured (e.g., age, sex, race, level of satisfaction, years teaching, etc.) and is thus ignorable. You as a researcher can simply select cases with data at both time points without worry that your data are biased—*if* all other important conditions are met (namely, a 100% response rate at all time points) and if you can afford to lose the power associated with all of those lost cases.

Now imagine that this surveying was part of the school district's initiative to keep teachers from leaving, and they wanted to focus on teachers with low satisfaction in September, perhaps with an intervention to help raise satisfaction of these low-satisfaction teachers. In this case, the missingness depends solely and completely on X, the initial score. Because the goal of the survey is to explore how these particular teachers fared, rather than all teachers in general, missingness is still considered *ignorable and MAR*.[3] If, on the other hand, other factors aside from initial satisfaction level were responsible (or partly responsible) for missingness, such that perhaps only teachers whose satisfaction had improved responded (the teachers who continued to be substantially dissatisfied may be less likely to return the survey), then the data are considered *MNAR* and are *not ignorable* (Rubin, 1976; Schafer & Graham, 2002) because they may substantially bias the results. In the case of MNAR, the average satisfaction of the follow-up group would be expected to be inflated if those who were most dissatisfied had stopped responding. If missingness were related to another external factor, such as if those teachers who were most dissatisfied were the most junior teachers (the teachers with least time in the profession), that would also qualify the missing data as MNAR.

---

3   As with many things in statistics, the implications of MAR missingness being *ignorable* has been misapplied. MAR missingness is a very particular type of missingness and is only ignorable for certain types of analyses (e.g., Heitjan & Basu, 1996). I think that most researchers are best served by not ignoring any type of missingness, even MCAR.

In other words, it is only legitimate to assume that your observed data are representative of the intended population if data are convincingly MAR or MCAR.[4] For simplicity, I will proceed through the rest of the chapter focusing on MCAR versus MNAR. MAR (ignorable missingness) is probably more common than MCAR but MNAR is probably most common, and thus MCAR is merely presented as a comparison point. In truth, best practices in handling missing data appear to be equally effective regardless of whether the data are MCAR, MAR, or MNAR.

How Do You Know If Your Data Are MCAR, MAR, or MNAR?

It is important that a researcher investigates why missing data are present when this is not expected. There are many methods suggested in the literature, but one of my favorite methods is to utilize binary logistic regression for the purpose of this investigation (regardless of what type of analysis I am ultimately going to perform for my hypothesis testing). One of the methods for testing whether missingness is MCAR, MAR, or MNAR is testing whether missingness is related to other variables.

Quantifying missingness is relatively easy. In most modern statistical programs, there is a system missing value that does not appear in analysis but can be manipulated. For example, we can recode system missing values into real values in SPSS. Using data from the National Health Interview Survey of 2010 (NHIS2010)[5] as an example, 952 cases (of 27,731) have missing values on the body mass index (BMI) variable we have previously used in this book. If I want to explore missingness on this variable, I can create a dummy variable that represents whether each case has missing data or complete data on that variable:

```
Recode BMI (sysmis = 1) (else = 0) into MISSING_BMI.
Execute.
```

In SPSS, SYSMIS allows us to manipulate data that have system missing values. Using this simple recode command, I created a new variable called MISSING_BMI that is coded 1 if a case had a missing value on BMI, and 0 if a case did not.[6]

Once I have a dummy variable that captures missingness on a particular variable, I can construct a logistic regression equation predicting missingness from other variables in the data set. My hope is that none are significantly related to missingness (strengthening my view that missingness might be random). The frequency

---

4   Once again, let us be clear that values that are "out of scope" or legitimately missing, such as nonsmokers who skip the question concerning how many cigarettes are smoked a day, are not considered "missing" and are not an issue (Schafer & Graham, 2002). In this example, let us imagine that nonclassroom teachers (e.g., guidance counselors, teacher assistants, or other personnel) who took the initial survey were not included in the follow-up because they are not the population of interest, classroom teachers. These would be legitimate missing data.

5   Public data are available from http://www.cdc.gov/nchs/nhis/nhis_2010_data_release.htm.

6   Other software packages have similar ways of dummy-coding missingness.

Table 14.1a   Frequency Distribution of MISSING_BMI

		Frequency	Percent	Valid Percent	Cumulative Percent
	.00	26,779	96.6	96.6	96.6
Valid	1.00	952	3.4	3.4	100.0
	Total	27,731	100.0	100.0	

SOURCE: National Health Interview Survey of 2010 (NHIS2010) from the National Center for Health Statistics (http://www.cdc.gov/nchs/nhis/nhis_2010_data_release.htm).

distribution of MISSING_BMI confirms that my recode worked, because I have the expected number of cases with 1 and 0 (Table 14.1a).

Once we have a dummy variable indicating missingness on a particular variable, we need to test three different things to evaluate the type of missingness mechanism:

a.   That missingness is not related to the dependent variable (DV) of interest;

b.   That missingness is not related to the independent variable (IV) or any other important variable in the data set—even if it is not being used in this particular analysis; and

c.   That when all variables are used to predict missingness, they collectively do not account for significant model fit improvement (as measured by the change in -2 log likelihood [$\Delta-2LL$] as we have done in previous chapters).

If all three of these conditions are met, and it is reasonable to argue that you have measured all important variables that *could be related to missingness on this variable*, then you can conclude that missingness is completely at random. Furthermore, it must be reasonable that there was enough power to detect a significant effect if indeed it existed. Because we have more than 27,000 cases, this is a reasonable assertion and non-significance can be taken seriously. However, if we only had 75 cases in our data, establishing no relationship also requires looking at the odds ratios (ORs) and regression coefficients and arguing they are small enough to be ignorable. A reasonably sized effect should have a reasonable probability of being detected for this type of analysis to matter.

The first test was conducted as a simple logistic regression analysis predicting MISSING_BMI from DIABETES, which was not significant. The other analyses can be combined into a single logistic regression equation, in which I include as many other variables with as little missing data as possible (fortunately, in this example, many of the other variables have mostly complete data). A logistic regression analysis predicted MISSING_BMI from other available variables I could think of that had no missing data: AGE, DIABETES (0 = no, 1 = yes), currently MARRIED (0 = no, 1 = yes), SEX (1 = male, 2 = female), and the original RACE variable (coded into six categories). Both the second and third conditions were failed in this example. The $\Delta-2LL$ was 217.05 ($p < .0001$) when these variables were entered into the equation, leading to the conclusion that there are significant relationships between one or more of these variables and missingness on BMI. As you can see in Table 14.1b, there is a slight trend with AGE, indicating that older respondents are more likely to be missing BMI than younger respondents. Overall, RACE was not significant, but one of the contrasts was

Table 14.1b   Variables Predicting Missingness in the BMI Variable

		B	SE	Wald	df	p	Exp(B)	95% CI for Exp(B)	
								Lower	Upper
	AGE	0.004	0.002	3.975	1	.046	1.004	1.000	1.007
	DIABETES	0.116	0.109	1.132	1	.287	1.123	0.907	1.390
	MARRIED	−0.013	0.068	0.038	1	.846	0.987	0.864	1.127
	SEX	−1.030	0.079	171.160	1	.000	0.357	0.306	0.416
	RACE			8.919	5	.112			
Step 1[a]	RACE(1)	−0.165	0.094	3.090	1	.079	0.848	0.705	1.019
	RACE(2)	0.265	0.312	0.718	1	.397	1.303	0.706	2.404
	RACE(3)	−0.375	0.165	5.172	1	.023	0.687	0.497	0.949
	RACE(4)	−0.620	1.012	0.375	1	.540	0.538	0.074	3.912
	RACE(5)	−0.062	0.276	0.051	1	.821	0.940	0.547	1.613
	Constant	−3.298	0.206	255.307	1	.000	0.037		

Variables in the Equation

[a]Variable(s) entered on step 1: AGE, DIABETES, MARRIED, SEX, and RACE. Race was entered as a categorical variable, dummy coded with the first category (White) as the reference group.

SOURCE: National Health Interview Survey of 2010 (NHIS2010) from the National Center for Health Statistics (http://www.cdc.gov/nchs/nhis/nhis_2010_data_release.htm).

significant. No other variable except SEX was significant, but the magnitude of the SEX effect is conclusive evidence that the data are not MCAR. The OR indicates that women are *substantially less likely* to be missing BMI than males.[7]

At times, these analyses can give insight into the reason data might be missing, but our primary motive is to establish whether the missingness mechanism can reasonably be considered random or not.

If the overall model (Δ−2LL) is significant, this is sufficient to conclude nonrandom missingness. A more sensitive measure of each variable is to look at the *variables not in the equation* statistics that some software provide. The results in Table 14.1b are unique effects, controlling for all other variables in the equation. The results in Table 14.1c are the simple relationships between missingness and each term in the model, which can be a more appropriate evaluation of whether data are MCAR or not.

As you can see, AGE, SEX, and one of the RACE contrasts were significant when considered without other variables entered into the model.

What Do We Do With Randomly Missing Data?

To illustrate some of the effects of missing data handling on linear models, I utilized data from the National Education Longitudinal Study of 1988 (NELS88) from the National Center for Education Statistics to provide an example. The sample of 5,550 students who had complete data on the relevant variables (smoking, achievement test scores, and other variables we will introduce later on) were defined, for our purposes, as our population with which we will compare the results of our examples.[8] These results are represented in the first row of Table 14.2.

---

7   Note that to be thorough, one could also model interactions as predictors of missingness.

8   Note that this sample is not the complete NELS88 sample; rather, it is a subsample of those with complete data for purposes of this example.

Table 14.1c Variables Not in the Equation

			Score	df	p
Step 0	Variables	AGE	8.848	1	.003
		DIABETES	2.145	1	.143
		MARRIED	1.079	1	.299
		SEX	186.998	1	.000
		RACE	7.860	5	.164
		RACE(1)	0.778	1	.378
		RACE(2)	0.808	1	.369
		RACE(3)	5.347	1	.021
		RACE(4)	0.383	1	.536
		RACE(5)	0.049	1	.825
	Overall statistics		203.185	9	.000

SOURCE: National Health Interview Survey of 2010 (NHIS2010) from the National Center for Health Statistics (http://www.cdc.gov/nchs/nhis/nhis_2010_data_release.htm).

Table 14.2 Comparison of "Population" With a Sample That Randomly Lost 25% to 75% of Data Predicting Smoking Status From Student Achievement

		B	SE	Wald	df	p	Exp(B)	95% CI for Exp(B)	
								Lower	Upper
Full sample	zACH	0.317	0.058	30.150	1	.000	1.374	1.226	1.538
N = 5,550	Constant	−2.722	0.058	2,199.538	1	.000	0.066		
25% missing	zACH	0.300	0.066	20.387	1	.000	1.350	1.185	1.538
N = 4,116	Constant	−2.705	0.067	1,634.233	1	.000	0.067		
50% missing	zACH	0.330	0.084	15.333	1	.000	1.391	1.179	1.641
N = 2,853	Constant	−2.822	0.085	1,104.232	1	.000	0.059		
75% missing	zACH	0.329	0.112	8.560	1	.003	1.389	1.115	1.731
N = 1,387	Constant	−2.686	0.114	553.251	1	.000	0.068		
75% missing Mean substitution	zACH	0.330	0.109	9.111	1	.003	1.390	1.123	1.722
	Constant	−2.690	0.057	2,237.091	1	.000	0.068		
75% missing Weak imputation	zACH	0.336	0.080	17.487	1	.000	1.400	1.196	1.638
	Constant	−2.704	0.058	2,201.363	1	.000	0.067		
75% missing Strong imputation	zACH	0.289	0.066	18.946	1	.000	1.336	1.172	1.522
	Constant	−2.701	0.057	2,244.825	1	.000	0.067		
75% missing Multiple imputation	zACH	0.306	0.074	17.205	1	.000	1.358	1.175	1.569
	Constant	−2.698	0.057	2,248.827	1	.000	0.067		

SOURCE: National Education Longitudinal Study of 1988 (NELS88) from the National Center for Education Statistics (http://nces.ed.gov/surveys/nels88/).

Data MCAR

To simulate MCAR situations, I created a random yes/no variable for each student in order to randomly assign them 0 or 1. To do this, I used a Mersenne Twister random number generator and a random variable Bernoulli probability generator with either a 0.25, 0.50, or 0.75 probability. This produced three dummy variables that identified 25%, 50%, or 75% of the sample for deletion from the analysis, thus simulating a situation no researcher wants to see—that a sizable portion of the sample has been potentially lost as missing data. As a confirmation of the randomness of the missingness, two analyses were performed. First, there was no relationship between the DV and the missingness variable (similar to that created in the NHIS2010 example above) in any of the three analyses. Second, there was no correlation between the missingness variable and the IV or any other substantive variable in the data set. Finally, a logistic regression predicting missingness (0 = not, 1 = missing) from SMOKING, zACH, and other predictors we will introduce later was not significant overall. These results confirm that the goal of completely random missing data was achieved.

The assertion in the literature is that data MCAR should have no substantial effect on the results (aside from loss of power and increased SE due to smaller sample size). Indeed, comparing the results in Table 14.2, this is confirmed. Loss of either 25% or 50% or even 75% of the sample in this random way resulted in no meaningful change in the results. The slope (and ORs) for student achievement is significant at the same level and the coefficients are reasonably similar. The SEs become predictably larger as the percentage of the sample that is missing increases because larger samples have smaller SEs and, correspondingly, the 95% confidence intervals (CIs) for the ORs become slightly wider as missingness increases. Wald statistics also scale with sample size, shrinking as the sample shrinks. I think one would be hard pressed to claim that loss of even 75% of the sample in a completely random way influenced these results appreciably for this analysis. Thus, in the case of data MCAR (and *only* in the case of MCAR), it seems that complete case analysis does not substantially bias the results, which is what authors have said in the literature for decades. But what if an enterprising researcher decided to attempt to recapture the missing data?

Mean Substitution

As mentioned above, this is a discredited method of dealing with missing data, for reasons you will quickly see. To examine the effects of this method of dealing with missing data, the sample of 75% missing was "recaptured" by substituting the mean of the existing 25% for each missing value. This had the effect of very slightly altering the mean (from 0.10 in the original variable to 0.09 in the mean-substituted variable); more important, it halved the standard deviation (*SD*; from 0.98 in the original variable to 0.49 in the mean-substituted variable). Interestingly, as you can see in Table 14.2, the problems introduced by the mean substitution do not substantially bias these results. Indeed, the primary issue that mean substitution introduces is creating a highly kurtotic variable. Because logistic regression is nonparametric, this does not lead to substantial bias in this example. However, the results can be more disruptive in other examples involving parametric assumptions.

Yet careful examination of this row in Table 14.2 shows that the mean substitution did not give the analysis much in the way of power or precision over the 75% missing analysis. Indeed, comparing this with the 75% missing analysis and the 0% missing analysis, the mean-substituted analysis is almost identical to the 75% missing analysis compared with the full sample analysis: the 95% CIs, Wald statistics, and *SEs* are all substantially different than the initial analysis with complete data. Thus, we cannot conclude that mean substitution did anything positive for the analysis in the case of data MCAR. Furthermore, remember that this is a best-case scenario in that the data are truly MAR, which is not often the case.

Strong and Weak Regression Imputation

Many authors have argued that using imputation (predicting scores via multiple regression modeling) is superior to mean substitution or complete case analysis, particularly when data are MAR. In the case of these data, which are MCAR, we will still introduce these two possibilities, which will become more useful later in the chapter. In this case, we will impute the missing cases using *strong imputation* (in this case, a highly correlated variable) or *weak imputation* (in this case, two variables that are more weakly related to the variable with the missing data, eighth-grade student achievement). The strong imputation will be performed by utilizing tenth-grade student achievement, which is correlated $r = 0.82$ with eighth-grade achievement. The weak imputation will use two less closely related variables, socioeconomic status (zSES) and grades (GRADES; 1 = mostly F's to 5 = mostly A's). These latter two variables have a multiple correlation $R = 0.58$. Thus, in the strong imputation example, we are using a variable that shares 67.24% of the variance of the target variable; and in the weak imputation case, we have variables that share 33.70% of the variance.

Looking first at weak regression imputation, an OLS regression predicting zACH with 75% MCAR from zSES and eighth-grade GRADES produced the following regression equation in Equation 14.1:

$$\hat{Y} = -2.954 + 0.264(\text{zSES}) + 0.719(\text{GRADES}) \tag{14.1}$$

Using this analysis to predict values in the DV when missing produced a variable that is strongly correlated with the original variable prior to random deletion of cases ($r = 0.71$, $p < .0001$).[9] As you can see in Table 14.2, weak imputation with 75% data MAR (admittedly a relatively esoteric case) produced results that slightly overestimated the

---

9   Some authors have argued that this overfits the data, because only values on the regression line are estimated, whereas values vary around the regression line in real data. Thus, some have suggested that adding a small random error to each prediction more closely simulates real-world data and prevents overfitting. However, how large an error to add is open to debate. I do not add this slight modification to this example, although it is something that should be considered, particularly if the predictors are very strongly related to the variable being imputed.

magnitude of the effect but had *SEs* and CIs that were closer to the full sample than the 50% MCAR sample, complete case analysis, or mean substitution. In other words, the weak imputation did not completely recapture the results when all 5,550 students had complete data, but it was superior to either of the other options, particularly in providing the power to detect an effect (e.g., Wald).

Looking at stronger imputation, predicting zACH with 75% MCAR from tenth-grade achievement test scores (ACH10) produced the regression equation in Equation 14.2:

$$\hat{Y} = -5.118 + 0.092(ACH10) \qquad (14.2)$$

Creating imputed values based on this equation for those without valid scores (the 75% of cases with randomly missing data) produced a variable that is strongly correlated with the original variable prior to random deletion of cases ($r = 0.86$, $p < .0001$). As you can see in Table 14.2, strong imputation with 75% data MAR produced results that were better than either complete case analysis or mean substitution or weak imputation—with statistics, *SEs*, and CIs that were much closer to the full sample than complete case analysis, mean substitution, or weak imputation. The point estimate was not closer, but all point estimates were reasonably close to begin with (all ORs with 0.04 of the full sample). In other words, the strong imputation did not completely recapture the results when all 5,550 students had complete data, but it is arguably the best outcome thus far, particularly in terms of power.

Both of these imputation procedures had laudable outcomes but obviously would be more efficient with a lower percentage of data missing.

Multiple Imputation (Bayesian)

Using AMOS software from IBM/SPSS (although many software packages, including SAS, can perform multiple imputation), I used Bayesian estimation to produce the recommended minimum of 20 parallel data sets using the *weak imputation* variables of eighth-grade grades and family SES. The workings of multiple imputation are explored elsewhere in detail, including in my book on data cleaning (Osborne, 2013), and are too lengthy to detail here. Briefly, multiple imputation creates multiple parallel data sets that impute different values for the missing values to give the researcher a theoretical range of outcomes, rather than a single outcome. This has certain benefits, including being able to estimate the stability or volatility of the results of imputation, as well as estimating a reasonable CI within which the actual results should fall.

The results of this analysis, however, are illustrative. As Table 14.2 shows in the final row, multiple imputation even with relatively modest predictors produced superior outcomes over regression imputation with weak predictors, or even regression imputation with a strong predictor if the totality of the results is considered. Most notably, the 95% CI for the multiple imputation analysis is closer to that of the full sample than either simple imputation with "weak" predictors or even (in this case) with a single strong predictor. The *SEs* are better than the original weak imputation as well, and the point estimate is also reasonably close to the full sample.

Summary

This first section explores a relatively rare situation in which data are truly MAR. In this case, even large portions of missing data can produce relatively accurate point estimates, although handling the missing data through regression imputation or multiple imputation seems to produce better results overall. Admittedly, this is splitting hairs between pretty good and very good results from an extremely bad situation. In the next section, we will explore what happens more commonly—when data are *not* MAR. In this case, I suspect you will see more profound benefits to dealing with missing data.

Data MNAR

To explore this more common issue of data MNAR, I created three types of nonrandom missingness: MNAR-low, MNAR-high, and MNAR-smoke. For the first two examples, students were grouped into low achieving (scoring at or below the 25th percentile on the standardized eighth-grade composite, BY2XCOMP), average achieving (between the 25th and 75th percentiles), and high achieving (above the 75th percentile). All analyses were predicting whether a student reported having smoked tobacco or not during high school.

A random Bernoulli generator individually identified individuals within each of those groups as missing or not.[10] Because the outcome was smoking, which was a small population, we randomly sampled at least 40% of students admitting to smoking across all achievement groups to ensure that the group of interest was not randomly removed from the data.

For MNAR-low (an imaginary scenario in which low-achieving students were least likely to respond), 95% of low-scoring students, 75% of average-scoring students, and 1% of high-scoring students assigned to be missing. For MNAR-high (an imaginary scenario in which high-achieving students were least likely to respond), 1% of low-scoring students, 75% of average-scoring students, and 95% of high-scoring students were assigned to be missing. This created highly skewed samples on achievement, representing perhaps a worst-case scenario in representative sampling. As a result of these schemes, about 59% of subjects were selected to be missing in the MNAR-low and MNAR-high conditions. Of course, there are infinite ways in which missingness can occur, and these are only intended to be examples, and rather extreme ones at that. We will use the first row of Table 14.3 as the "gold standard" or target value of the "population" we are exploring, and see how nonrandom missingness can influence conclusions. More important, we will explore whether any of these methodologies for dealing with missingness can at least partially repair the damage done by nonrandom missingness.

Example 1. Nonrandom Missingness Reverses the Effect

In this first example, nonrandom sampling leads to a conclusion that is significant and in the *opposite direction* of the effect when all data are present ("population"

---

10   In the case of MNAR-low and MNAR-high, smokers were privileged with lower missingness rates as they were such a small portion of this fictitious population to begin with (6.6%).

Table 14.3 Effectiveness of Various Missing Data Handling Strategies When Data Are Missing Not at Random

		B	SE	Wald	df	p	Exp(B)	95% CI for Exp(B) Lower	Upper
Full sample N = 5,550	zACH	0.317	0.058	30.150	1	.000	1.374	1.226	1.538
Missing not at random: low-scoring students more likely to be missing									
MNAR-low	zACH	−0.307	0.081	14.189	1	.000	0.736	0.627	0.863
Mean substitution	zACH	−0.441	0.094	22.098	1	.000	0.643	0.535	0.773
Reg-weak	zACH	−0.009	0.086	0.012	1	.914	0.991	0.838	1.172
Reg-strong	zACH	0.178	0.067	6.981	1	.008	1.195	1.047	1.363
Multiple imputation	zACH	0.039	0.083	0.223	1	.637	1.040	0.883	1.225
Missing not at random: high-scoring students more likely to be missing									
MNAR-high	zACH	1.148	0.088	168.749	1	.000	3.153	2.652	3.750
Mean substitution	zACH	1.113	0.074	228.284	1	.000	3.045	2.635	3.518
Reg-weak	zACH	0.860	0.080	116.522	1	.000	2.364	2.022	2.763
Reg-strong	zACH	0.635	0.077	68.056	1	.000	1.887	1.623	2.194
Multiple imputation	zACH	0.832	0.078	113.145	1	.000	2.298	1.971	2.678

SOURCE: National Education Longitudinal Study of 1988 (NELS88) from the National Center for Education Statistics (http://nces.ed.gov/surveys/nels88/).

NOTE: Mean substitution, missing data replaced by mean score; Reg-weak, regression imputation with two weak predictors; Reg-strong, regression imputation with a strong predictor.

coefficient: $b = 0.317$, $SE = 0.058$, Wald $= 30.15$, OR $= 1.37$ [CI of 1.23, 1.54]; in MNAR-low: $b = -0.307$, $SE = 0.081$, Wald $= 14.89$, OR $= 0.7437$ [CI of 0.63, 0.86]). Not only is this a significant finding in the incorrect direction (leading an unsuspecting researcher to an error of inference), but it is so wildly inaccurate that the 95% CI *does not include the population OR*. This is important, because when data were MCAR, even massive missingness still kept the population OR within the 95% CIs, thus somewhat protecting the researcher who correctly utilizes CIs to avoid serious mistakes in inference. In this situation, where data are MNAR, the researcher is now vulnerable to serious errors of inference.

This first example shows how powerful nonrandom missingness can be in biasing results.[11] This is not surprising, but it does give one pause when considering what these examples imply for the state of the results reported in our journals.

Missing data literature would suggest that addressing missingness could help the problem. However, the effects of mean substitution are that this enhances the error, increasing the magnitude of the effect in the wrong direction. It also increases the *SE* over complete case analysis. Neither of these is desirable. Regression imputation with moderately weak predictors (the same zSES and GRADES predictors as in the prior example) helps the situation, moving the effect back toward the original value, but leaves the effect nonsignificant. In this case, the researcher would find a neutral, rather than inverse, result. Strong imputation (again using the tenth-grade achievement variable for strong imputation) was the best at repairing the damage. After this repair strategy, we are left with a much weaker effect (but in the correct direction and significant) and with a *SE* much closer to the original analysis. Thus, the researcher would be drawing conclusions that match the underlying "population" parameters. Multiple imputation with the moderately weak predictors was about as good as regression imputation in this example, but it was still highly preferable to complete case analysis or mean substitution.

Example 2. Nonrandom Missingness Dramatically Inflates the Effect

Similar results were observed for the MNAR-high example. This scenario produced a strongly inflated effect in the same direction as the "population" ($b = 1.15$, $SE = 0.088$, Wald = 168.75, OR = 3.15 [CI of 2.65, 3.75]). Again in this case, the CI does not come remotely close to including the population OR of 1.37. In this example, the researcher would be drawing a conclusion that matches the population direction, but the effect size is wildly exaggerated. In the population, a 1 *SD* increase in achievement is roughly equal to a 37% increase in the odds of smoking. Under MNAR-high, that effect is approximately 8 times larger. As with the previous example, mean substitution did little to repair the damage, and regression imputation and multiple imputation moved the effect sizes back closer to the original "population" effect sizes. No repair strategy, however, fixed a missing data problem as severe as presented in this example.

Summary

These examples show the risk of using complete case analysis when data are not MCAR (for more examples of this effect, see Shafer & Graham, 2002, Table 2).

---

11   I think we should propose that this type of error of inference be called a Type III error. It is an error that involves correctly rejecting the null hypothesis but asserting that the effect is in the exact opposite direction from the population due to sample bias, nonrandom missingness, or other methodological issues. This type of error can lead to decades of debate in the literature and can be destructive.

The important thing to remember is that as a researcher, one does not generally have access to the true population values and thus has no idea if missingness is leading to a serious (Type III) error of inference (as in MNAR-low) or a wild misestimation of effect size (as in MNAR-high). Thus, I and many others have argued that it is important to deal responsibly with missing data.

The lessons to take away are that complete case analysis is only an innocuous practice when (a) the number of cases with missing data is a small percentage of the overall sample, (b) the data are *demonstrably* MAR, and (c) power in the remaining cases is sufficient to detect a reasonable range of effects. Mean substitution does not appear to be a defensible repair strategy, and imputation is only as effective as the predictors are good (i.e., closely correlated with the variable experiencing missingness).

These examples represent extremely challenging missingness issues often not faced by average researchers, but it should be comforting to know that appropriately handling missing data even in extremely unfortunate cases can still produce desirable (accurate, reproducible) outcomes. Unfortunately, no technique can completely recapture the population parameters when there are such high rates of missingness, and in such a dramatically biased fashion. However, these techniques would at least keep you, as a researcher, on safe ground concerning the goodness of inferences you would draw from the results.

How Missingness Can Be an Interesting Variable in and of Itself

Missing data are often viewed as lost, an unfilled gap, although as I have demonstrated in this chapter, they are not always completely lost, given the availability of other strongly correlated variables. Going one step further, missingness can be considered an outcome in itself and can be an interesting variable to explore in some cases. There is information in missingness. The act of refusing to respond or responding in and of itself might be of interest to researchers, just as quitting a job or remaining at a job can be an interesting variable. Aside from attempting to determine whether the data are MCAR, MAR, or MNAR, these data could yield important information. Imagine two educational interventions designed to improve student achievement, and further imagine that there is much higher dropout in one condition than in the other condition, and further that the students dropping out are those with the poorest performance. Not only is that important information for interpreting the results (because the differential dropout would artificially bias the results), but it might give insight into the intervention itself. Is it possible that the intervention with a strong dropout rate among those most at risk indicates that the intervention is not supporting those students well enough? Is it possible that intervention is alienating the students in some way, or might be inappropriate for struggling students? All of this could be important information for researchers and policymakers, but many researchers discard this potentially important information. Remember, you (or someone) worked hard to obtain your data. Don't discard anything that might be useful!

Summing Up: Benefits of Appropriately Handling Missing Data

There are some very good books by some very smart people dealing solely with missing data (e.g., Little & Rubin, 1987; Schafer, 1997), and I had no wish to replicate that work here. The goal of this chapter was to convince you, the researcher, that this is a topic worthy of attention, that there are good, simple ways to deal with this issue, and that effectively dealing with the issue makes your logistic regression results better. Logistic regression is not immune from the effects of nonrandomly missing data that can bias your sample and results. Complete case analysis and mean substitution are not defensibly considered best practices anymore. Regression or multiple imputation can minimize the damage missing data can do to inference and estimates of population parameters, and in the age of modern computing, neither is tremendously difficult to attempt. The examples in this chapter are entirely relating to simple binary logistic regression, but the general conclusions should apply to most areas of the general linear model. Examples relating to univariate inference and simple OLS regression (simple correlation) can be found in Chapter 6 of Osborne (2013), along with an extended discussion of multiple imputation.

Because we often gather multiple related variables, we often know (or can estimate) a good deal about the missing values. Aside from examining missingness as an outcome in itself (which I strongly recommend), modern computing affords us the opportunity to fill in many of the gaps with high-quality data. This is not merely "making up data" as some fear; rather, it is the act of utilizing the vast amount of data you do have. As these limited examples show, the act of estimating values and retaining cases in your analyses most often leads to more replicable findings because they are generally closer to the actual population values than analyses that discard those with missing data (or worse, substitute means for the missing values). Thus, using best practices in handling missing data makes the results a better estimate of the population in which you are interested. It is also surprisingly easy to do, once you know how.

Thus, it is my belief that best practices in handling missing data include the following:

- First, do no harm! Use best practices and careful methodology to minimize missingness. There is no substitute for complete data,[12] and some careful forethought and planning can often save a good deal of frustration in the data analysis phase of research.
- Be transparent! Report any incidences of missing data (rates, by variable, and reasons for missingness, if possible). This can be important information to reviewers and consumers of your research and is the first step in thinking about how to effectively deal with missingness in your analyses.
- Explicitly discuss whether data are MAR or not (i.e., if there are differences between individuals with incomplete and complete data). Using analyses similar

---

12 Except in certain specialized circumstances in which researchers purposely administer selected questions to participants or use other advanced sampling techniques that have been advocated for in the researching of very sensitive topics.

to those I model in this chapter, you can give yourself and the reader a good sense of why data might be missing and whether it is at random or not. That allows you, and your audience, to think carefully about whether missingness may have introduced bias into the results. I would advocate that all authors report this information in the methods section of formal research reports.

- Discuss how you as a researcher dealt with the issue of incomplete data and the results of your intervention. A clear statement concerning this issue is simple to add to a manuscript and can be valuable for future consumers as they interpret your work. Be specific—if you used imputation, how was it done and what were the results? If you deleted the data (complete case analysis), justify why.

ENRICHMENT

1. Play with the MNAR data examples and see if you can replicate or improve on the missing data handling.

REFERENCES

Cole, J. C. (2008). How to deal with missing data. In J. W. Osborne (Ed.), *Best practices in quantitative methods.* Thousand Oaks, CA: SAGE.

Greenland, S., & Finkle, W. D. (1995). A critical look at methods for handling missing covariates in epidemiologic regression analyses. *American Journal of Epidemiology, 142*(12), 1255–1264.

Heitjan, D. F., & Basu, S. (1996). Distinguishing "missing at random" and "missing completely at random." *American Statistician, 50*(3), 207–213. doi: 10.2307/2684656

Little, R., & Rubin, D. (1987). *Statistical analysis with missing data* (1st ed.). New York, NY: Wiley.

Osborne, J. W. (2013). *Best practices in data cleaning: A complete guide to everything you need to do before and after collecting your data.* Thousand Oaks, CA: SAGE.

Osborne, J. W., Kocher, B., & Tillman, D. (2012, February). *Sweating the small stuff: Do authors in APA journals clean data or test assumptions (and should anyone care if they do)?* Paper presented at the annual meeting of the Eastern Educational Research Association, Hilton Head, SC.

Rubin, D. B. (1976). Inference and missing data. *Biometrika, 63*(3), 581–592. doi: 10.1093/biomet/63.3.581

Schafer, J. L. (1997). *Analysis of incomplete multivariate data.* Boca Raton, FL: Chapman & Hall/CRC.

Schafer, J. L., & Graham, J. W. (2002). Missing data: Our view of the state of the art. *Psychological Methods, 7*(2), 147–177. doi: 10.1037/1082-989X.7.2.147

Vach, W., & Blettner, M. (1995). Logistic regression with incompletely observed categorical covariates—investigating the sensitivity against violation of the missing at random assumption. *Statistics in Medicine, 14*(12), 1315–1329. doi: 10.1002/sim.4780141205

15

TRUSTWORTHY SCIENCE

Improving Statistical Reporting

Advance Organizer

We live in an era of unprecedented technology and science. Anyone alive during the space race and scientific advances of the 1960s and 1970s imagined a 21st century where we were living on the moon or in floating cloud cities on Earth, talking with robots, and living lengthy, healthy and happy lives, largely without problems or worry.[1] Instead, we live in a 21st century world where people question whether vaccination of children is good, whether humans have impacted the environment, and whether evolution is a "belief" or a well-supported theory.

Trust of science is low among the general population and is decreasing precipitously in some segments (Gauchat, 2012). This is ironic given that there are more cell phones (brought to us by the aforementioned amazing science that people now often mistrust) in the United States than people,[2] each of which has more computing power than a room-sized mainframe from the 1970s, and scientific advances have benefitted almost every area of our lives in recent decades.

People much more expert than I have discussed the possible reasons for this. They include highly publicized scandals involving unethical behavior by scientists, conflicts of interest and profit motives, political agendas, the publication of bizarre findings in prominent outlets,[3] and probably poor communication about the limits of real scientific inquiry within the popular media. Regardless of the source, both within and outside the scientific community, we need to move toward open and trustworthy science or we risk losing the tremendous gains we have made in recent decades. To do this, I think we need to remember the original goals of the endeavor we call "science" or "research."

---

1 For some reason, these futuristic depictions also had us all wearing unisex and unflattering jump suits. I don't think science has an explanation for that one yet.

2 As reported by Cecilia Kang in the *Washington Post* on October 11, 2011 (https://www.washingtonpost.com/blogs/post-tech/post/number-of-cell-phones-exceeds-us-population-ctia-trade-group/2011/10/11/gIQARNcEcL_blog.html).

3 One example is studies by Bem (2011) asserting a scientific observation of extrasensory perception (for a report on the failure to replicate Bem, 2011, see Ritchie, Wiseman, & French, 2012).

In my mind, our goal is to enable us to say something meaningful (and/or useful) about a given population at a given time. The purpose of research ought not to be the self-aggrandizement of the researcher, or awards of tenure or degrees. Those are important personal milestones, but they should be won in the pursuit of excellent and impactful (and valid) research. They should not be the sole purpose of performing the research.

One of the principles of science is the principle of replication—that another scientist, performing similar tasks on a similar sample using similar methods, should see similar results. Another principle is objectivity—that anyone, seeing the same phenomena, would draw the same conclusions. Yet replication and objectivity seem in short supply in many areas of science today, as I lamented briefly in Chapter 1 and as many other scholars have discussed recently (for example, see Asendorpf et al., 2013; Horton, 2015; Ioannidis, 2005; Simmons, Nelson, & Simonsohn, 2011).

A particular subcategory of replication is the issue of reproducibility—the idea that another person, given the *same data* another researcher has already analyzed, should be able to independently produce the same results (Peng, 2015). This issue has been discussed recently in computationally intensive disciplines, such as genomics (Stodden, 2011), economics (J. Ioannidis & Doucouliagos, 2013), and computational chemistry (Couchman, 2014), but this most basic test has been raised in medicine as well (Kraus, 2014). This new discussion has led to conversations about publishing null findings of high rigor and quality, new models for peer review, and open access to data and computational models that lead to published results.

One of the purposes of writing my previous book on data cleaning and testing of assumptions was to demonstrate that there are simple and defensible aspects of statistical methods that can help improve the trustworthiness (e.g., generalizability and replicability) of findings from a particular study. In presenting examples of data cleaning in parts of this book, I have attempted to encourage you as a statistician and researcher to consider these basic issues of data quality and testing of assumptions as a path to improving trustworthiness of our results. In my mind, it does no good if the data you analyze are biased or flawed to such an extent that the results fail to generalize to the population you sought to understand, or if others (perhaps with cleaner or better data) fail to replicate your results. Too many "controversies" in the sciences seem to arise from poor data or, perhaps, poor power (e.g., Tressoldi, 2012). Indeed, in Galton's (1907) seminal work on the wisdom of crowds, he saw the commonsense desirability of removing indefensible data points to improve point estimates.

Precision is another aspect of statistical methods that is often overlooked in the dissemination of findings. We often focus on *point estimates* (e.g., the slope is 0.50) without discussion of the precision of that estimate (e.g., the 95% confidence interval [CI] for the slope could be broad: 0.10, 0.90) or the volatility of that estimate (similar analyses of similar samples find wildly different results). Standard errors (*SE*s) are informative of how precise our estimates are, in that the smaller they are, the more precise our estimates are (and the narrower our CIs are). If you have very broad CIs spanning different magnitudes of effects, it is tough to assert that you have anything useful to say about the population you seek to discuss; furthermore, if you attempt to replicate your findings and get wildly different results, one has to question the usefulness of your findings from a scientific perspective.

This book cannot help you design better studies, but the goal of this chapter is to provide some tools to allow you to more honestly communicate the precision and expected replicability of your results in a larger discussion of what your results mean. In this chapter, I will briefly introduce the concept of bootstrap resampling analysis, and show how it can be used to inform you (and those you disseminate your results to) about the precision and replicability of your results.

In this chapter, we will cover

- A basic overview of power
- A basic introduction to bootstrap analysis
- Assumptions of bootstrap analysis
- How sample size can influence replicability
- Exploring whether effect size influences replicability
- Examples of bootstrap analysis of replicability of simple and complex effects

Guidance on how to perform these analyses in various statistical packages will be available online at study.sagepub.com/osbornerlm.

What Is Power, and Why Is It Important?

If your goal is to report effects that are likely to be valuable to others in your field, one of the first things we need to worry about is whether the study you are engaged in has enough power to detect the effects you expect to observe. All other things being equal, increases in sample size should increase precision and decrease the volatility (which influences replicability and/or generalizability) of parameter estimates in linear modeling. Because sample size is a continuous variable, there is no magic cutoff point (contrary to rules of thumb that many authors propose): $N = 499$ is not inadequate and $N = 501$ is not perfectly adequate. The honest advice you will glean from this chapter is that larger, representative (cleaned) samples are better than smaller or biased samples,[4] and that researchers should give readers information not only about effect sizes and hypothesis tests but also the trustworthiness of the findings.

Statistical power is the ability to correctly reject a false null hypothesis (in other words, to detect effects when indeed there are effects present) and is calculated based on a particular effect size, alpha level, sample size, and in the context of a particular analytic strategy. Authors have been discussing the issue of power for more than 60 years (see Deemer, 1947). Jacob Cohen (1962, 1988, 1992) spent many years encouraging the use of power analysis in planning research, reporting research, and interpreting results (particularly where null hypotheses are not rejected). Although it is unclear how many researchers actually calculate *a priori* power or required sample size (power calculated *before* conducting a study, contrasted with power calculated during the statistical analysis, *a posteriori* power), few report having calculated power, and

---

4   This epiphany wins the award for "most obvious piece of advice that nobody wants" because it is simultaneously obvious and self-evident while being maximally unhelpful in planning a study.

a relatively low percentage of studies meet Cohen's criterion for "acceptable" power. Only 29% of randomized experimental studies and only 44% of nonexperimental (or quasi-experimental) studies in prominent psychology journals met the criterion of having calculated power of 0.80 or higher (Osborne, 2008; Osborne, Kocher, & Tillman, 2012).[5]

Correctly Rejecting a Null Hypothesis

Power is the probability of correctly rejecting a false null hypothesis, and it is obviously important. If there really is an effect in a population (in the unknowable "reality"), a study with greater power will be more likely to reject the null hypothesis, leading the researcher to the correct conclusion—that there is a relationship between two variables (when in fact there is a relationship between the variables). Therefore, if you are testing a new drug and the drug is really having a beneficial effect on patients, power is the probability that you will detect that effect and correctly reject the null hypothesis.

Theoretically, if your power is 0.80, you will correctly reject the null hypothesis on average 80% of the time (given a particular effect size, sample size, and alpha level). Even in situations in which there is a real effect in the population, a researcher with a power level of 0.80 will fail to reject the null hypothesis 20% of the time. You may have a wonderfully effective drug that can save people from misery, disease, and death, but under this hypothetical scenario, 20% of the time you will not realize it. This is a Type II error—the failure to reject a null hypothesis when there is an effect in the unknowable "reality."

As you can imagine, this is an undesirable outcome. Although we want to be sure to avoid Type I errors, it seems to me equally troubling to fail to see effects when they are present. Fortunately, there is a simple way to minimize the probability of Type II errors—ensure that you have sufficient power to detect the expected effects. Researchers who fail to do *a priori* power analyses risk wasting resources gathering either less or more data than might be needed to test their hypotheses. If a power analysis indicates that $N = 100$ subjects would be sufficient to reliably detect a particular effect, gathering a sample of $N = 400$ is a substantial waste of resources. Likewise, if $N = 1,000$ is required to reliably detect an effect, a sample of $N = 400$ is woefully inadequate and an equal waste of effort. Complicating this is that researchers are not often able to anticipate the exact magnitude of the expected effect, particularly when complex effects such as curvilinear or interaction effects might be present.

Informing Null Results

The second way power can inform an analysis is in determining the meaning of null results through post hoc power analyses. For example, if a study that fails to reject

---

5 By the way, I personally find the "common wisdom" that power of 0.80 is adequate to be wholly *inadequate*. I doubt few of us would aspire to failing to see an important effect 20% of the time when it is there to be seen. Few of us would aspire to medical doctors incorrectly diagnosing our children 20% of the time, or for structural engineers to build bridges that stay up 80% of the time. In the modern era of research, it seems to me we can do better, and it often doesn't take a great deal of extra effort to do so.

the null hypothesis had power of 0.90 to detect anticipated or reasonable effect sizes, one can be relatively confident that failing to reject the null was the correct decision[6] and the researcher can be more confident in asserting that the null hypothesis is an accurate description of the state of affairs in the population. However, in the context of poor power, failure to detect a null hypothesis leaves ambiguity rather than clarity.

Null results are routinely underappreciated in scientific literature, yet they can be valuable if they are truly able to assert equality across groups or lack of effect. Imagine, for example, you were researching two educational interventions, both designed to achieve the same goal. One is very expensive, in either human resources or financial cost, and could be disruptive to how most classrooms run. Another is cost-effective, easy to implement, and minimally disruptive. How would you interpret null results? The fact that your innovative, expensive intervention did not outperform the cost-effective and easy intervention could be disappointing. On the other hand, there is really good news here from a societal standpoint: the cost-effective and easy intervention is as effective as the difficult and expensive intervention! If you had a good-enough sample to detect reasonable effects reliably and did not detect any, one can be more confident in asserting that both interventions are equally effective.

Is Power an Ethical Issue?

Aside from Type II error rates, low power has implications for Type I error rates in the overall body of research. The social sciences have been accused of having low power as a field for many decades, despite reviews of the literature that find good power to detect effects in most studies (e.g., Osborne, 2008). Because science relies on more than just one study, Rossi (1990) argued that the *power of a group of studies* can influence their reliability and usefulness as a whole. The argument Rossi proposed was that low power in a large group of studies can increase the proportion of Type I errors in a field.

To understand this novel perspective on the importance of power to Type I error rates, let us assume that there is a strong bias toward journal editors publishing only studies that report statistically significant findings (Dickersin & Min, 1993; Hopewell, Loudon, Clarke, Oxman, & Dickersin, 2009). Let us further imagine a field where there is very poor power, such that the average study in the field had a power of 0.20. With a standard p value of .05, 5 of every 100 studies will end up with Type I errors and get published with erroneous results. With power of 0.20, 20 of 100 studies will detect real effects that exist, whereas 80% will miss existing effects and be discarded or remain unpublished. Thus, taking this extreme situation, 20 studies with true effects will be published for every 5 that are published with false effects (Type I errors). Following Rossi's argument, that leads us to a 20% Type I error rate (5 of 25 total articles published in the field), rather than the 5% we assume from setting alpha at 0.05.

This example represents an extreme and unrealistic situation. Or does it? Cohen's (1962) survey of the top-tier psychological literature concluded these studies had power of 0.48 to detect medium-sized effects. Although that is substantially better

---

6   Particularly if you have done appropriate data cleaning and testing of assumptions.

than the 0.20 in the example above, it does produce inflated Type I error rates for the field. Again, assuming a very strong publication bias, we would have approximately 50 true effects published for every 5 false effects, which leads to a 10% Type I error rate in the field—double the rate we think it is. Although this might not sound like a serious problem, if you were being tested for a life-threatening disease, would you rather the test have a 5% chance of a false positive or a 10% chance?

Furthermore, we often have controversies in fields that lead to confusion and tremendous expenditure of resources when in fact the conflicting results may be more methodological than substantive—marginal power producing conflicting results, noncomparable samples or methods, and so forth. When power is low in a field, there can be wide variation in the replicability of a real, moderately strong effect in the population. In a simulation presented later in this chapter, small, low-power samples taken from a "population" produced wildly fluctuating estimates of the effect, leading to a Type II error almost half the time. Imagine this was the study of the effect of homework on student achievement (or exercise on lifespan). If there was a real relationship between the two variables in the population but researchers repeatedly used convenience samples that had poor power, one could easily see how controversy could develop. Half of the researchers would assert that there is no significant relationship between the two variables, and the other half would find powerful, significant effects, leaving practitioners unable to draw valid conclusions.

Thus, we are left with a situation in which statistical power is a very important concept, but reviews of power in many disciplines are discouraging. Cohen's (1962) initial survey of the *Journal of Applied Social Psychology*, a top-tier psychology journal at the time, found that power to detect a small effect in this literature was 0.18, a medium effect was 0.48, and a large effect was 0.83. In other words, unless researchers in psychology were studying phenomena with large effect sizes (which is assumed to be relatively rare in the social sciences), researchers generally had less than a 50:50 chance of detecting effects that existed—the exact situation described in the previous paragraph. Reviews of other areas (Rossi, 1990; Sedlmeier & Gigerenzer, 1989) paint a similarly bleak picture for more recent research. These reviews indicate that by the end of the 1980s, little had changed from the early 1960s regarding power.[7]

Power in Linear Models

Power in different models is calculated differently. Many statistical computing packages include methods to compute *a priori* power, and there are also stand-alone packages such as the free G*POWER.[8] Power calculations for some models (e.g., analysis of variance or ordinary least squares [OLS] regression) are relatively straightforward, whereas other models (e.g., logistic regression models) are a bit more complicated. For most cases, the primary variable that a researcher has control over is sample size

---

7 Although my review of the educational psychology literature (Osborne, 2008) indicated that observed power in some branches of the social sciences may be much better than generally assumed.

8 Downloads and documentation are available at http://www.gpower.hhu.de/en.html.

(data cleaning can also reduce error variance, increasing power). Other aspects of power analysis (e.g., as the criterion for significance [α]) and effect size are not generally under your control in the modern era.[9] In regression models, the number of the predictors in the model is a factor in calculating power; in logistic models, power calculations also take into account the relative proportions of individuals that fall into each group in the dependent variable (DV).[10]

Using G*POWER to do some simple calculations, you can see in Table 15.1 that power and sample size are closely related when other factors are held constant. This is valuable to know because this is one of the few factors under the researcher's control.

Table 15.1   Sample Sizes Needed to Achieve Power of 0.80 and 0.95 Given Small, Medium, and Large Effect Sizes

	Sample Size Needed to Achieve Power = 0.80	*Sample Size Needed to Achieve Power = 0.95*
Simple correlation/ordinary least squares regression[1]		
$\rho = 0.10$	779	1,289
$\rho = 0.30$	82	134
$\rho = 0.50$	26	42
Independent-groups t test[2]		
$d = 0.20$	788	1,302
$d = 0.50$	128	210
$d = 0.80$	52	84
Simple logistic regression[3]		
Odds ratio = 1.50	308	503
Odds ratio = 3.00	53	80
Odds ratio = 8.00	26	35

NOTE: Effect size conventions taken from Cohen (1988) where available.

[1]Calculated at $\alpha = 0.05$, two-tailed test.
[2]Calculated at $\alpha = 0.05$, two-tailed test, equal cell sizes, total sample reported.
[3]Odds ratios of 3.00 are considered important in epidemiological literature (Kraemer, 1992); thus, this was selected as a medium effect size. Odds ratios inflate quickly (and not linearly); therefore, 8.00 was selected as a large effect size (Hsieh, Bloch, & Larsen, 1998) and 1.50 was selected as a small effect size. Unfortunately, there is not agreement about what constitutes large, medium, and small effects in odds ratios. Calculations assumed two-tailed tests; no other independent variables were included in the analysis with an independent variable that was assumed to be normal.

9   During earlier parts of the 20th century, the α criterion for significance was considered on a case-by-case basis. However, as we moved into the latter part of the 20th century, it was generally considered to be fixed at 0.05. Now, of course, there are movements to do away with significance testing altogether. In situations in which hypothesis testing is used, it is generally considered fixed, which leads me to consider it not generally under the control of the researcher.

10 Power and sample size in multinomial logistic regression is a bit more complicated still.

You can also see that when effects are reasonable in magnitude, the difference in sample size needed to achieve power of 0.80 versus 0.95 is not often dramatic (e.g., 82 versus 134 for a population correlation of 0.30, 128 versus 210 in the case of a t test where the effect is one-half of a standard deviation, etc.).

OLS Regression With Multiple Predictors

OLS regression is relatively straightforward in terms of power, being primarily influenced by the effect size, sample size, and number of predictors in the equation. In this section, we will explore the power of something less often discussed in textbooks: the power to detect more complex effects (e.g., an interaction or curvilinear effect) after the main effects of the variables are in the equation. Let us take one of our examples from Chapter 7, the curvilinear relationship between institution size and faculty salary (see Table 7.2 for the relevant statistics). In this example, we had the linear, quadratic, and cubic effects in the equation, entering the quartic effect ($\Delta R^2 = 0.029$). There was strong power in that data set ($n > 1,000$), but we can ask the question as to how large a sample we would have needed to reliably detect an effect of that nature. In Figure 15.1, I present power calculations for three different scenarios. The top line represents the power for the analysis. As you can see, even with a much smaller sample, that effect was easily detectable, with power exceeding 0.90 with only a sample of $N = 150$.

However, what if that effect was less robust? Often, complex effects are smaller in effect size due to collinearity with other effects in the model. The middle line is the power at different sample sizes if the quartic effect had been more moderate ($\Delta R^2 = 0.015$). As you can see in those calculations, even with a sample of $N = 200$, power reaches 0.80. Finally, with an even more modest effect ($\Delta R^2 = 0.0075$), a sample of $N = 200$ never reaches power of 0.50. In fact, under that scenario, the sample would have to be over $N = 700$ to reach power of 0.95. Some of our interaction examples in Chapters 9 and 10 have effect sizes in that range. This reinforces the fact that complex effects require larger samples to detect and researchers should use appropriately large samples to enable detection of these types of effects.

Binary Logistic Regression

Figure 15.2 presents three examples of the relationship between power and sample size in a simple binary logistic regression analysis with one predictor, in which we anticipate various effect sizes (odds ratio [OR] = 1.5, 2.0, and 3.0) but hold the proportion of the population constant at 80%/20% (it does not matter which is Y = 1 or Y = 0). As you can see from Figure 15.2, with a population OR of 2.0 and these parameters, an $N = 40$ gives power of about 35%. In other words, 65% of the time, the analysis will not find an effect that is in fact real. In this case, failure to reject the null is probably not strong evidence for equality between groups because there is a relatively small chance that we would have seen the effect if it were there. With an expected OR of 1.5, the situation becomes more questionable. With $N = 40$, power is less than 0.20; even with an expected OR of 3.0, power is less than 0.70.

Figure 15.1   An Example of Power to Detect Complex Effects in Ordinary Least Squares Regression

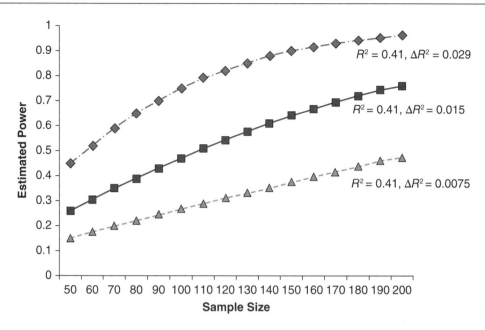

There are some general lessons we can learn from Figure 15.2. First, as one would expect, as effect size increases, power at any given sample size increases. Second, despite an OR of 1.50 being a reasonable effect size, a sample size of $N = 300$ is required to attain power approaching 0.80, whereas a sample of approximately $N = 120$ accomplishes the same goal for an OR of 2.0, and an OR of 3.0 surpasses power of 0.90 with that sample size (again, keeping in mind the other parameters—20% of the population in one category and only one predictor in the equation).

Sample size is one of the factors in power that is easily controlled or manipulated by the researcher. As we see in Figure 15.3, the distribution of the population across the outcome variable groups also has a substantial effect on power for logistic regression. With only one predictor in the equation, power tends to increase as the proportion of the population in each group becomes more equal, and power is lowest when dealing with rare outcomes. However, larger effect sizes and sample sizes ameliorate this issue. For example, when we expect an OR of 2.0 and have $N = 250$ in our sample, any population with at least 10% in the outcome group has power greater than 0.90; in comparison, a smaller expected OR of 1.50 and sample size of $N = 100$ does not have power exceeding 0.50 even if the distribution of the population is 50%/50% across the two outcome groups.[11] Conversely, with the same OR of 1.50 and a sample of $N = 250$, your power reaches 0.80 if about 30% of the population is in the $Y = 1$ group. With larger effect sizes (OR = 2.0), smaller sample sizes ($N = 100$) allow us to reach power of 0.80 with 25% of the population in $Y = 1$, and with $N = 250$ at less than 10% of the population.

---

11   Note that these curves are symmetrical, so that power with a proportion of 0.60 is equal to that of 0.40, 0.90 is equal to that of 0.10, and so on.

Figure 15.2   A Priori Power as a Function of Total Sample Size, Assuming the Probability of Being in the "1" Category of 0.20

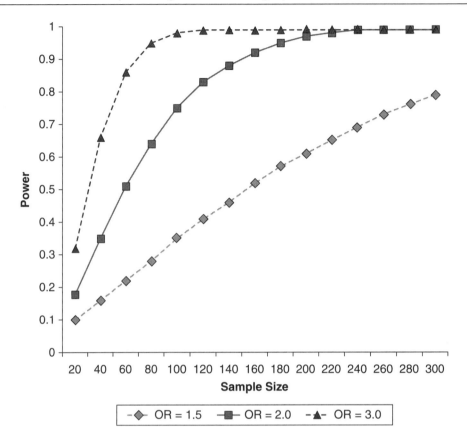

Summary of Points Thus Far

The complexities of linear models prohibit the effective formulation of simple, universal rules of thumb regarding sample size, particularly when dealing with logistic regression analyses. Many studies published in top-tier journals are not large enough to have power to detect complex effects reliably, nor to have enough precision to give a high expectation of replication of reported effects. In the next section, we will explore how you can explore the replicability and stability or volatility of your analyses.

Who Cares as Long as $p < .05$? Volatility in Linear Models

In this section, we will explore how volatile linear models can be, even when analyses have adequate power. To demonstrate this, let us examine the 16,610 participants in the National Education Longitudinal Study of 1988 (NELS88)[12] who had complete data on variables of interest. This is similar to the simple logistic regression analysis

---

12   For more information on NELS88, visit the National Center for Education Statistics website (http://nces. ed.gov/surveys/nels88/).

Figure 15.3   How the Proportion in Y = 1 Affects Power

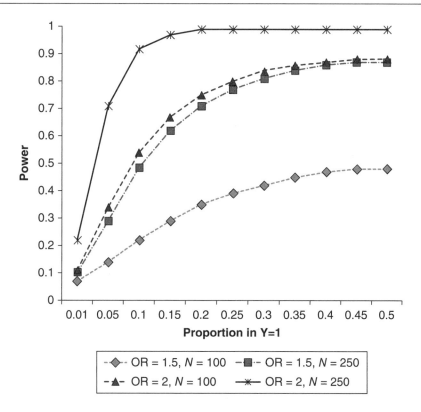

reported in Chapter 5, except that we will use GRADUATE (whether a student in the initial eighth-grade sample was retained through graduation in twelfth grade, coded 0 if the student did not graduate and 1 if the student graduated high school) as our DV and socioeconomic status (zSES) as our independent variable. In this sample, 1,477 of 16,610 students (8.9%) failed to graduate.

As discussed above, one key question in science is whether the findings of any given study are an accurate estimate of the population parameters. In the examples below, we are going to explore the volatility of two different analyses: one based on small ($N = 100$) samples and another based on much larger ($N = 500$) samples. In both, we will use a variable we have used repeatedly through prior chapters: SES (converted to z-scores). In the full sample of $N = 16,610$, which we will refer to as the known "population," SES has an OR of 2.83 when predicting retention and graduation in high school. We will use this as the "population estimate" or gold standard for comparison with other analyses. A priori power analyses for these upcoming examples should be 0.85 when $N = 100$, and 0.999 with $N = 500$. To demonstrate how volatile results can be when samples are small, even when power is relatively good, I drew 50 random samples (with replacement between each sample) from the larger sample and analyzed each separately.

Given that most samples were relatively small compared with the size of the "population," the probability that a single individual influenced more than a single analysis was relatively small, even in the $N = 500$ condition. Thus, the samples are plausibly considered independent. This procedure simulates how real scientists pull samples from real populations and informs us about how accurate we can expect our analyses to be. The advantage we have that real scientists do not have is that we have the "population" parameters with which we can compare our results.

The results of these analyses are presented in Table 15.2 and Figures 15.4a and 15.4b. With a population OR of 2.83 (included as a dashed black line on the graph), we would expect most samples to cluster around that point.

Small Samples Versus Large Samples

As you can see, the 50 smaller samples produced startlingly inaccurate point estimates of the population parameter. The ORs from each sample (represented as squares on the graph) ranged from 1.59 to 11.28. CIs for these analyses are included, but many extended beyond the upper limit of 14.0 on the graph (although they are accurately reported in Table 15.2, upper bounds ranged up to 213.51). All CIs in these analyses contained the population OR, and 40 of the 50 analyses (80%) showed that the OR was significant, which was in line with our *a priori* power estimate of 0.85. Furthermore, at the bottom of Table 15.2, you can see that the average of all of these estimates produces a reasonable approximation of the population parameter, despite the high volatility across the individual samples.

Table 15.2 Predicting the Odds of High School Graduation by Family Socioeconomic Status ($N = 100$)

							95% CI	
Sample	*β*	*SE*	*Wald*	*df*	*p*	*Exp(β)*	*Lower*	*Upper*
"Population"	1.041	0.033				2.831	2.656	3.018
1	1.383	0.464	8.870	1	.003*	3.986	1.605	9.904
2	0.855	0.365	5.473	1	.019*	2.351	1.149	4.813
3	1.816	0.728	6.223	1	.013*	6.149	1.476	25.621
4	1.107	0.369	8.976	1	.003*	3.024	1.466	6.238
5	1.421	0.645	4.851	1	.028*	4.140	1.169	14.657
6	1.116	0.429	6.785	1	.009*	3.053	1.318	7.072
7	1.513	0.539	7.864	1	.005*	4.540	1.577	13.070
8	0.681	0.380	3.206	1	.073	1.976	0.938	4.163
9	1.675	0.544	9.481	1	.002*	5.337	1.838	15.499
10	0.986	0.493	4.009	1	.045*	2.682	1.021	7.043
11	2.190	0.835	6.886	1	.009*	8.940	1.741	45.906
12	1.020	0.358	8.131	1	.004*	2.774	1.376	5.593

Sample	β	SE	Wald	df	p	Exp(β)	95% CI Lower	95% CI Upper
13	1.164	0.458	6.448	1	.011*	3.202	1.304	7.862
14	0.435	0.409	1.128	1	.288	1.545	0.692	3.447
15	1.817	0.627	8.406	1	.004*	6.156	1.802	21.031
16	0.964	0.444	4.722	1	.030*	2.623	1.099	6.259
17	0.984	0.422	5.436	1	.020*	2.674	1.170	6.114
18	1.444	0.535	7.279	1	.007*	4.236	1.484	12.091
19	1.111	0.394	7.943	1	.005*	3.038	1.403	6.578
20	0.670	0.371	3.252	1	.071	1.954	0.944	4.045
21	0.569	0.352	2.623	1	.105	1.767	0.887	3.519
22	1.882	0.616	9.318	1	.002*	6.565	1.961	21.979
23	0.741	0.401	3.408	1	.065	2.098	0.955	4.608
24	0.429	0.317	1.836	1	.175	1.536	0.826	2.858
25	1.826	0.599	9.404	1	.002*	6.272	1.940	20.278
26	0.971	0.476	4.152	1	.042*	2.639	1.038	6.714
27	0.883	0.441	4.000	1	.045*	2.418	1.018	5.744
28	1.150	0.568	4.102	1	.043*	3.159	1.038	9.619
29	0.978	0.496	3.899	1	.048*	2.660	1.007	7.026
30	0.541	0.454	1.417	1	.234	1.718	0.705	4.184
31	1.952	0.566	11.908	1	.001*	7.042	2.324	21.336
32	0.961	0.441	4.741	1	.029*	2.614	1.101	6.207
33	1.051	0.396	7.046	1	.008*	2.860	1.316	6.214
34	1.234	0.454	7.399	1	.007*	3.435	1.412	8.359
35	1.140	0.383	8.880	1	.003*	3.126	1.477	6.617
36	0.897	0.326	7.591	1	.006*	2.452	1.295	4.641
37	0.931	0.402	5.366	1	.021*	2.538	1.154	5.582
38	0.635	0.349	3.301	1	.069*	1.887	0.951	3.743
39	0.972	0.480	4.091	1	.043*	2.642	1.031	6.773
40	2.423	1.501	2.607	1	.106	11.276	0.596	213.505
41	0.971	0.480	4.095	1	.043*	2.641	1.031	6.765
42	1.624	0.473	11.787	1	.001*	5.075	2.008	12.827
43	0.949	0.409	5.378	1	.020*	2.583	1.158	5.759

(Continued)

Table 15.2   (Continued)

Sample	β	SE	Wald	df	p	Exp(β)	95% CI Lower	95% CI Upper
44	1.105	0.698	2.504	1	.114	3.019	0.768	11.863
45	1.349	0.398	11.476	1	.001*	3.852	1.765	8.404
46	1.146	0.424	7.303	1	.007*	3.146	1.370	7.223
47	0.941	0.530	3.150	1	.076	2.563	0.907	7.244
48	1.663	0.484	11.801	1	.001*	5.275	2.043	13.625
49	1.008	0.449	5.032	1	.025*	2.740	1.136	6.609
50	0.701	0.336	4.338	1	.037*	2.015	1.042	3.896
Average	1.16					3.19		

SOURCE: National Education Longitudinal Study of 1988 (NELS88) from the National Center for Education Statistics (http://nces.ed.gov/surveys/nels88/).

NOTE: *p* values less than 0.05 are highlighted with an asterisk.

Figure 15.4a   Volatility in Odds Ratios With a Sample Size of $N = 100$

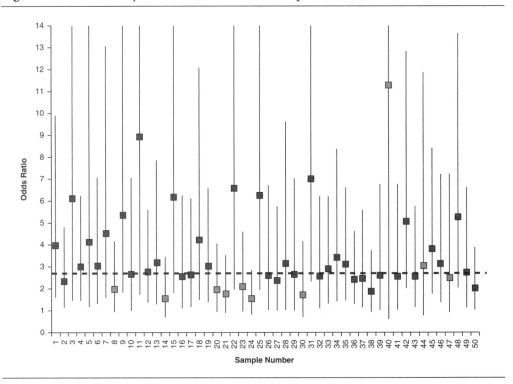

SOURCE: National Education Longitudinal Study of 1988 (NELS88) from the National Center for Education Statistics (http://nces.ed.gov/surveys/nels88/).

Figure 15.4b   Volatility in Odds Ratios With a Sample Size of $N = 500$

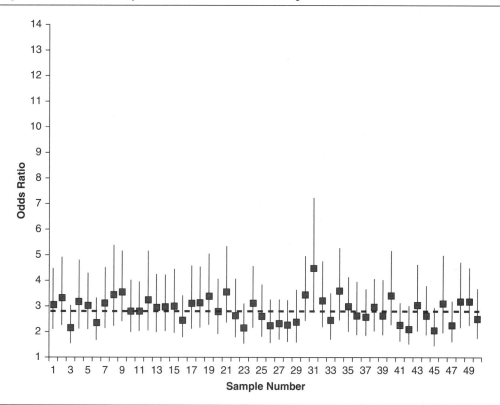

SOURCE: National Education Longitudinal Study of 1988 (NELS88) from the National Center for Education Statistics (http://nces.ed.gov/surveys/nels88/).

In comparison, when you examine the graphical summary of the larger samples analyzing the same data, you can see that these estimates are much more consistent, accurate, and less volatile estimates of the "population" parameters.[13] Unfortunately, most researchers do not have information about the true population parameters, so few researchers are aware of how volatile their population estimates could be. Even in the case of the second set of analyses ($N = 500$), the ORs varied relatively significantly, ranging from just over 2.00 to over 4.40.

Meta-analysis is of course designed to help address the problem that individual studies can be highly inaccurate estimates of population parameters, but aggregated together, they tend to produce much more reliable estimates of these unknowable values. CIs can help describe the precision of the point estimates, but CIs require certain assumptions that may or may not be met, and for some statistics, CIs may not be available.

Bootstrap resampling can be a tool that can help researchers understand and contextualize their findings from a single sample in which population values are

---

13 To conserve space, I declined to report the statistics from all 50 of the analyses in the $N = 500$ set. However, they are available in Chapter 10 of my logistic regression book (Osborne, 2015).

unknown. Almost all modern statistical software packages contain the capability to perform these analyses. The next section briefly reviews the basics of bootstrap resampling and provides some examples of how it can be used to bring context to findings.

A Brief Introduction to Bootstrap Resampling

Bootstrap analysis is a "resampling" methodology that is becoming more available as computers become more powerful and statistical computing software becomes more able to perform these types of analyses. In general, bootstrapping takes a particular group of cases and makes a large number of related samples from that single initial sample. It does this by sampling *with replacement,* meaning that any given case can be selected zero, one, or more times into each new sample. Thus, by using this simple concept, proponents of this methodology argue that researchers can essentially "pull themselves up by their own bootstraps" and make more effective use of what might be an inadequate sample. However, inadequate samples tend to produce inadequate estimates from this type of analysis.

In theory, if you use bootstrapping to create a large enough number of related samples and then analyze all of those samples, bootstrapping should be able to give a more robust estimate of the population parameters as well as potentially eliminate the influence of highly influential outliers or data points. This is because, in theory, if a relatively small number of cases are highly influential, resampling the same data set thousands of times should yield a large number of samples in which these cases are *not* influential, and thus are more accurate estimates of the population.

Bootstrapping can therefore be used to estimate the replicability (or volatility) of a result, because you can theoretically examine thousands of potential permutations of a data set. You can also use this methodology to compute CIs for statistics (like some effect sizes) that are highly desirable (Thompson, 2002; Wilkinson, 1999) but not easily estimated (Osborne, 2014; Smith & Osborne, in press). For example, I recently demonstrated bootstrap replication applied to both Cronbach's alpha and factor loadings in exploratory factor analysis, both of which could help researchers get a better estimate of how volatile their analyses are (Osborne, 2014).

In the following sections, we will use a particular method of CI estimation: percentile interval. This application of bootstrap analysis is relatively simple and assumes that the bootstrap distribution is reasonably unbiased (the mean of the bootstrapped values is reasonably close to the initial estimate from the original analysis) and symmetrical. Unless otherwise stated, the analyses meet these assumptions.

What you *cannot* do with bootstrapping is get estimates from data that are not in your data set (obviously). Thus, bootstrapping is probably not a substitute for a good, large, representative, unbiased original sample (or a second sample to test whether results replicate); however, bootstrapping in the context of a good sample can be fun and informative. In particular, I argue that you can get a much better estimate of the precision of your estimates, as well as a good estimate of the volatility (which can inform the generalizability or replicability) of your results.

All other things being equal, we are going to explore the following principles:

1. Results from larger samples will be less volatile (more likely to replicate) than results from smaller samples.

2. Effect size magnitude should not substantially influence the volatility of the results.

3. Simple results (e.g., main effects) should be less volatile (more likely to replicate) than more complex results.

We will revisit some analyses from prior chapters to explore these simple and intuitive principles.

Principle 1. Results From Larger Samples Will Be Less Volatile Than Results From Smaller Samples

I will not belabor this point. It is not only intuitive, but we also demonstrated it convincingly in the prior section using binary logistic regression as our example. Results based on small samples were much more volatile, and thus less likely to replicate, than those in larger samples.

Principle 2. Effect Sizes Should Not Affect the Replicability of the Results

To explore this point, we will return to simple OLS regression analyses from Chapter 3. The relationship between SES and achievement test scores was about $\beta = 0.505$ without any data cleaning in the full sample of $N = 16,609$, and the relationship between family income and body mass index (BMI; Enrichment Exercise 3 in Chapter 3) was $\beta = 0.023$.

For the purposes of this exercise, I randomly selected 2% of the cases for the SES analysis, yielding a sample of $N = 331$, which produced results very similar to that of the full sample ($\beta = 0.497$), as you can see in Table 15.3.

After 5,000 bootstrap samples were analyzed, all replications of the zSES effect (100%) were significant at $p < .05$,[14] and there was only moderate variability in the

Table 15.3 Predicting Student Achievement From Family Socioeconomic Status, Both Converted to z-Scores ($N = 331$)

Coefficients[a]

Model		Unstandardized Coefficients		Standardized Coefficients	t	p	95% CI for B	
		B	SE	Beta			Lower Bound	Upper Bound
1	(Constant)	0.087	0.046		1.892	.059	−0.003	0.178
	zSES	0.486	0.047	0.497	10.411	.000	0.394	0.578

[a]Dependent variable: zACH.

---

14 In fact, none of the p values exceeded .001.

Table 15.4 Predicting Body Mass Index From Family Income Category ($N = 338$)

Model		Unstandardized Coefficients		Standardized Coefficients	t	p	95% CI for B	
		B	SE	Beta			Lower Bound	Upper Bound
1	(Constant)	23.478	0.759		30.917	.000	21.984	24.972
	FAM_INC	0.046	0.105	0.024	0.442	.659	−0.160	0.253

Coefficients[a]

[a]Dependent variable: BMI.

magnitude of the effects. The mean of the betas was 0.497, identical to the point estimate in Table 15.3, and the skewness of the distribution of betas was −0.16 and symmetrical, as you can see in Figure 15.5. Furthermore, the 95% CI was [0.419, 0.570], which is reasonably close to the calculated 95% CIs in this example. These results indicate that another relatively small sample from the same population is highly likely to have similar conclusions regarding the relationship between family SES and student achievement.

For the contrasting analysis, I replicated the analysis from Chapter 3, removing highly influential cases, leaving a small but significant effect in the large sample of 8,589 ($\beta = 0.025$, $p < .019$). A 4% random sample of that data set ($N = 338$) had very similar characteristics to that of the original sample, as you can see in Table 15.4, except that given the much smaller sample size, a regression coefficient this small is not significantly different than zero.

This effect is so close to zero that in the context of a reasonably sized sample, we should not usually reject the null hypothesis. In this case, even in such a small sample, 6.7% of the bootstrap replications rejected the null hypothesis, a Type I error rate slightly higher than the goal of $\alpha = 0.05$. However, with β that ranged from −0.170 to 0.216, a sample of this size would have the power to detect some of the more aberrant effects. Overall, however, the vast majority of the effects were null, matching Table 15.4, and leading us to believe that this null effect would be likely to replicate.

In terms of volatility of the effect, the distribution of the results were symmetrical, as you can see in Figure 15.6, with empirical 95% CIs derived from the bootstrap analyses of [−0.079, 0.125]. Note that it is not the magnitude of the effect size that will determine the width of the empirical or calculated CIs. In this case, the range of the empirical CIs is just over 0.20, and the range was just over 0.15 in the previous example. The quality of the data and size of the sample (which impacts the SE) are two of the primary factors influencing precision of the estimate (and hence, potential replicability).[15]

---

15 You can see that this effect has a larger SE and thus a larger set of empirical and calculated CIs. In fact, we can even produce empirical CIs from the 5,000 bootstrap samples for statistics in which there are not routinely ways to report CIs. In this example, we can present empirical CIs for the SEs themselves (the 95% CI for SE(b) is [0.096, 0.114].

Figure 15.5   Distribution of β Across 5,000 Bootstrap Samples Where $N = 331$ and the Original $\beta = 0.497$

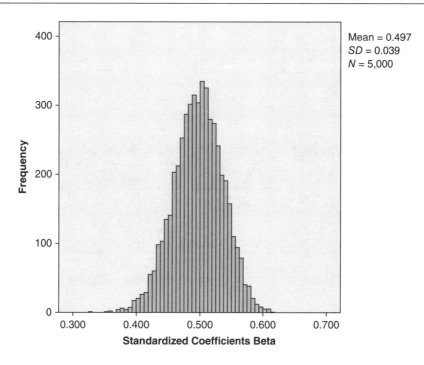

Figure 15.6   Distribution of β Across 5,000 Bootstrap Samples Where $N = 338$ and the Original $\beta = 0.024$

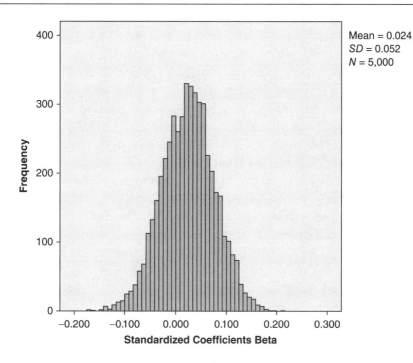

Table 15.5 Poisson Regression Analysis Predicting Number of Female Sex Partners From Biological Sex and Age of First Intercourse, Including Interaction, Reduced Sample After Data Cleaning (Reproduced From Table 11.7)

Omnibus Test[a]

Likelihood Ratio Chi-Square	df	p
70.625	3	.000

Dependent variable: SEXFNUM.
Model: (intercept), AGESEX_c, SEX, AGESEX_c*SEX.
[a]Compares the fitted model against the intercept-only model.

Parameter Estimates

Parameter	B	SE	95% Wald CI		Hypothesis Test		
			Lower	Upper	Wald Chi-Square	df	p
(Intercept)	−0.709	0.1610	−1.025	−0.394	19.407	1	.000
AGESEX_c	0.390	0.1222	0.151	0.630	10.205	1	.001
SEX	0.616	0.1638	0.295	0.937	14.154	1	.000
AGESEX_c*SEX	−0.486	0.1241	−0.730	−0.243	15.371	1	.000
(Scale)	1[a]						

Dependent variable: SEXFNUM.
Model: (intercept), AGESEX_c, SEX, AGESEX_c*SEX.
[a]Fixed at the displayed value.

Principle 3. Complex Effects Are Less Likely to Replicate Than Simple Effects, Particularly in Smaller Samples

In this example we can explore the examples from Chapter 11, wherein there was an interaction between AGESEX_c (age of first intercourse, centered at 4, age 17–18 years) and SEX (0 = female, 1 = male). Following data cleaning, the interaction was significant, with a regression coefficient of −0.486 ($SE = 0.12$, 95% CI = [−0.730, −0.243], $p < .001$; results of this analysis were presented in Figure 11.7 and in Table 11.7 [reproduced here as Table 15.5.]). With a final sample of 1,558, we would expect these results to replicate well and to be relatively stable. To test this, I performed bootstrap resampling analyses on the final sample of 1,558, requesting 5,000 replications and saving the coefficients and associated statistics for further analysis.[16] The results of these replications show rather large variability in the regression coefficients for the interaction term, despite the relatively strong sample size. The coefficients ranged from −0.107 to −0.915, with a mean of −0.498 and an empirical 95% CI of [−0.326, −0.717]. The results are presented in Figure 15.7.

One basic question in any type of replication analysis is whether the basic hypothesis test (e.g., rejection of the null hypothesis) is likely to be replicated in a similar analysis on a similar sample. In this case, only 32 of the 5,000 replications (0.6%) produced a significance test that would not result in rejection of the null hypothesis. This is a good sign for the robustness of our results. However, the next question, how likely the magnitude of the effect is to be replicated, is less clear. The CIs around this effect have a broad range, indicating that the estimate might not be as precise

---

16 The SPSS macro and data files will be available on the book website if you wish to adapt them for your own use.

Figure 15.7   Distribution of Poisson Regression Coefficient for SEX*AGESEX_c From
Bootstrap Analysis

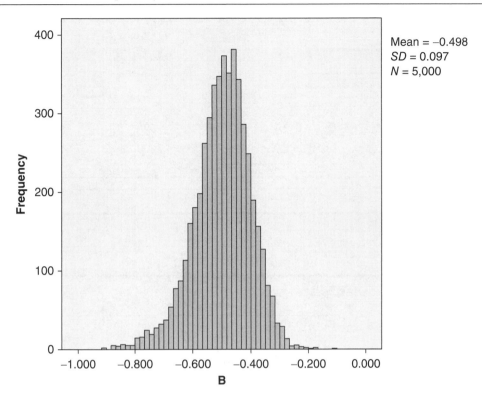

as one would like. On average, the results were very close. However, there is no way
of knowing whether an independent replication would have a b of −0.33 or −0.71.
Indeed, we cannot estimate the population parameter very precisely; thus, if we were
being honest, we can be confident in estimating that there really is an interaction and
that this coefficient is negative, but not much more.

The reality is a bit different with a smaller sample. To demonstrate this point, I
randomly sampled approximately 30% of the original sample, leaving $N = 442$. In this
sample, the overall results were similar to that in Table 15.5, as you can see in Table
15.6. The omnibus test is less strong, but the effect for the interaction is similar and
significant. If you, as a researcher, had this sample, would your conclusions be substantially different than in the prior analysis?

Unfortunately for many, even a sample this large ($N = 481$) can produce results that
are highly volatile. For example, repeating 5,000 bootstrap replications based on this
sample, only 58.3% of the replications produced a significant interaction ($p < .05$).
Thus, there is a much lower probability of replication given a sample similar to this
one. Furthermore, the range of coefficients was much broader. Given a relatively unbiased estimate ($b = −0.367$ versus −0.489 in the original analysis, skewness of −0.394), it
is legitimate to explore the empirical 95% CIs from this analysis. The calculated CIs in
Table 15.6 are not substantially different from the original (−0.70, −0.03). However, in

Table 15.6   Poisson Regression Analysis Predicting Number of Female Sex Partners
From SEX and AGESEX_c, Including Interaction (Reduced Sample of
$N = 481$)

Omnibus Test[a]

Likelihood Ratio Chi-Square	df	p
23.132	3	.000

Dependent variable: SEXFNUM.
Model: (intercept), AGESEX_c, SEX,
AGESEX_c*SEX.
[a]Compares the fitted model against the
intercept-only model.

Parameter Estimates

Parameter	B	SE	95% Wald CI		Hypothesis Test		
			Lower	Upper	Wald Chi-Square	df	p
(Intercept)	−0.608	0.2402	−1.078	−0.137	6.400	1	.011
AGESEX_c	0.228	0.1660	−0.098	0.553	1.883	1	.170
SEX	0.461	0.2465	−0.023	0.944	3.492	1	.062
AGESEX_c*SEX	−0.367	0.1707	−0.701	−0.032	4.612	1	.032
(Scale)	1[a]						

Dependent variable: SEXFNUM.
Model: (intercept), AGESEX_c, SEX, AGESEX_c*SEX.
[a]Fixed at the displayed value.

the bootstrap replication, we have coefficients ranging from −1.50 to 0.378 (as you can see in Figure 15.8), although the empirical CIs (−0.66, −0.13) were reasonably similar to the previous analysis.

As we have discussed throughout this book, more complex effects deserve careful interpretation and attention, because they are least likely to be observed (as they are often lower in power than simpler effects) and thus are least likely to be replicated. As a final examination of the analyses from Chapter 11, let us examine briefly the interesting and complex curvilinear interaction between biological sex (SEX) and the cubed effect of age of first sexual experience (AGESEX_c^3), which was presented in Table 11.8 and Figure 11.8. The original b of 0.119 ($SE = 0.0485$, 95% CI [0.024, 0.214], $p < .014$) left us with an interesting effect. But the more interesting and complex effects leave me wondering whether the effect was just a quirk of the data (in fact, different data cleaning in that data set can lead to different results) or whether the effect was robust and replicable. Of course, absent a new study with similar data, it is difficult to know. However, a simple bootstrap resampling can give us some information about the likelihood that a similar sample would produce similar results.

Analyses of 5,000 bootstrap samples gives us some context for this interesting effect. First, 65.8% of the bootstrap samples produced a significant curvilinear interaction effect (leaving about one-third of the samples not resulting in significant effects). Thus, it is likely that another similar sample would replicate the results, but it is far from guaranteed. Second, the magnitude of the results varied around a mean of 0.118, as you can see in Figure 15.9, but the empirical 95% CI was [0.032, 0.203], indicating that a replication is likely to find an effect in the same general range as the original observed effect.

Figure 15.8   Distribution of Poisson Regression Coefficient for SEX*AGESEX_c From Bootstrap Analysis With a Smaller Sample ($N = 481$)

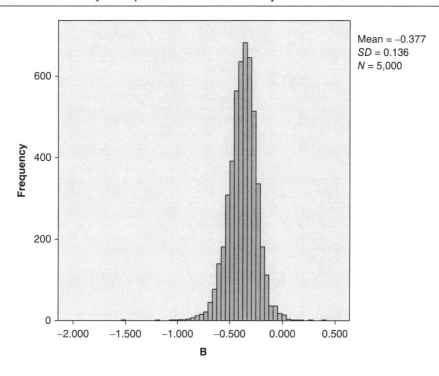

Figure 15.9   Distribution of Poisson Regression Coefficient for SEX*AGESEX3 From Bootstrap Analysis in the Full Sample ($N = 1,567$)

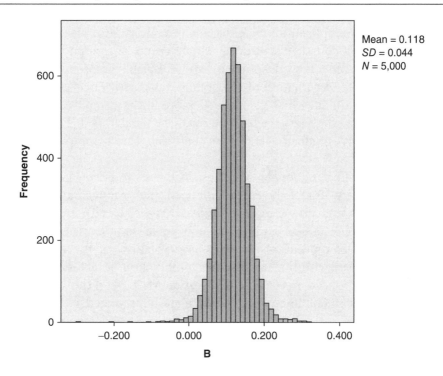

Table 15.7   Do Curvilinear Interaction Shapes Replicate?

Parameter	Original B	Bootstrap Replication Number 1,462	Bootstrap Replication Number 4,226	Bootstrap Replication Number 3,891
(Intercept)	−0.492	−0.752	−0.634	−0.638
AGESEX_c	0.665	0.961	0.644	0.916
SEX	0.372	0.617	0.503	0.518
AGESEX^2	−0.151	−0.125	−0.012	−0.221
AGESEX^3	−0.095	−0.130	−0.075	−0.158
AGESEX_c*SEX	−0.895	−1.204	−0.843	−1.176
SEX*AGESEX^2	0.163	0.130	0.033	0.235
SEX*AGESEX^3	0.119	0.156	0.091	0.190
(Scale)	1	1	1	1

One of the other critical questions in complex effects such as curvilinear effects and interactions is whether the shape or nature of the effect will replicate. To test this, I randomly[17] selected three examples from the 5,000 bootstrap replications to graph. As you can see in Table 15.7, the coefficients vary somewhat across the three examples. Furthermore, in the second example, the cubic interaction term was significant at $p < .064$, thus representing one of the cases in which that effect would not have been significant. However, as you can see in Figure 15.10, the original effect and the three examples of bootstrap replications were reasonably consistent in the nature and shape of the interaction. Of course, because the female group was much smaller, it is not surprising that those effects were somewhat more variable across replications. Even within this group, the basic nature of the graphs was mostly consistent and thus we could have more confidence that the effect is more likely to replicate than not, that the magnitude of the effects are likely to be relatively similar, and that the end result is a reasonably consistent effect even where coefficients are somewhat more volatile (as with the effect of SEX).

To examine the effect of smaller samples on the same effect, I again randomly selected approximately 30% of the cases ($N = 456$) that produced a similar result to the prior analysis, as you can see in Table 15.8 and Figure 15.11.

Combining complex effects and (relatively) small sample size should lead to reduced probability of replication and increased volatility of the results. The first and most obvious issue with this undesirable combination of factors was that 112 of 5,000 replications produced an error in the Hessian Matrix and thus were not able to be completed. Of the 4,888 analyses that successfully produced results, only 45.9%

---

17   Using a random number generator at http://random.org.

Figure 15.10 Comparison of Original SEX*AGESEX^3 Curvilinear Interaction and Three Replication Examples

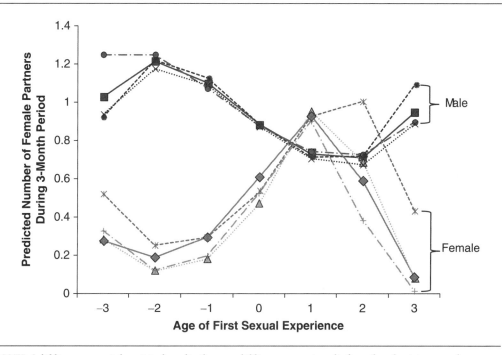

NOTE: Solid lines represent the original results; the nonsolid lines represent results from three bootstrap samples.

Table 15.8 Curvilinear Interaction With a Smaller Sample ($N = 456$)

Omnibus Test[a]

Likelihood Ratio Chi-Square	df	p
35.332	7	.000

Dependent variable: SEXFNUM.
Model: (intercept), AGESEX_c, SEX, AGESEX^2, AGESEX^3, AGESEX_c*SEX, SEX*AGESEX^2, SEX*AGESEX^3.
[a]Compares the fitted model against the intercept-only model.

Parameter Estimates

Parameter	B	SE	95% Wald CI		Hypothesis Test		
			Lower	Upper	Wald Chi-Square	df	p
(Intercept)	−0.566	0.3539	−1.260	0.128	2.559	1	.110
AGESEX_c	0.956	0.4041	0.164	1.748	5.601	1	.018
SEX	0.370	0.3609	−0.337	1.078	1.053	1	.305
AGESEX^2	−0.096	0.1343	−0.359	0.168	0.508	1	.476
AGESEX^3	−0.140	0.0754	−0.287	0.008	3.423	1	.064
AGESEX_c*SEX	−1.231	0.4110	−2.037	−0.426	8.977	1	.003
SEX*AGESEX^2	0.131	0.1367	−0.137	0.398	0.914	1	.339
SEX*AGESEX^3	0.176	0.0766	0.026	0.326	5.259	1	.022
(Scale)	1[a]						

Dependent variable: SEXFNUM.
Model: (intercept), AGESEX_c, SEX, AGESEX^2, AGESEX^3, AGESEX_c*SEX, SEX*AGESEX^2, SEX*AGESEX^3.
[a]Fixed at the displayed value.

Figure 15.11   Comparison of Original SEX*AGESEX³ Curvilinear Interaction and the Same Effect From a Smaller Sample ($N = 456$)

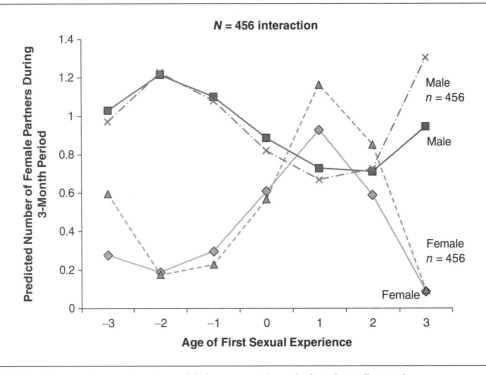

NOTE: Solid lines are the original analysis, and dashes represent the results from the smaller sample.

produced significant results for this interaction term (SEX*AGESEX³). If we included the samples that did not produce results due to an error, we would have an even less positive outcome. Thus, we immediately see the drawbacks to relatively small samples.

Aside from that issue, there was much greater variability in this example regression coefficient. As you can see in Figure 15.12, this statistic ranged from −0.55 to 1.29 across the 5,000 bootstrap samples, with an empirical 95% CI of [−0.01, 0.417], a much broader range than in the previous bootstrap analyses when the sample was larger. Given that the CI includes zero, this would be a problematic finding in which to put a good deal of faith. However, absent this bootstrap analysis, the results in Table 15.8 would be publishable. The effect is significant and interesting (at least to me, someone who is completely ignorant of the literature around this issue). Given that this sample of many hundreds is larger than many journals see in high-quality articles, there probably would not be much hesitation to publish the results. Without the information that this effect is not likely to replicate with high confidence, and without acknowledging the high volatility of the result (which is underestimated in the computed CIs in Table 15.8), there would be no reason for concern. Given the information you now possess, however, would you be confident in publishing this result, and do you think it is desirable to publish this result without the additional context of how likely it is to replicate?

Figure 15.12   Variability in the Regression Coefficient of SEX*AGESEX³ Across 5,000
Bootstrap Replication Analyses

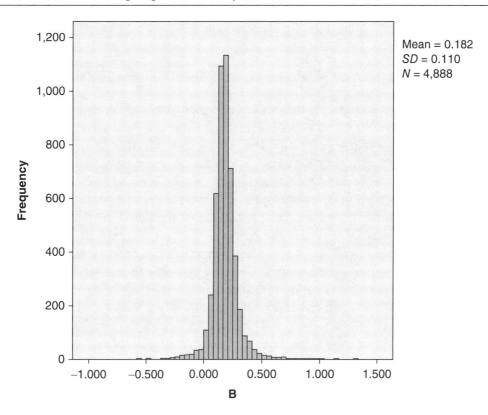

SUMMARY AND CONCLUSIONS

When thinking about science, a key question is whether the observed effects are reproducible and/or replicable. My assertion is that more reproducible and replicable results are more meaningful, in part because they might be more likely to reflect population parameters. Effects that are not reproducible or replicable should not be of interest to the scientific community.

In thinking about replication, there are two main questions that mirror much of the discussion throughout this book:

1. Is the effect statistically significant (in other words, do we reject the null hypothesis)?

2. If so, is the effect important (in other words, is the effect of a size that would lead anyone to care about it)?

When examining replication of effects with significance tests, we must then have two different pieces of information:

1. How likely is a replication with another sample to match the initial null hypothesis statistical test?

2. How likely is the replication with another sample to produce a similar effect size?

The lesson in this chapter is that we need to report more than significance levels, and even effect sizes or computed CIs (although those are great starts!). We saw, for example, that large samples are much less volatile in replication than small samples. We further saw that with simple analyses and reasonable sample sizes for those analyses, both the hypothesis testing and the effect sizes were likely to replicate.

However, those types of effects are rarely of interest in science. Moderation (interaction effects) is more interesting and also less simply replicated, particularly complex effects like curvilinear interactions, and particularly when those effects are in the context of inappropriately small samples. When a simple linear interaction was examined in the context of a Poisson regression analysis with a sample size of more than 1,500 participants, over 99% of the replications produced similar rejection of the null hypothesis. Yet when the sample was reduced to $N = 481$ (maintaining a similar effect size, etc.), the rejection of the null hypothesis was observed in only 58% of the samples. Thus, we can see that complex effects require far larger samples than is typical in many fields to have reasonable expectations of replication of anything more than the most basic effects.

The situation was more serious in the case of the curvilinear interaction. With $N = 1,500$, we saw only about two-thirds of the samples rejecting the null hypothesis and less than half when the sample was smaller. Similarly, the breadth of the CIs tracked (inversely) with the sample sizes. There was much more volatility in the magnitude of the effect with small samples than large samples, leaving us again with less confidence that effects will replicate when samples are small (small being relative because these samples were in the range of $N = 400$–500).

Because our goal in science is to say something about a population that is generalizable and potentially replicable, this information might provide important context. I would argue that researchers should begin exploring the robustness of their results in this fashion and reporting the results of these explorations to provide important context. Although no researcher would want to find a significant effect and then report that it is highly unlikely that the effect would replicate, that is exactly what many researchers are in essence doing when they report results based on small samples, small effect sizes, using data that have not been cleaned, and so forth. Although it is not always within reason for every researcher to replicate results with independent samples (although that would be ideal), bootstrap replication analysis can provide some information concerning the robustness of the results, and bootstrap resampling methods are now widely available in most statistical computing packages.[18] It does a

---

18 For SPSS users, examples of the macro I used will be available on the website. You do not have to purchase the extra bootstrapping module in order to perform these types of analyses!

field no good to report results that are not likely to replicate, serving only to pad a researcher's curriculum vitae. If we care about science, journal editors and researchers should begin to embrace the call for replication and reproducibility... the basic building blocks of trustworthy science.

ENRICHMENT

1. Using data from the book's website, along with your own software or SPSS macros (also available on the book's website), explore bootstrap analyses from these or other examples within the book.

2. Ask a colleague or adviser to share a data set that was recently published (or better yet, one soon to be published) and perform bootstrap resampling analyses on that data. Discuss the results with the owner of the data set. What context does this analysis add to the original analyses? Do they change the interpretation(s)?

REFERENCES

Asendorpf, J. B., Conner, M., De Fruyt, F., De Houwer, J., Denissen, J. J., Fiedler, K., . . . Nosek, B. A. (2013). Recommendations for increasing replicability in psychology. *European Journal of Personality*, *27*(2), 108–119. doi: 10.1002/per.1919

Bem, D. J. (2011). Feeling the future: Experimental evidence for anomalous retroactive influences on cognition and affect. *Journal of Personality and Social Psychology*, *100*(3), 407–425. doi: 10.1037/a0021524.

Cohen, J. (1962). The statistical power of abnormal-social psychological research: A review. *Journal of Abnormal and Social Psychology*, *65*, 145–153. doi: 10.1037/h0045186

Cohen, J. (1988). *Statistical power analysis for the behavioral sciences* (2nd ed.). Hillsdale, NJ: Lawrence Erlbaum.

Cohen, J. (1992). A power primer. *Psychological Bulletin*, *112*(1), 155–159. doi: 10.1037/0033-2909.112.1.155

Couchman, J. R. (2014). Peer review and reproducibility: Crisis or time for course correction? *Journal of Histochemistry and Cytochemistry*, *62*(1), 9–10. doi: 10.1369/0022155413513462

Deemer, W. L., Jr. (1947). The power of the t test and the estimation of required sample size. *Journal of Educational Psychology*, *38*(6), 329–342. doi: 10.1037/h0060014

Dickersin, K., & Min, Y. (1993). Publication bias: The problem that won't go away. *Annals of the New York Academy of Sciences*, *703*(1), 135–148. doi: 10.1111/j.1749-6632.1993.tb26343.x

Galton, F. (1907). Vox populi (the wisdom of crowds). *Nature*, *75*, 450–451.

Gauchat, G. (2012). Politicization of science in the public sphere: A study of public trust in the United States, 1974 to 2010. *American Sociological Review*, *77*(2), 167–187. doi: doi: 10.1177/0003122412438225

Hopewell, S., Loudon, K., Clarke, M. J., Oxman, A. D., & Dickersin, K. (2009). Publication bias in clinical trials due to statistical significance or direction of trial results. *Cochrane Database Syst Rev*, *1*(1), MR000006. doi: 10.1002/14651858.MR000006.pub3

Horton, R. (2015). Offline: What is medicine's 5 sigma? *Lancet*, *385*(9976), 1380. doi: 10.1016/S0140-6736(15)60696-1

Hsieh, F. Y., Bloch, D. A., & Larsen, M. D. (1998). A simple method of sample size calculation for linear and logistic regression. *Statistics in Medicine*, *17*(14), 1623–1634. doi: 10.1002/(SICI)1097-0258 (19980730)17:14<1623::AID-SIM871>3.0.CO;2-S

Ioannidis, J. P. (2005). Why most published research findings are false. *Chance, 18*(4), 40–47.

Ioannidis, J., & Doucouliagos, C. (2013). What's to know about the credibility of empirical economics? *Journal of Economic Surveys, 27*(5), 997–1004. doi: 10.1111/joes.12032

Kraemer, H. C. (1992). *Evaluating medical tests: Objective and quantitative guidelines.* Newbury Park, CA: SAGE.

Kraus, W. L. (2014). Editorial: Do you see what I see? Quality, reliability, and reproducibility in biomedical research. *Molecular Endocrinology, 28*(3), 277–280. doi: 10.1210/me.2014-1036

Osborne, J. W. (2008). Sweating the small stuff in educational psychology: How effect size and power reporting failed to change from 1969 to 1999, and what that means for the future of changing practices. *Educational Psychology, 28*(2), 151–160. doi: 10.1080/01443410701491718

Osborne, J. W. (2014). *Best practices in exploratory factor analysis.* Seattle, WA: CreateSpace Publishing.

Osborne, J. W. (2015). *Best practices in logistic regression.* Thousand Oaks, CA: SAGE.

Osborne, J. W., Kocher, B., & Tillman, D. (2012, February). *Sweating the small stuff: Do authors in APA journals clean data or test assumptions (and should anyone care if they do)?* Paper presented at the annual meeting of the Eastern Educational Research Association, Hilton Head, SC.

Peng, R. (2015). The reproducibility crisis in science: A statistical counterattack. *Significance, 12*(3), 30–32. doi: 10.1111/j.1740-9713.2015.00827.x

Ritchie, S. J., Wiseman, R., & French, C. C. (2012). Failing the future: Three unsuccessful attempts to replicate Bem's "retroactive facilitation of recall" effect. *PLoS One, 7*(3), e33423. doi: 10.1371/journal.pone.0033423

Rossi, J. S. (1990). Statistical power of psychological research: What have we gained in 20 years? *Journal of Counseling and Clinical Psychology, 58*(5), 646–656. doi: 10.1037/0022-006X.58.5.646

Sedlmeier, P., & Gigerenzer, G. (1989). Do studies of statistical power have an effect on the power of studies? *Psychological Bulletin, 105*(2), 309–316. doi: 10.1037/0033-2909.105.2.309

Simmons, J. P., Nelson, L. D., & Simonsohn, U. (2011). False-positive psychology undisclosed flexibility in data collection and analysis allows presenting anything as significant. *Psychological Science, 22*(11), 1359–1366. doi: 10.1177/0956797611417632

Smith, E. A., & Osborne, J. W. (in press). Realizing the dream of confidence intervals for effect sizes through bootstrap resampling. *Practical Assessment, Research and Evaluation.*

Stodden, V. C. (2011). Trust your science? Open your data and code. *Amstat News, 409,* 21–22.

Thompson, B. (2002). What future quantitative social science research could look like: Confidence intervals for effect sizes. *Educational Researcher, 31*(3), 25–32. doi: 10.3102/0013189X031003025

Tressoldi, P. E. (2012). Replication unreliability in psychology: Elusive phenomena or "elusive" statistical power? *Frontiers in Psychology, 3,* 218. doi: 10.3389/fpsyg.2012.00218

Wilkinson, L., & Task Force on Statistical Inference, APA Board of Scientific Affairs. (1999). Statistical methods in psychology journals: Guidelines and explanations. *American Psychologist, 54*(8), 594–604. doi: 10.1037/0003-066X.54.8.594

RELIABLE
MEASUREMENT MATTERS

16

Advance Organizer

One of the basic assumptions of quantitative research is that we are measuring the variables we think we are measuring with a high degree of reliability (the actual assumption is that of perfectly reliable measurement, which is not attainable). Our variables are often difficult to measure, making measurement error a concern, particularly in the social/behavioral and health sciences.

Despite impressive advancements in measurement in recent years (particularly the broad dissemination of structural equation modeling, Rasch measurement methodologies, and item response theory [IRT], to name but a few), the reliability of measurement remains an issue. In simple analyses in which there is only a single predictor, unreliable measurement causes relationships or effects to be *underestimated* (or attenuated), increasing the risk of Type II errors. In the case of multiple regression (or analyses with multiple predictors), effect sizes of variables with poor reliability can be underestimated while causing other variables in the analysis to be simultaneously *overestimated*, because the full effect of the variable with poor measurement qualities might not be removed.

This is a significant concern if the goal of research is to accurately model the "real" relationships evident in the population. Although most authors assume that reliability estimates (e.g., Cronbach's alpha [α] internal consistency estimates) of 0.70 and above are acceptable (e.g., Nunnally & Bernstein, 1994), and Osborne (2008) reported that the average alpha reported in top educational psychology journals was 0.83,[1] many scientists fail to understand that measurement of this quality contains enough measurement error to significantly alter the results of their research.

Many authors seem not terribly concerned with reliable measurement (by virtue of failing to report any facts relating to the quality of their measurement, again as reported by Osborne, 2008). Others seem to believe that moderately good reliability is "good enough" to accurately model the population relationships and produce generalizable, accurate results. Thus, the goal of this chapter is to briefly explore

---

1 These findings are among those articles actually reporting internal consistency estimates, which may be a biased sample.

some issues related to suboptimal measurement, which potentially impact all linear models, and to encourage researchers to more rigorously assess the quality of their measurement, more fully understanding the effect of imperfect reliability on their results, and, where possible, take action to correct the issue.

In this chapter, we will cover

- A brief overview of the concept of reliability
- A brief overview of Cronbach's alpha
- Factors that influence alpha
- Effects of poor reliability on
  - Simple regression
  - Multiple regression
  - Interactions
  - Logistic regression

A More Modern View of Reliability

When we discuss reliability, we are essentially discussing error in measurement. The ideal is error-free measurement, which is unattainable, but good measurement should not be. Reliability and measurement could be the topic of an entire book (e.g., the classic psychometrics text by Nunnally & Bernstein, 1994); therefore, in this brief chapter, we merely highlight some issues relevant to linear modeling.

There are several methods of estimating reliability, depending on whether your measure is a single-item measure or observation (e.g., diastolic blood pressure, whether a behavior occurred or not) or a scale (e.g., a depression inventory). Much of our discussion is grounded in *classical test theory*, which posits that any observed score is a function of the true value (unknowable) and error (both random and nonrandom). Modern measurement theories have different models of thinking about measurement (e.g., estimating person or item parameters, as in IRT, Rasch measurement, etc.), although we do not have space to cover all of these topics in depth (but some excellent books on these topics are available; see Bond & Fox, 2006; Edelen & Reeve, 2007; Hambleton, Swaminathan, & Rogers, 1991; Smith & Smith, 2004; Wilson, 2005). For our purposes in this chapter, we will consider reliable measurement to be a situation in which error is minimized, and we will use the example of Cronbach's alpha as our example indicator of reliability (again knowing there are many different ways to measure reliability).

Furthermore, in the modern era, we do not talk about "reliable" or "unreliable" measures. Rather, modern psychometrics acknowledges that things such as factor structure, reliability, and validity are joint properties of the measure and of the particular sample data being used (Fan & Thompson, 2001; Wilkinson, 1999). It should be self-evident to modern researchers that a scale needs to be well developed in order to be useful, but that we evaluate the scale in the context of a particular sample (or series of samples, as we recommended when discussing replication). Thus, those of us interested in measurement must hold two somewhat bifurcated ideas in mind simultaneously: that a scale can be stronger or weaker, and that scales are only strong or weak in the context of the particular sample being used. This can lead to a nihilistic mindset if carried to an extreme, so I recommend we take a moderate position in this discussion: that scales can be more or less strong, but that all scales need to be evaluated in the particular populations or data in which they reside.

What Is Cronbach's Alpha (and What Is It Not)?

Cronbach's alpha (Cronbach, 1951) is one of the most widely reported indicators of scale reliability in the social sciences. It has some conveniences over other methods of indicating reliability, and it also has some drawbacks. There are also misconceptions about what alpha represents.

Some scholars have attempted to expand inference around alpha: such as development of a test of equivalence across two samples (Feldt, 1980) and computation of confidence intervals (CIs) for alpha (see Barnette, 2005).[2]

Let us start with the original goal for alpha. Prior to Cronbach's seminal work in this area, the reliability of a scale in a particular sample was evaluated through methods such as test-retest correlations. This type of reliability is still discussed today in psychometrics textbooks but has serious drawbacks, including the difficulty of convening the same group of individuals to retake instruments, memory effects, and attenuation due to real change between administrations. Another serious drawback to test-retest reliability is an inability to evaluate the reliability of unstable constructs (e.g., mood, content knowledge) that are expected to change over time. Thus, as Cronbach himself put it, test-retest reliability is generally best considered as an index of *stability* rather than reliability *per se*.

The split-half reliability estimate was also developed early in the 20th century. This reliability estimate requires scale items to be divided into two groups (most commonly, even- and odd-numbered items) and scored. Those two scores are then compared as a proxy for an immediate test-retest correlation. This too has drawbacks—the number of items is halved, there is some doubt as to whether the two groups of items are parallel, and different splits of items can yield different correlation coefficients. The Spearman-Brown correction was developed to help correct for the reduction in item number and to give a coefficient intended to be similar to the test-retest coefficient. As Cronbach (1951) pointed out, this coefficient is best characterized as an indicator of equivalence between two forms, much as we today also talk about parallel forms reliability.

Alpha and the Kuder-Richardson Coefficient of Equivalence

The Kuder-Richardson Formula 20 (KR-20) was developed to address some of the concerns over other forms of reliability, particularly split half, and preceded alpha. The KR-20 is specific to items scored ether "0" or "1" as in many academic tests or scales, and it is simpler to calculate by hand. Alpha is more general than KR-20, but KR-20 and alpha will arrive at the same solution if items are binary. It does not therefore appear that KR-20 is necessary.

---

2   However, neither practice seems to have been adopted widely in the literature I am familiar with.

The Correct Interpretation of Alpha

Cronbach (1951) himself wrote and provided proofs for several assertions about alpha. These include the following:

- Alpha is ($n/n - 1$) times the ratio of inter-item covariance to total variance—in other words, a direct assessment of the ratio of error (unexplained) variance in the measure.
- The average of all possible split-half coefficients for a given test.[3]
- The coefficient of equivalence from two tests composed of items randomly sampled (without replacement) from a universe of items with the mean covariance as the test or scale in question.
- A lower-bound estimate of the coefficient of precision (accuracy of the test with these particular items) and coefficient of equivalency (simultaneous administration of two tests with matching items).
- The proportion (lower bound) of the test variance due to all common factors among the items.

Given these points, I conclude that the standards we use for alpha, and the average alpha found in strong journals, are often not good enough. As Nunnally and Bernstein (1994, p. 235) distill from all of this, alpha is an expected correlation between one test and an alternative form of the test containing the same number of items. The square root of alpha is also, as they point out, the correlation between the score on a scale and errorless "true scores." Let us unpack this for a moment.

This means that if one has an $\alpha = 0.80$ for a scale, this is interpreted as the expected correlation between that scale and another scale sampled from the same domain of items with the same covariance and number of items. The square root of 0.80 is 0.89, which represents an estimate of the correlation between that score and the "true scores" for that construct. As you probably know by this point in the book, the square of a correlation is an estimate of shared variance, so squaring this number brings us back to the proportion of "true score" in the measurement (and $1 - \alpha$ is the proportion of error variance in the measurement).

A review of educational psychology literature from 1969 and 1999 indicated average (reported) α values of 0.86 and 0.83, respectively (Osborne, 2008). This is not bad but still leaves almost 20% of the measurement as error, which seems suboptimal. Also keep in mind that even in modern, high-quality journals, only 26% of articles reported this basic data quality information, meaning that this estimate is probably an overestimate of the routine measurement quality in journals. Despite these optimistic averages, it is not difficult to find journal articles in almost any discipline reporting analyses with much lower alphas. Poor measurement can have profound (and often unpredictable) effects on outcomes.

---

3   This implies that there is a distribution of split-half coefficients based on different splits, and that alpha is the mean of all of these splits. This is an interesting idea that many of us miss, because we focus just on the one number we calculate.

What Alpha Is Not

Note that alpha is *not* a measure of unidimensionality (an indicator that a scale is measuring a single construct rather than multiple related constructs) as is often thought (Cortina, 1993; Schmitt, 1996). Unidimensionality is an important assumption of alpha, in that scales that are multidimensional will cause alpha to be underestimated if not assessed separately for each dimension, but high values for alpha are not necessarily indicators of unidimensionality (Cortina, 1993; Schmitt, 1996).

Factors That Influence Alpha

Length of the Scale

All other things being equal, longer scales will have higher alpha values.

Average Inter-Item Correlation

All other things being equal, alpha is higher when the average correlation between items is higher. However, Cronbach specifically pointed out that when inter-item correlations are low, alpha can be high when a scale contains a large number of items with low intercorrelations. This is one of the chief drawbacks to interpretability of alpha—that with enough mostly unrelated items, alpha will move into the "reasonable" range that most researchers use as a rule of thumb.

Reverse-Coded Items (Negative Item-Total Correlations)

Many scales are constructed with reverse-coded items. However, alpha cannot provide accurate estimates when the analysis includes items with negative item-total correlations. Thus, any item that is expected to have a negative item-total correlation (e.g., if the factor loading is negative when most others are positive) should be reversed prior to analysis.

Random Responding or Response Sets

Random responding tends to attenuate all of these estimates because it primarily adds random error. Thus, failure to identify this issue in your data will lead to underestimation of the internal consistency of the data. Response sets can have a variety of effects, depending on the response set. Some types of response sets will inflate alpha estimates and some can attenuate alpha (for an overview of response sets and how one can identify them, you might see Osborne & Blanchard, 2011).

Multidimensionality

A key assumption of alpha is that all items within a particular analysis represent a single dimension, or factor. To the extent that assumption is violated, alpha will be misestimated. Thus, the factor structure of the scale should be considered before submitting items to this type of analysis. Note that a high alpha is not proof of unidimensionality.

Outliers

Outliers (inappropriate values) usually have the effect of increasing error variance, which would have the effect of attenuating the estimate of alpha. Thus, data should be carefully screened prior to computing alpha.

Other Assumptions of Alpha

Alpha was built for a time when researchers often summed or averaged items on scales. Many researchers do this today despite the existence of more modern measurement options. Of course, when summing or averaging items in a scale, you are making an assumption that all items contribute equally to the scale—that the weighting of each item is identical. Alpha also assumes that all items contribute equally, but this might not be a valid assumption.

For example, when scales are submitted to factor analysis, items often have different factor loadings, indicating that items vary in the extent to which they measure the construct in question. There are similar indicators in Rasch measurement and other measurement models. Latent variable modeling (e.g., structural equation modeling) also directly addresses this issue when estimating individual scores on latent variables.

There is not currently a good way to deal with this issue (in my opinion). Rasch and IRT have different methods of estimating reliability of measures; in confirmatory factor analysis/structural equation modeling, there are also ways to assess goodness of fit, but those interested in estimating internal consistency via alpha must accept that most scales violate this assumption.

What Is "Good Enough" for Alpha?

Many authors have asserted that an alpha of 0.70 or 0.80 represents "adequate" or "good" reliability, respectively (Nunnally & Bernstein, 1994). Let us just say for the present that reliability is a continuous variable, and thus, higher is better, probably with diminishing returns once one exceeds 0.90 (which still represents about 10% error variance in the measurement).

What constitutes "good enough" also depends on the purpose of the data and the method of analysis. Better data are always better, of course, but using the data to choose children for an educational program (or decide if teachers are doing a good job) is different from evaluating correlations between constructs for a dissertation.

In the next sections, we will explore examples of the effect of poor reliability on some linear models, and you can judge for yourself what range of reliability (and measurement error) you are comfortable with.

Reliability and Simple Correlation or Regression

Because "the presence of measurement errors in behavioral research is the rule rather than the exception" and the "reliabilities of many measures used in the behavioral sciences are, at best, moderate" (Pedhazur, 1997, p. 172), it is important that researchers be aware of accepted methods of dealing with this issue. For simple correlation, Equation 16.1 (Cohen, Cohen, West, & Aiken, 2003) provides an estimate of the "true" relationship between the independent variable (IV) and dependent variable (DV) in the population:

$$r_{12}^* = \frac{r_{12}}{\sqrt{r_{11}r_{22}}} \tag{16.1}$$

In this equation, r_{12} is the observed correlation, and r_{11} and r_{22} are the reliability estimates of the variables.[4] To illustrate use of this formula, I take an example from my own research, which involved two scales measuring how psychologically invested students were in their education. The sample included 214 high school students in an urban Midwestern high school. The two scales, the School Perceptions Questionnaire (SPQ; Osborne, 1997) and the Identification With School scale (IWS; Voelkl, 1996, 1997), should be strongly correlated, tend to have good internal consistency, and serve our purposes as a good example of what can happen when measurement quality is suboptimal.

First, internal consistency was estimated to be $\alpha = 0.88$ for the SPQ and $\alpha = 0.81$ for the IWS, both of which are considered by traditional benchmarks to be good levels of internal consistency. After averaging the items to create composite scores and testing assumptions, the scales were correlated at $r = 0.75$, which translates to a coefficient of determination (percent variance accounted for) of 0.56, a relatively strong correlation for the social sciences.

Using Equation 16.1, we can correct this correlation to estimate the disattenuated, or "true" population correlation, resulting in an estimated $\rho = 0.89$. Looking at variance accounted for, we see a sizable improvement from 0.56 to 0.79 from observed correlation to the disattenuated "population" correlation. This represents a 40.29% increase in effect size or variance accounted for, or a 28.72% *underestimation* of the true correlation attributable to measurement error. Many are surprised that variables having such "good" reliability could still show such a dramatic misestimation of the "true" population effect. Some might be skeptical that this classic formula overcorrects for measurement error. We can validate this calculation using structural

---

4   Note again that many estimates of reliability, such as alpha, are "lower bound" estimates that can lead to *overcorrection*. Thus, conservativeness is advised when correcting in this manner. Specifically, one of my advisers in graduate school advocated correcting reliability to 0.95, rather than 1.00, to reduce the probability of overcorrection when using structural equation modeling. In this case, to be conservative and limit overcorrection, I think a good idea is to add 0.05 or so to any reliability estimate prior to using it in a formula such as this.

equation modeling to estimate correlations between latent (error-free) constructs. Without going into detail (because structural equation modeling is beyond the scope of this book), the estimated correlation between latent constructs was estimated to be $r = 0.90$, very close to the $\rho = 0.89$ estimated above. Thus, we can have a good level of confidence that this formula indeed corrects appropriately for attenuation due to imperfect measurement.

Other examples of the effects of attenuation due to poor reliability are presented in Table 16.1.

For example, even when reliability is 0.80 (row highlighted in Table 16.1), it appears that the observed correlation is attenuated substantially (about 33%) from what we would calculate if the assumption of perfect measurement was met. When reliability drops to 0.70 or below, we can see that about half the variance in these simple effects is lost, potentially leading to Type II errors. How many articles in respected journals have reliability estimates at or below that important conceptual threshold, and how many studies would find significant results if their effect sizes were not substantially attenuated due to poor measurement?

Reliability and Multiple IVs

With each IV added to a regression equation, the effects of less than perfect reliability become more complex and the results of the analysis more questionable. With the addition of one IV with less than perfect reliability, each succeeding variable entered has the opportunity to claim part of the shared variance that should have been accounted for by the unreliable variable(s). The apportionment of the explained variance among the IVs will thus be incorrect and reflect a misestimation of the true population effect.

Table 16.1   Example Disattenuation of Simple Correlation Coefficients

Reliability Estimate (α)	Observed Correlation Coefficient					
	$r = 0.10$ (0.01)	$r = 0.20$ (0.04)	$r = 0.30$ (0.09)	$r = 0.40$ (0.16)	$r = 0.50$ (0.25)	$r = 0.60$ (0.36)
0.95	0.11 (0.01)	0.21 (0.04)	0.32 (0.10)	0.42 (0.18)	0.53 (0.28)	0.63 (0.40)
0.90	0.11 (0.01)	0.22 (0.05)	0.33 (0.11)	0.44 (0.19)	0.56 (0.31)	0.67 (0.45)
0.85	0.12 (0.01)	0.24 (0.06)	0.35 (0.12)	0.47 (0.22)	0.59 (0.35)	0.71 (0.50)
0.80	0.13 (0.02)	0.25 (0.06)	0.38 (0.14)	0.50 (0.25)	0.63 (0.39)	0.75 (0.56)
0.75	0.13 (0.02)	0.27 (0.07)	0.40 (0.16)	0.53 (0.28)	0.67 (0.45)	0.80 (0.64)
0.70	0.14 (0.02)	0.29 (0.08)	0.43 (0.18)	0.57 (0.32)	0.71 (0.50)	0.86 (0.74)
0.65	0.15 (0.02)	0.31 (0.10)	0.46 (0.21)	0.62 (0.38)	0.77 (0.59)	0.92 (0.85)
0.60	0.17 (0.03)	0.33 (0.11)	0.50 (0.25)	0.67 (0.45)	0.83 (0.69)	—

NOTE: Reliability estimates for this example assume the same reliability for both variables. Percent variance accounted for (shared variance – coefficient of determination) is given in parentheses.

In essence, low reliability in one variable can lead to substantial overestimation of the effect of another related variable.

As more IVs with low levels of reliability are added to the equation, the likelihood that the variance accounted for is not apportioned correctly will increase. This can lead to erroneous findings and increased potential for Type II errors for the variables with poor reliability and Type I errors for the other variables in the equation. Obviously, this gets increasingly complex as the number of variables in the equation grows and becomes increasingly unacceptable in terms of replicability and confidence in results. In these cases, structural equation modeling could be considered a best practice.

Reliability and Interactions in Multiple Regression

Up to this point, the discussion has been confined to the relatively simple issue of the effects of imperfect reliability on simple effects. However, many interesting hypotheses involve curvilinear or interaction effects, as we have seen in prior chapters. Of course, poor reliability in main effects is compounded dramatically when those effects are used in cross-products, such as squared or cubed terms, or interaction terms. Aiken and West (1991) present a good discussion on the issue. An illustration of this effect is presented in Table 16.2.

As Table 16.2 shows, even at relatively high reliabilities, the reliability of cross-products is often substantially weaker (except when there are strong correlations between the two variables). This, of course, has deleterious effects on power and inference. According to Aiken and West (1991), there are two avenues for dealing with this: correcting the correlation or covariance matrix for low reliability, and then using the corrected matrix for the subsequent regression analyses,[5] or using structural

Table 16.2 The Reliabilities of Interaction Terms as a Function of Reliabilities of Individual Variables

Reliability of X and Z	Correlation Between X and Z			
	$r = 0$	$r = 0.20$	$r = 0.40$	$r = 0.60$
0.90	0.81	0.82	0.86	0.96
0.80	0.64	0.66	0.71	0.83
0.70	0.49	0.51	0.58	0.72
0.60	0.36	0.39	0.47	0.62

NOTE: These calculations assume that both variables are centered at 0 and that both X and Z have equal reliabilities. Numbers reported are cross-product reliabilities.

---

5 In theory, you could correct an entire correlation matrix for poor reliability, and then most statistical computing packages allow you to use correlation matrices as the data entered rather than raw data. This is both cumbersome and potentially subject to error.

equation modeling to model the relationships in an error-free fashion (which would be my recommendation).

Protecting Against Overcorrecting During Disattenuation

The goal of disattenuation is to be simultaneously accurate (in estimating the "true" relationships) and conservative in preventing overcorrecting. Overcorrection serves to further our understanding no more than leaving relationships attenuated.

There are several scenarios that might lead to inappropriate inflation of estimates, even to the point of impossible values. A substantial underestimation of the reliability of a variable would lead to substantial overcorrection, and potentially impossible values. This can happen when reliability estimates are biased downward by heterogeneous scales, for example. Researchers need to seek precision in reliability estimation in order to avoid this problem.

Given accurate reliability estimates, however, it is possible that sampling error, well-placed outliers, or even suppressor variables could inflate relationships artificially and thus, when combined with correction for low reliability, produce inappropriately high or impossible corrected values. In light of this, I would suggest that researchers make sure they have checked for these issues prior to attempting a correction of this nature (researchers should check for these issues regularly anyway).

Other (Better) Solutions to the Issue of Measurement Error

Fortunately, as the field of measurement and statistics advances, other options to these difficult issues emerge. One obvious solution (already mentioned above) to the problem posed by measurement error is to use structural equation modeling to estimate the relationship between constructs (which can provide estimates of error-free results given the right conditions), rather than utilizing our traditional methods of assessing the relationship between measures. This eliminates the issue of potential overcorrection or undercorrection from other methods of dealing with measurement error, as well as eliminating the question as to which estimate of reliability to use (since there are many, and they often fail to agree). Given the easy access to structural equation modeling software and a proliferation of structural equation modeling manuals and texts, this approach is more accessible to researchers now than ever before. Having said that, structural equation modeling is still a complex process and should not be undertaken without proper training and mentoring (of course, that is true of all statistical procedures).

Another emerging technology that can potentially address this issue is the use of modern measurement methodologies such as Rasch or IRT modeling. Rasch measurement utilizes a fundamentally different approach to measurement than classical test theory, which many of us were trained in. Use of Rasch measurement not only provides more sophisticated, and probably accurate, measurement of constructs but also gives more sophisticated information on the reliability of items and individual scores. Even an introductory treatise on Rasch measurement is outside the limits of this book, but individuals interested in exploring more sophisticated measurement models are encouraged to refer to Bond and Fox (2001) for an excellent primer on Rasch measurement.

Does Reliability Influence Other Analyses, Such as Analysis of Variance?

Analysis of variance (ANOVA) uses the same ordinary least squares (OLS) estimation as regression and produces the same results as OLS regression when performed appropriately (as we demonstrated in Chapter 4). Therefore, poor reliability in the DV should affect ANOVA in the same way. We often fail to talk about measurement error in categorical variables, but that can certainly happen as well (e.g., miscategorization of an individual).

In some simulations I ran while preparing for this section of the chapter, I often saw a 15% to 20% reduction in effect sizes between the ANOVA analyses with average reliability and the ones with poor reliability. In the example below, we use the SPQ with original reliability ($\alpha = 0.88$) and reduced reliability ($\alpha = 0.64$) created by random substitution of numbers. In this analysis, we will look at identification with academics (SPQ scores) as a function of whether the student withdrew from school or not.[6] Conceptually, those who stay in school (do not withdraw/drop out) should have higher scores on the SPQ than those who do withdraw/drop out. This simple ANOVA[7] performed on the SPQ with normal reliability ($\alpha = 0.88$) showed a strong difference between those who stayed in school (mean = 4.35 of 5.00, with a standard deviation [SD] = 0.45) and those who withdrew from school (mean = 3.50 of 5.00, with a $SD = 0.61$). This effect was significant ($F_{(1, 166)} = 94.96$, $p < .0001$, $\eta^2 = 0.36$).

When the DV was the SPQ with poor reliability ($\alpha = 0.64$), the effect was still significant and in the same direction (mean = 3.87 versus 3.22). However, the effect size was reduced ($F_{(1, 166)} = 58.29$, $p < .0001$, $\eta^2 = 0.26$) by 27.78% (which is approximately what would be expected given that in the parallel analysis with two variables of poor reliability the effect size was reduced by about 54%: $r = 0.75$ versus $r = 0.51$). This effect of poor reliability is significant enough that researchers are more likely to experience a Type II error in an analysis with reduced sample size (and hence, reduced power). This highlights the importance of paying attention to the quality of measurement in research and emphasizes the point that ANOVA and other linear models are susceptible to attenuation of effects due to poor measurement.[8]

---

6 Those of you interested in this topic can refer to a recent theoretical piece by myself and my colleague Brett Jones (Osborne & Jones, 2011) explaining how identification with academics is theoretically related to outcomes like dropping out of school. For purposes of this example, you have to trust that it should be.

7 Again, all assumptions were met except perfect reliability.

8 This is not a perfect simulation, of course, as you can see from the change in means between the original and less reliable SPQ scales. Because of the strong social desirability of the scale, responses are biased toward the upper end of the Likert scale, and random substitution of values tended to lower mean scores. However, the mean difference is still relatively large, the SDs were similar (0.57 versus 0.52), and the result is hopefully intuitive enough to drive the point home.

Reliability in Logistic Models

In my experience, reliability is more often discussed in the context of OLS regression–type models than in logistic-type models. However, there is no reason to expect that reliability is irrelevant in *any linear model*.

To provide an example of the effect of poor measurement on simple logistic regression, we will revisit our simple example predicting GRADUATE (coded 0 = no, 1 = yes) from family socioeconomic status (zSES) from earlier chapters. Recall that using the original data, students from families with higher zSES have increased probabilities of graduating (odds ratio [OR] = 2.83).

We will assume that SES and graduation are both perfectly measured and that this OR = 2.83 represents the gold standard against which we will compare the effects of poorer measurement. To simulate poorer measurement, I randomly selected 25%, 50%, or 75% of the cases to have zSES replaced with a random number generated from a standard normal distribution with a mean of 0 and *SD* of 1.0.[9] As confirmation, after these scores were randomly replaced by random numbers, the correlation between the original SES variable and the new "less reliably measured" SES variable was $r = 0.75$, 0.52, and 0.27 for the 25%, 50%, and 75% replacement conditions, respectively.

As you can see in Table 16.3, when 25% of SES values are replaced by random numbers (simulating poorer reliability), the OR decreases from 2.83 to 2.11, and the logit decreases from 1.04 to 0.75. Similarly, less reliable measurement (50% or 75% of values substituted with random values) results in more attenuation. Thus, it is clear that poor reliability in logistic regression has the same deleterious effect on accuracy of the estimates as it does in other common linear models.

Table 16.3 Family Socioeconomic Status and Odds of Graduating High School in Models With Different Levels of Measurement Error

% Replaced by Random Numbers		B	SE	Wald	df	p	Exp(B)
Original	zSES	1.041	0.033	1,018.355	1	.000	2.831
	Constant	2.740	0.037	5,409.564	1	.000	15.480
25%	SES_25	0.748	0.030	624.180	1	.000	2.112
	Constant	2.544	0.033	6,115.089	1	.000	12.737
50%	SES_50	0.491	0.029	296.872	1	.000	1.634
	Constant	2.423	0.030	6,688.529	1	.000	11.282
75%	SES_75	0.235	0.028	73.123	1	.000	1.265
	Constant	2.348	0.028	7,138.306	1	.000	10.463

SOURCE: National Education Longitudinal Study of 1988 (NELS88) from the National Center for Education Statistics (http://nces.ed.gov/surveys/nels88/). Adapted from Osborne (2015).

---

9 The correlation between the original SES variable and the randomly generated variable was $r = 0.015$, meaning I was replacing mostly good data with unrelated "noise" or random error.

But Other Authors Have Argued That
Poor Reliability Isn't That Important. Who Is Right?

I am, of course!

Even if you are not completely convinced of that last statement, I hope I have persuaded you that this is an issue worthy of attention. There are other articles that espouse the opposite view, just as there are authors who argue it is irresponsible to remove outliers or transform non-normal data.

One example of such an author is Schmitt (1996), who argued that even relatively low alphas (e.g., $\alpha = 0.50$) do not seriously attenuate validity coefficients, which are correlations. In his example, he cites the maximum validity coefficient under alpha as the square root of alpha (in this case, $\alpha = 0.50$ would yield a maximum validity coefficient of 0.71). Following this logic, you have now set the ceiling for the highest possible effect size you can report in your research at 0.50, rather than 1.00. This should be a serious concern for any researcher. Even Schmitt's own results show profound effects of poor reliability. In one example (Schmitt, 1996, Table 4), two variables with $\alpha = 0.49$ and 0.36, respectively, show an observed correlation of $r = 0.28$. A simple calculation from Equation 16.1 shows a corrected correlation of $r = 0.67$; not only does this look substantially larger in this metric, but when one examines the coefficients of determination ($r^2 = 0.08$ and 0.45, respectively), one can truly see how misguided this argument is. In this case, the true effect size is attenuated by 82.18%—or *the correct effect size is almost six times that of the observed effect size*! Furthermore, if this example were extended to a multiple regression model, it is likely that other related predictors would have their unique effects substantially overestimated due to the very poor measurement of this example predictor.

I hope you are of the mind that you are not happy to misestimate your effect sizes to such a degree in your research.

Sample Size and the Precision/Stability of Alpha-Empirical CIs

Fan and Thompson (2001) point out that few authors provide context in their articles as to the precision of their effect size point estimates. Alpha is merely a point estimate like any other statistic that we have talked about thus far; alpha is also sample dependent, with representative samples better than biased samples, and larger samples better than smaller samples. Thus, when authors report alpha without the context of CIs, readers have no way to understand how precise that estimate is and how likely that estimate is to replicate.

Although there have been attempts in the literature to construct methods to calculate CIs for alpha, these have not gained traction in routine practice. With bootstrap resampling, we can easily provide empirical estimates that are valuable for readers. Reliability estimates of $\alpha = 0.80$ that have 95% CIs of [0.78, 0.82] are much more precise than those with 95% CIs of [0.50, 0.90], and this information provides important context. CIs around statistics such as alpha are desirable (and relatively easily calculated, now that you are expert in bootstrap analysis; for an example of bootstrap analysis and an associated SPSS macro for alpha, see Osborne, 2014, Chapter 10).

SUMMARY AND CONCLUSIONS

Throughout this book, we have talked about the importance of meeting assumptions to ensure basic quality of the results. One consistent assumption throughout all linear models was that of perfect measurement. As one deviates from that assumption, we can assume that the results of the models are increasingly less useful, often in unpredictable ways.

Modern statistical methods provide ways to help remedy unreliable measurement, such as the use of structural equation modeling to model error-free latent variables; however, even structural equation modeling models are only as good as the data they use. We must, as scientists, end the fallacy that poor measurement is forgivable and not harmful to our results, and we must focus on basing our research on high-quality measurement.

This chapter is more conceptual, and as such, I am not including exercises. My book on exploratory factor analysis (Osborne, 2014) has more in-depth information on using SPSS to explore measurement issues such as reliability. A forthcoming book (Osborne & Smith, in press) will explore the same content using SAS. Both include hands-on exercises with real data.

REFERENCES

Aiken, L. S., & West, S. (1991). *Multiple regression: Testing and interpreting interactions.* Thousand Oaks, CA: SAGE.

Barnette, J. J. (2005). ScoreRel CI: An Excel program for computing confidence intervals for commonly used score reliability coefficients. *Educational and Psychological Measurement, 65*(6), 980–983. doi: 10.1177/0013164405278577

Bond, T. G., & Fox, C. M. (2006). *Applying the Rasch model: Fundamental measurement in the human sciences.* London, England: Psychology Press.

Cohen, J., Cohen, P., West, S., & Aiken, L. S. (2003). *Applied multiple regression/correlation analysis for the behavioral sciences.* Mahwah, NJ: Lawrence Erlbaum.

Cortina, J. M. (1993). What is coefficient alpha? An examination of theory and applications. *Journal of Applied Psychology, 78*(1), 98–104. doi: 10.1037/0021-9010.78.1.98

Cronbach, L. J. (1951). Coefficient alpha and the internal structure of tests. *Psychometrika, 16*(3), 297–334. doi: 10.1007/BF02310555

Edelen, M. O., & Reeve, B. B. (2007). Applying item response theory (IRT) modeling to questionnaire development, evaluation, and refinement. *Quality of Life Research, 16*(Suppl. 1), 5–18. doi: 10.1007/s11136-007-9198-0

Fan, X., & Thompson, B. (2001). Confidence intervals for effect sizes confidence intervals about score reliability coefficients, please: An EPM guidelines editorial. *Educational and Psychological Measurement, 61*(4), 517–531. doi: 10.1177/0013164401614001

Feldt, L. S. (1980). A test of the hypothesis that Cronbach's alpha reliability coefficient is the same for two tests administered to the same sample. *Psychometrika, 45*(1), 99–105. doi: 10.1007/BF02293600

Hambleton, R. K., Swaminathan, H., & Rogers, H. J. (1991). *Fundamentals of item response theory.* Thousand Oaks, CA: SAGE.

Nunnally, J. C., & Bernstein, I. H. (1994). *Psychometric Theory* (3rd ed.). New York, NY: McGraw Hill.

Osborne, J. W. (1997). Identification with academics and academic success among community college students. *Community College Review, 25*(1), 59–67. doi: 10.1177/009155219702500105

Osborne, J. W. (2008). Sweating the small stuff in educational psychology: How effect size and power reporting failed to change from 1969 to 1999, and what that means for the future of changing practices. *Educational Psychology, 28*(2), 151–160. doi: 10.1080/01443410701491718

Osborne, J. W. (2014). *Best practices in exploratory factor analysis.* Seattle, WA: CreateSpace Publishing.

Osborne, J. W. (2015). *Best practices in logistic regression.* Thousand Oaks, CA: SAGE.

Osborne, J. W., & Blanchard, M. R. (2011). Random responding from participants is a threat to the validity of social science research results. *Frontiers in Psychology, 1*, 220. doi: 10.3389/fpsyg.2010.00220

Osborne, J.W., & Jones, B. (2011). Identification with academics and motivation to achieve in school: How the structure of the self influences academic outcomes. *Educational Psychology Review, 23*, 131–158. doi: 10.1007/s10648-011-9151-1

Osborne, J. W., & Smith, E. A. (in press). *Exploratory factor analysis using SAS: Best practices and modern methods.* Cary, NC: SAS Publishing.

Pedhazur, E. J. (1997). *Multiple regression in behavioral research: Explanation and prediction.* Fort Worth, TX: Harcourt Brace College Publishers.

Schmitt, N. (1996). Uses and abuses of coefficient alpha. *Psychological Assessment, 8*(4), 350–353. doi: 10.1037/1040-3590.8.4.350

Smith, E. V., Jr., & Smith, R. M. (2004). *Introduction to Rasch measurement.* Maple Grove, MN: JAM Press.

Voelkl, K. E. (1996). Measuring students' identification with school. *Educational and Psychological Measurement, 56*(5), 760–770. doi: 10.1177/0013164496056005003

Voelkl, K. E. (1997). Identification with school. *American Journal of Education, 105*(3), 294–318.

Wilkinson, L., & Task Force on Statistical Inference, APA Board of Scientific Affairs. (1999). Statistical methods in psychology journals: Guidelines and explanations. *American Psychologist, 54*(8), 594–604. doi: 10.1037/0003-066X.54.8.594

Wilson, M. (2005). *Constructing measures: An item response modeling approach.* Mahwah, NJ: Lawrence Erlbaum.

17

PREDICTION IN THE GENERALIZED LINEAR MODEL

Advance Organizer

Imagine you have developed a great new intervention to, say, help students with reading difficulties or to help obese people lose weight. Your initial research shows that it works well, but it is extremely expensive and so only those who are most likely to benefit from your intervention should have access to it. How are you going to decide who should get the intervention?

Or imagine you are in charge of admissions for your college or university. You know that certain things predict success in your school, at least partly. How do you decide who gets admitted?

Most research we hear about is *explanatory,* meaning that the goal of the research is attempting to understand a phenomenon. For example, studies explore what variables explain success in your college, or what variables help ameliorate the negative effects of obesity, or which intervention seems to produce the best outcomes among students having difficulty reading. Almost every published research study has a section at the end where the authors tell us why we should care about their particular findings. In essence, these summary statements (e.g., people with height-to-waist ratios of less than 1.5 might benefit most from this intervention, or students scoring above the 80th percentile on this particular measure are three times as likely to succeed in our college as students scoring lower) are predictions of efficacy in the future.

But how do we know that these results will generalize to your patients, your students, or your community partners? You can try replicating the results in another sample, and you can keep replicating the results *ad nauseam*, which will give you more confidence if you keep getting the same results. However, there are traditional ways of evaluating how well a regression equation will generalize to a new sample—something we addressed more generally in Chapter 15.

This chapter seeks to present a process of validation, and an example of how best to do this. Authors have been writing about this process for decades, yet it is rarely covered in depth in statistics textbooks. Where prediction is covered, it is most often (in my experience) covered in the context of ordinary least squares (OLS) regression, although the concepts apply equally well to other aspects of the generalized linear model (GLM; e.g., logistic models).

In this chapter, we will cover some basic concepts around prediction:

- Prediction versus explanation
- OLS regression concepts such as
  - Shrinkage
  - Cross-validation of prediction equations
  - Calculating confidence intervals (CIs) around individual predictions
- Analogues for logistic models
- Applications of bootstrap resampling

Guidance on how to perform these analyses in various statistical packages will be available online at study.sagepub.com/osbornerlm.

Prediction Versus Explanation

There are two general applications for linear models: prediction and explanation.[1] These applications roughly correspond to two differing goals in research: being able to make valid projections concerning an outcome for a particular individual (prediction), or attempting to understand a phenomenon by examining a variable's correlates on a group level (explanation). There has been substantial, often contentious debate as to whether these two applications of multiple regression are grossly different, as authors such as Scriven (1959) and Anderson and Shanteau (1977) assert, or whether they are necessarily part and parcel of the same process (DeGroot, 1969; Kaplan, 1964; for an overview of this discussion, see Pedhazur, 1997, pp. 195–198). I believe both are necessarily part of the scientific process. We seek to take our understanding of phenomena from our research and apply it to people who were not specifically involved in our research (a process called generalization). Although both processes are, in my mind, part of the same goal and process, there are some nuances to the recommended analytic procedures involved with the two different types of analyses.

When one uses linear models for explanatory purposes, this person is exploring relationships between multiple variables in a sample to shed light on a phenomenon, with a goal of generalizing this new understanding to a general or specific population. When one uses linear models for prediction, this researcher is using a sample to create a regression equation that would optimally predict a particular phenomenon within a particular population. The difference is that because the equations are going to be applied to cases not yet studied or measured (i.e., individuals *not in the sample used in the analysis*), we need to have a reasonable level of confidence

---

1 Some readers may be uncomfortable with the term *explanation* when referring to multiple regression because these data are often correlational in nature, whereas the word *explanation* often implies causal inference. However, *explanation* will be used in this chapter because (a) it is the convention in the field, (b) here I am talking about regression with the *goal* of explanation, and (c) one can come to an understanding of phenomena by understanding associations without positing or testing strict causal orderings.

that what we saw in our particular sample will generalize. Let us imagine research-ers creating a linear model to predict twelfth-grade achievement test scores from eighth-grade variables, such as family socioeconomic status, race, sex, educational plans, parental education, grade point average (GPA), and participation in school-based extracurricular activities. The goal is not to understand why students achieve at a certain level, because we already have abundant research on these issues; rather, we aim to create the best equation so that, for example, guidance counselors could predict future achievement scores for their students, and (hopefully) intervene with those students identified as at risk for poor performance, or select students into programs based on their projected scores. Although theory is useful for identifying what variables should be in a prediction equation, the variables do not necessarily need to make conceptual sense. If the single greatest predictor of future achieve-ment scores was how high that student could jump, or the number of hamburgers a student eats, it should be in the prediction equation regardless of whether it makes sense (although researchers paying attention might be interested enough in this type of finding to pursue some explanatory research on the topic, again demonstrating the relatedness of these two processes . . .).

For those of you who are not completely comfortable with this assertion, let us acknowledge that the basic aspects of the mathematics are the same for both. You will get the same intercepts and regression weights given a data set regardless of whether your intent is explanation or prediction. This conversation reminds me of another conversation people get passionate about—causal modeling. Using these applications to create causal inferences as opposed to relational inferences has nothing to do with the mathematics of the analysis, but rather the intent, and the mechanisms to generate the data. If you generated data causally, you may be able to make causal inferences. If not, you may not.

How Is a Prediction Equation Created?

The general process for creating a prediction equation involves gathering relevant data from a *large, representative sample* from the population you wish to generalize to. What constitutes "large" is open to debate. Although guidelines for general appli-cations of regression are as small as $50 + 8 \times$ the number of predictors (Tabachnick & Fidell, 1996),[2] guidelines for prediction equations are more stringent due to the need to generalize beyond a given sample. Whereas some authors have suggested that 15 subjects per predictor is sufficient (Park & Dudycha, 1974; Pedhazur, 1997), others have suggested a minimum total sample (e.g., 400; see Pedhazur, 1997) and still others have suggested a minimum of 40 subjects per predictor (Cohen, Cohen, West, & Aiken, 2003; Tabachnick & Fidell, 1996). Of course, because the goal is a stable regression equation that is representative of the population regression equa-tion, more is better, but only to the extent that it increases the representativeness of the sample.

---

2   Personally, I find rules of thumb less useful than processes such as power calculations (e.g., see Chapter 15). Please take all such rules of thumb with a large grain of salt and skepticism.

Let me be excruciatingly clear on this next point: getting a truly representative sample of the population you wish to generalize to is the key ingredient in this type of research. You can do all of the fancy analyses you want, but if you validate a prediction equation on a sample that does not mirror the intended use, you've wasted your time. True representativeness is one of the holy grails of research, and it is something that is not talked about quite enough. It is not usually something you can test for, so most assume their sampling strategies are adequate. Yet this is probably not the case. Make sure you clearly define your population of interest, and sample in such a way as to maximize the probability of getting a representative sample(s).

Methods for Entering Variables Into the Equation

There are many ways to enter predictors into the regression equation. This is something we have not discussed deeply in this book because most modern applications of linear modeling for scientific research prefer forced (researcher-controlled) entry. However, stepwise-entry methods might be defensible if your application of regression modeling is *solely and entirely* to predict an outcome, and particularly if there are many hundreds of variables from which to choose.

Several entry methods rely on the statistical properties of the variables to determine order of entry (e.g., forward selection, backward elimination, stepwise). Others rely on the experimenter to specify order of entry (hierarchical, blockwise) or have no order of entry (simultaneous). Almost all statisticians today eschew stepwise methods in favor of analyst-controlled entry, and they discourage entry based on the statistical properties of the variables because it is atheoretical.[3] I agree with this point of view for the majority of research, which is often explanatory. However, when it comes to creating prediction equations, the only goal is creating the best prediction equation. If you examine the processes most authors recommend for creating prediction equations, they essentially involve manual stepwise entry. I see no compelling reason not to use stepwise methods provided that (a) your *only* goal is creating a prediction equation and (b) you are knowledgeable of stepwise methods and can make educated decisions about how to manage the process.

Regardless of the entry method ultimately chosen by the researcher, it is critical that the researcher examine individual variables to ensure that only variables contributing significantly to the variance accounted for by the regression equation are included. Variables not accounting for significant portions of variance should be deleted from the equation, and the equation should be re-estimated (which is essentially what many stepwise procedures do automatically). Furthermore, researchers might want to examine excluded variables to see whether their entry would significantly improve prediction (a significant increase in R^2; again, what some stepwise methods do).

---

3   A thorough discussion of this issue is beyond the scope of this book, so the reader is referred to Cohen, Cohen, West, and Aiken (2003) and Pedhazur (1997) for overviews of the various techniques, and Thompson (1989, 1995) and Schafer (1991, 1992) for more discussions of the issues.

There are downsides to stepwise-entry methods, which should not be ignored. For example, they cannot evaluate nonlinear effects easily, nor can they easily automatically model interaction terms. They do not commonly evaluate influential data points and do not check assumptions. Any one of these factors could adversely influence the quality of your outcome.

Shrinkage and Evaluating the Quality of Prediction Equations

Any linear model produced based on a particular sample is optimized for the sample at hand. Because this process capitalizes on chance and error in the sample (a phenomenon many authors refer to as "overfitting"), the equation produced in one sample will not generally fare as well in another sample (i.e., R^2 in a subsequent sample using the same equation will not be as large as R^2 from the original sample), a phenomenon called *shrinkage*. The most desirable outcome in this process is for minimal shrinkage, indicating that the prediction equation will generalize well to new samples or individuals from the population examined. Although there are equations that can estimate shrinkage, the best way to estimate shrinkage and test the prediction equation is through cross-validation or double cross-validation.

Cross-Validation

To perform cross-validation, a researcher will gather either two large samples or one very large sample that will be split into two samples via random selection procedures. The prediction equation is created in the first sample. That equation is then used to create predicted scores for the members of the second sample. The predicted scores are then correlated with the observed scores on the dependent variable ($r_{y\hat{y}}$). This is called the *cross-validity coefficient*. The difference between the original model R^2 and $r_{y\hat{y}}^2$ is the shrinkage estimate. The smaller the shrinkage, the more confidence we can have in the generalizability of the equation.

Double Cross-Validation

In double cross-validation, prediction equations are created in both samples and each is then used to create predicted scores and cross-validity coefficients in the other sample. This procedure involves little work beyond cross-validation and produces a more informative and rigorous test of the generalizability of the regression equation(s). In addition, because two equations are produced, one can look at the stability of the actual regression line equations.

An Example Using Real Data

We will use a small ($N = 700$) subset of the National Education Longitudinal Survey of 1988 from the National Center for Education Statistics (NELS88) data to provide a realistic example of prediction, shrinkage, and cross-validation. We will predict twelfth-grade self-reported GPA (GPA12; measured on a scale from 1 to 12) from eighth-grade variables such as GPA (GPA8; measured on a scale from 1 to 5), RACE (0 = Caucasian, 1 = African American), participation in school activities (PART; composite of extracurricular activities ranging from 1 to 2.75), and parent education (PARED; ranges from 1 = did not finish high school to 6 = Ph.D., M.D., or other terminal degree). A random sample of 700 students from NELS88 (more accurately reflecting typical samples researchers collect) was randomly split into two groups: a prediction and a validation sample (samples 1 and 2, respectively, in the data set). The results from the first sample of $N = 350$ revealed the following regression line equation:

$$\hat{Y} = -2.45 + 1.83(GPA) - 0.77(RACE) + 1.03(PART) + 0.38(PARED)$$

In the first group, this analysis produced an overall $R^2 = 0.55$. To compute the shrinkage estimate, we can take this equation and predict scores for the second sample and correlate the predicted scores with the observed scores. Those predicted scores correlated $r_{y\hat{y}} = 0.73$ with observed scores. With a $r_{y\hat{y}}^2$ of 0.53 (cross-validity coefficient), shrinkage was 2%, a good outcome.

Double Cross-Validation

In double cross-validation, prediction equations are created in both samples and then each is used to create predicted scores and cross-validity coefficients in the other sample. This procedure involves little work beyond cross-validation and produces a more informative and rigorous test of the generalizability of the regression equation(s). In addition, because two equations are produced, one can look at the stability of the actual regression line equations.

The following regression equation emerged from analyses of the second sample:

$$\hat{Y} = -4.03 + 2.16(GPA) - 1.90(RACE) + 1.43(PART) + 0.28(PARED)$$

This analysis produced an $R^2 = 0.60$. This equation was used in the first group to create predicted scores in the first group, which correlated 0.73 with observed scores, for a cross-validity coefficient of 0.53. Note that (a) the second analysis revealed larger shrinkage than the first, (b) the two cross-validation coefficients were identical (0.53), and (c) the two regression equations are markedly different, even though the samples had large subject to predictor ratios (over 80:1).

So How Much Shrinkage Is Too Much Shrinkage?

There are no clear guidelines concerning how to evaluate shrinkage, except the general agreement that less is always better. But is 3% acceptable? What about 5% or 10%? Or should it be a proportion of the original R^2 (so that 5% shrinkage on an R^2 of 0.50 would be fine, but 5% shrinkage on an R^2 of 0.30 would not be)? There are no guidelines in the literature. However, Pedhazur has suggested that one of the advantages of double cross-validation is that one can compare the two cross-validity coefficients; if they are similar, one can be fairly confident in the generalizability of the equation.

The Final Step

If you are satisfied with your shrinkage statistics, the final step in this sort of analysis is to combine both samples (assuming shrinkage is minimal) and create a final prediction equation based on the larger sample. In our data set, the combined sample produced the following regression line equation:

$$\hat{Y} = -3.23 + 2.00(\text{GPA8}) - 1.29(\text{RACE}) + 1.24(\text{PART}) + 0.32(\text{PARED})$$

How Does Sample Size Affect the Shrinkage and Stability of a Prediction Equation?

As discussed above, there are many different opinions as to the minimum sample size one should use in prediction research. As an illustration of the effects of different subject to predictor ratios on shrinkage and stability of a regression equation, data from NELS88 were used to construct prediction equations identical to our running example. Two samples representing each of the sample sizes that correspond to ratios of 5, 15, 40, 100, and 400 subjects per predictor were randomly selected from the full data set sample (randomly selecting from the full sample for each new pair of a different size). Following selection of the samples, prediction equations were calculated, and double cross-validation was performed. The results are presented in Table 17.1.

The first observation from Table 17.1 is that, by comparing regression line equations, the very small samples can have wildly fluctuating parameter estimates (both

Table 17.1 Comparison of Double Cross-Validation Results With Differing Subject to Predictor Ratios

Sample Ratio	Obtained Prediction Equation	R^2	$r^2_{y\hat{y}}$	Shrinkage
Population	$\hat{Y} = -1.71 + 2.08(\text{GPA8}) - 0.73(\text{RACE}) -$ $0.60(\text{PART}) + 0.32(\text{PARED})$	0.48		
5:1				
Sample 1	$\hat{Y} = -8.47 + 1.87(\text{GPA8}) - 0.32(\text{RACE}) +$ $5.71(\text{PART}) + 0.28(\text{PARED})$	0.62	0.53	0.09
Sample 2	$\hat{Y} = -6.92 + 3.03(\text{GPA8}) + 0.34(\text{RACE}) +$ $2.49(\text{PART}) - 0.32(\text{PARED})$	0.81	0.67	0.14
15:1				
Sample 1	$\hat{Y} = -4.46 + 2.62(\text{GPA8}) - 0.31(\text{RACE}) +$ $0.30(\text{PART}) + 0.32(\text{PARED})$	0.69	0.24	0.45
Sample 2	$\hat{Y} = -1.99 + 1.55(\text{GPA8}) + 0.34(\text{RACE}) +$ $1.04(\text{PART}) - 0.58(\text{PARED})$	0.53	0.49	0.04

(Continued)

Table 17.1   (Continued)

Sample Ratio	Obtained Prediction Equation	R^2	$r^2_{y\hat{y}}$	Shrinkage
40:1				
Sample 1	$\hat{Y} = -0.49 + 2.34(\text{GPA8}) - 0.79(\text{RACE}) -$ $1.51(\text{PART}) + 0.08(\text{PARED})$	0.55	0.50	0.05
Sample 2	$\hat{Y} = -2.05 + 2.03(\text{GPA8}) - 0.61(\text{RACE}) -$ $0.37(\text{PART}) + 0.51(\text{PARED})$	0.58	0.53	0.05
100:1				
Sample 1	$\hat{Y} = -1.89 + 2.05(\text{GPA8}) - 0.52(\text{RACE}) -$ $0.17(\text{PART}) + 0.35(\text{PARED})$	0.46	0.45	0.01
Sample 2	$\hat{Y} = -2.04 + 1.92(\text{GPA8}) - 0.01(\text{RACE}) +$ $0.32(\text{PART}) + 0.37(\text{PARED})$	0.46	0.45	0.01
400:1				
Sample 1	$\hat{Y} = -1.26 + 1.95(\text{GPA8}) - 0.70(\text{RACE}) -$ $0.41(\text{PART}) + 0.37(\text{PARED})$	0.47	0.46	0.01
Sample 2	$\hat{Y} = -1.10 + 1.94(\text{GPA8}) - 0.45(\text{RACE}) -$ $0.56(\text{PART}) + 0.35(\text{PARED})$	0.42	0.41	0.01

intercept and regression coefficients). Even the 40:1 ratio samples have impressive fluctuations across the two random samples. Although the fluctuations in the 100:1 sample are fairly small in magnitude, some coefficients reverse direction or are far off of the population regression line. As expected, it is only in the largest samples in which the equations stabilize and remain close to the population equation.

Variance accounted for (R^2) is also misestimated in many of the analyses of samples below a 100:1 ratio. Cross-validity coefficients vary a great deal across samples until a 40:1 ratio is reached, where they appear to stabilize. Finally, it appears that shrinkage appears to minimize as a 40:1 ratio is reached. If one compares cross-validity coefficients to determine whether an equation is stable, from these data one would advise that a sample size of 40:1 ratio or better is needed. If the goal is to get an accurate, stable estimate of the population regression equation (which it should be if that equation is going to be widely used outside the original sample), it appears desirable to have at least 100 subjects per predictor, given these data.

Improving on Prediction Models

There are many ideas in the literature about how to create better prediction models. Some of the ideas involve "tuning," "recalibration," or incrementally improving models according to parameters such as error terms or using techniques that are in the same general family as bootstrap resampling to create more robust and general models. For example, there is currently discussion in the literature about multisample tuning, such as repeatedly sampling 90% of the data and using the remaining 10% to evaluate the quality of the model. This process can be repeated multiple times, tuning the model along the way until a stable model is created.

Leave-one-out models are the most extreme, in which the entire sample minus one case is used to create a prediction model (similar to "jackknife" resampling methodology). This process is repeated until a stable regression equation is created.

Authors have also talked about using shrunken coefficients, in which each coefficient is modified (i.e., shrunk) to create more accurate prediction equations. For example, if you are using OLS estimation, one general proposal for a shrinkage factor ($\hat{c}$) could be calculated using Equation 17.1:

$$\hat{c} = 1 - \frac{ps^2}{SS_{model}} \quad (17.1)$$

where SS_{model} is the explained sum of squares, p is the number of predictors, and s^2 is the residual variance (all of which are available via standard output in most statistical packages). This factor can then be used as a weight for each regression coefficient, thus attempting to create a more generalizable equation (e.g., van Houwelingen & Sauerbrei, 2013).

Bootstrap resampling analysis can also be used to simplify and improve the above-described processes. Using resampling, we can create large numbers of derivative samples based on the full original sample, and models from these bootstrap samples can be tested in the original sample or on cases not included in the bootstrapped sample (e.g., a modification of the 90%/10% methodology above).

I am not sure there is a general answer for an improved methodology aside from having a large sample (or several large samples) and using generalizable, clean data. The more high stakes the prediction, the more care you should feel obligated to take in order to ensure that you are using the best possible prediction equation, and the more care that should be used when using predictions to make decisions about individuals.

Unfortunately, prediction is never straightforward, even with recalibrated models. As Steyerberg, Borsboom, van Houwelingen, Eijkemans, and Habbema (2004) note, extensive recalibration of models can lead to overly optimistic estimates of prediction effectiveness. If this is a particular interest to you, I would encourage you to review some of the recent literature in this area and use prediction with caution, particularly where there may be heterogeneity within samples (in which case, interactions might be appropriate for the model).

Calculating a Predicted Score, and CIs Around That Score

There are two categories of predicted scores relevant here: scores predicted for the original sample, and scores that can be predicted for individuals outside the original sample.

Individual predicted scores and CIs for the original sample are available in the output available from most common statistical packages. Thus, the latter will be addressed here.

Once an analysis is completed and the final regression line equation is formed, it is possible to create predictions for individuals who were not part of the original sample that generated the regression line (one of the attractive features of regression). Calculating a new prediction based on an existing regression line is a simple matter of substitution and algebra. However, no such prediction should be presented without CIs, especially in light of the observed volatility in regression lines in Table 17.1. The only practical way to do this is through the following formula (Equation 17.2):

$$\hat{Y} \pm t_{(\alpha/2,\, df)} (s_{\hat{y}}) \tag{17.2}$$

$s_{\hat{y}}$ is calculated as

$$s_{\hat{y}} = \sqrt{s_{\hat{\mu}}^2 + MS_{\text{residual}}}$$

where $S_{\hat{\mu}}^2$ is the squared standard error of mean predicted scores (standard error of the estimate, squared), and the mean square residual, both of which can be obtained from typical regression output.[4]

Prediction (Prognostication) in Logistic Regression (and Other) Models

One of the recurrent themes of this book is that OLS regression is not an island unto itself; as such, prediction equations and parallel methods are available for other aspects of the GLM (e.g., logistic, multinomial, or Poisson models). These are sometimes called "prognostic" models rather than predictive models, but they follow many of the same general principles as those already discussed.

Because of the different estimation and mechanics of logistic-type models (which should include Poisson models), there are some different metrics that one can use to evaluate the performance of a prediction model. However, themes already developed apply throughout the GLM: overfitting of a model to a sample is likely to take place, and thus models from one sample are unlikely to have as good a fit with different samples, particularly with smaller samples or samples from different demographic groups.

Overall Performance

To evaluate the overall performance or average prediction error, we can calculate the Brier score as in Equation 17.3, where y_i is the observed outcome (observed category, 0 or 1) for subject i, and p_i is the prediction (predicted probability, which

---

4 It is often the case that one will want to use standard error of the predicted score when calculating an individual CI. However, because that statistic is only available from statistical program output, and only for individuals in the original data set, it is of limited value for this discussion. Here I suggest using the standard error of the mean predicted scores, because it is the best estimate of the standard error of the predicted score, knowing it is not completely ideal but lacking any other alternative.

most modern statistical software packages will save) for subject i in the data set of n subjects. The Brier score is 0 for perfect models, and models that diverge substantially from zero (Brier scores range up to 0.25) are less optimal. Because the Brier score requires observed probabilities, for our purposes, it is most easily applied to models with categorical predictors only.

$$\text{Brier score} = \Sigma \frac{\left(y_i - p_i\right)^2}{n} \tag{17.3}$$

Calibration refers to the extent of bias in the prediction equation. A basic metric for calibration is the intercept, which is also referred to as the "case mix" because it represents the average probability of experiencing an outcome when all other variables are zero. If, on average, 9% of a group is diagnosed with diabetes, then the prediction equation should have average predicted probabilities of 0.09 for that group. Minor deviations from the expected probability are routine; however, as predicted probabilities move farther from the observed probabilities, the calibration of the prognostic equation is called into question (e.g., Harrell, Lee, & Mark, 1996). The other aspect of calibration is the slope of each individual variable, which can fluctuate significantly across samples, as you can see in Table 17.1. Some authors have recommended that prediction equation coefficients be scaled by the slope of the linear predictor (also referred to as the prognostic index [PI], which is merely a predicted value based on the regression equation) predicting the outcome. This is also referred to as the "calibration slope" (e.g., Steyerberg, Harrell, Borsboom, Eijkemans, Vergouwe, & Habbema, 2001b), something we will explore below.

Concordance or Discrimination

For binary outcomes, concordance (c) is the typical measure of discrimination, or the ability to distinguish high-risk from low-risk cases. Put another way, discrimination is the ability of a variable to correctly separate cases that experienced one outcome over another (e.g., body mass index helps separate those diagnosed with diabetes versus those not diagnosed with diabetes). This statistic is identical to the area under the receiver operating characteristic (ROC) curve and varies between 0.50 and 1.00 for reasonable models, with higher numbers indicating better models (Royston, Moons, Altman, & Vergouwe, 2009; Steyerberg et al., 2001b).[5] For binary outcomes, we can conceptualize concordance as pairs of cases, in which predicted probabilities were in line with actual outcomes. If we were looking at DIABETES, a concordant model would assign a higher probability to a case that reported being diagnosed with diabetes as compared with one not diagnosed.

When a model has good discrimination but poor calibration, improvement in the model is possible. When models have poor discrimination, further attempts at calibration are likely to be fruitless (Harrell et al., 1996).

---

5 Calculating the area under the ROC curve requires either some advanced mathematical skills or a modern statistical software package. For example, in SPSS, you can save the predicted probabilities from a logistic regression analysis and select the ROC curve function under ANALYZE to compute the area under the ROC curve. Similar options are likely available in other packages.

Estimated Shrinkage in Logistic Models

According to Harrell et al. (1996), a good way to estimate the expected shrinkage from a logistic model (Equation 17.4) is to calculate the ratio of model χ^2 (the total likelihood ratio statistic when all p parameters are in the equation) to χ^2.

Expected shrinkage

$$(\hat{\gamma}) = \frac{\chi^2_{model} - p}{\chi^2_{model}} \qquad (17.4)$$

This statistic can roughly indicate the amount of overfitting in the model. Some authors have argued that a good estimate of shrinkage or miscalibration could then be used to weight regression terms to better calibrate a model; however, there are questions in the literature because the calibration coefficients are also rooted in specific samples.

Other Proposed Methods of Estimating Shrinkage

There are other methods for estimating shrinkage aside from Equation 17.4. These include ridge regression (Hoerl & Kennard, 1970), penalized maximum likelihood (Harrell et al., 1996), the least absolute shrinkage and selection operator (LASSO; Tibshirani, 1996), and bootstrap resampling (Steyerberg, Eijkemans, & Habbema, 2001a), which we will explore below. As Styerberg et al. (2001a) reported, many of the internal validation methods converged at similar points as sample sizes (particularly events per variable) increased, but bootstrap resampling resulted in stable and the most unbiased estimates of performance of the prognostic equation compared with other methods (e.g., split half, 90%/10%, etc.), which is sensible because it takes advantage of the full sample.

An Example of External Validation of a
Prognostic Equation Using Real Data

To explore applications in prognostic equations, we will return to a relatively simple example from Chapter 9, Table 9.6b, predicting graduation from high school (GRADUATE) from eight-grade achievement (zACH), family socioeconomic status (zSES), and the interaction of the two predictors (SESACH). With just these three variables in the equation, and removing 88 cases with Cook's Distance 5 standard deviations or more above the mean[6] and entering all three terms simultaneously into the equation,[7] we are left with the following overall statistics, in Table 17.2.

---

6 In Chapter 9, this cutoff removed only 86 cases; this is because SEX and the interactions related to SEX slightly altered the statistics for two cases.

7 We do not need to do blockwise entry here, because we already established the importance of the interaction term. We simply want the final equation.

Table 17.2   Regression Results for the Full Sample ($N = 16,520$)

Omnibus Tests of Model Coefficients

		Chi-Square	df	p
	Step	2,174.146	3	.000
Step 1	Block	2,174.146	3	.000
	Model	2,174.146	3	.000

Variables in the Equation

		B	SE	Wald	df	p	Exp(B)	95% CI for Exp(B)	
								Lower	Upper
Step 1[a]	zACH	1.453	0.065	494.237	1	.000	4.277	3.763	4.862
	zSES	0.972	0.062	244.831	1	.000	2.643	2.340	2.985
	zACH by zSES	0.292	0.061	22.589	1	.000	1.339	1.187	1.510
	Constant	3.386	0.063	2,851.816	1	.000	29.561		

[a]Variable(s) entered on step 1: ZACH, ZSES, ZACH*ZSES.

External Validation of a Prediction Equation

The real challenge of prediction is testing the prediction equation on a completely independent sample. External validity is also more relevant, because the application of prediction or prognostic models is to create predictions (prognostications) about individuals not part of the original sample. Incidence of the outcome in external samples is likely to be different, as is the relationship of predictors to outcomes (discrimination). External samples can be made by nonrandom splits of large data sets (e.g., data from different locations or years/waves). As you can see in Table 17.3, data from NELS88 are categorized into four regions. To explore external validation of our model, let us leave the smallest region (the Northeast) out of the development and internal validation process. We will then use a random sample of the same number of cases in the Northeast region to explore how robust (and generalizable) the shrunken prediction equation is.

To make this exercise more realistic, we will work with a 10% sample of the non-Northeast cases, leaving us with 1,337 cases, 114 of which did not graduate (a 38:1 ratio of events to predictors, which is not bad according to prior research such as

Table 17.3   Regions Within NELS88

g8regon

		Frequency	Percent	Valid Percent	Cumulative Percent
	1 NORTHEAST	3,127	18.9	18.9	18.9
	2 NORTH CENTRAL	4,435	26.8	26.8	45.8
	3 SOUTH	5,743	34.8	34.8	80.5
Valid	4 WEST	3,184	19.3	19.3	99.8
	8 {MISSING}	31	0.2	0.2	100.0
	Total	16,520	100.0	100.0	

Table 17.4   Logistic Regression in the Reduced ($n = 1,337$) Sample

Omnibus Tests of Model Coefficients

		Chi-Square	df	p
	Step	159.495	3	.000
Step 1	Block	159.495	3	.000
	Model	159.495	3	.000

Variables in the Equation

		B	SE	Wald	df	p	Exp(B)	95% CI for Exp(B)	
								Lower	Upper
	zACH	1.458	0.240	36.997	1	.000	4.295	2.686	6.870
Step 1[a]	zSES	1.203	0.216	31.147	1	.000	3.330	2.183	5.081
	SESACH	0.581	0.213	7.430	1	.006	1.787	1.177	2.713
	Constant	3.401	0.232	215.376	1	.000	29.988		

[a]Variable(s) entered on step 1: ZACH, ZSES, SESACH.

Steyerberg et al., 2001a, 2001b). I also sampled 1,337 cases from the Northeast sample so we can directly compare model statistics to observe shrinkage.

The analysis of this original sample is presented briefly in Table 17.4. As you can see in Table 17.4, this random subsample of the overall data set already has drifted from the original regression coefficients. Although the intercept and effect of zACH are relatively similar, the effects of zSES and the interaction have drifted substantially.

Overall Performance (Brier Score)

To calculate the Brier score for this example, I used Equation 17.4 to create prediction/prognostication scores and then used Equation 17.3, which resulted in a score of 0.047, which is on the low end of the performance scale (recall that 0.00 would be perfect prognostication/prediction).

Estimated Shrinkage

Based on Equation 17.4, with our initial model $\chi^2 = 159.495$ and three predictors, we would expect shrinkage of 0.981, which is very small (expected overall model fit shrunken to $\chi^2 = 156.495$). We could then weight each coefficient by that factor and supposedly get a more replicable analysis. However, as you can see in Table 17.5, this shrinkage estimate is a bit conservative.

If you were performing simple prognostication (prediction) based on this single sample, you could create a prediction equation (as shown in Equation 17.5) and observe how closely the prediction explains the results.

$$\hat{Y} = 3.401 + 1.458(zACH) + 1.203(zSES) + 0.581(SESACH) \qquad (17.5)$$

Taking scores generated by Equation 17.5 and entering these scores as the sole predictor produced the results in Table 17.5, which represents an indicator of calibration. There are a few aspects to this analysis to note. First, because we have the same number of cases, we can compare the model chi-square to the original model (147.641 versus the original chi-square of 159.495) as an example of shrinkage, an analogue to the indicators that we get from OLS regression.

Table 17.5   Prognostication Index and Shrinkage in Logistic Regression

Omnibus Tests of Model Coefficients

		Chi-Square	df	p
Step 1	Step	147.641	1	.000
	Block	147.641	1	.000
	Model	147.641	1	.000

Variables in the Equation

		B	SE	Wald	df	p	Exp(B)
Step 1[a]	Predicted	1.158	0.147	61.888	1	.000	3.182
	Constant	−0.095	0.280	0.116	1	.733	0.909

[a]Variable(s) entered on step 1: PRED.

Second, the slope of the PI can serve as the "calibration slope"—a constant (1.158) that can be used to modify the original regression weights to create a "shrunken" equation that is supposed to be less subject to overfitting. This modified equation is presented in Equation 17.6:

$$\hat{Y} = 3.401 + 1.458*1.158(zACH) + 1.203*1.158(zSES) +$$
$$0.581*1.158(SESACH) \tag{17.6}$$

or

$$\hat{Y} = 3.401 + 1.688(zACH) + 1.393(zSES) + 0.673(SESACH)$$

Concordance and Discrimination

As mentioned above, the area under the ROC curve for this analysis is also the estimate of discrimination, or the ability to discriminate between cases who have and have not experienced an outcome correctly.[8] You can see in Table 17.6 and Figure 17.1 that discrimination is not bad in this example (recall that this ranges from 0.50 to 1.00, with higher numbers being better). In this example, the area under the curve is significantly different from 0.50 (null hypothesis), meaning that the equation categorized individuals at a rate significantly better than chance.

It is likely more significant predictors in the equation would help improve the model discrimination.

Table 17.6   Discrimination (Area Under the Receiver Operating Characteristic Curve)

Area Under the Curve

Test Result Variable(s): PRE_1

Area	SE[a]	Asymptotic Sig.[b]	Asymptotic 95% CI	
			Lower Bound	Upper Bound
0.869	0.015	.000	0.841	0.898

[a]Under the nonparametric assumption.
[b]Null hypothesis: true area = 0.5.

---

8   This is the same concept in measurement and psychometrics where we want items on a test to discriminate between high- and low-scoring individuals.

Figure 17.1   Receiver Operating Characteristic Curve

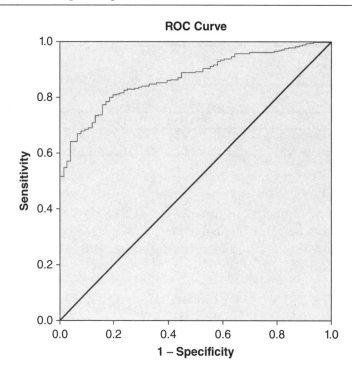

**Using Bootstrap Analysis to
Estimate a More Robust Prognostic Equation**

**General Bootstrap Methodology
for Internal Validation of a Prognostic Model**

Several authors have talked about using bootstrap methodologies for internal validation purposes, keeping in mind that the model is only as good as the sample. In general, starting with your internal validation sample, perform the following steps:

1. Perform bootstrap resampling to produce a derivative sample of the same size.

2. Estimate the regression equation in the bootstrap sample.

3. Using the equation from step 2, calculate the PI (another term for predicted values) within the *original sample.*

4. Create a model with the binary outcome as the dependent variable and the PI as the only predictor in the original sample.

5. Record *s*, the unstandardized regression coefficient for PI.

6. Repeat steps 1–4 multiple times to create a large number of *ss* that can be averaged to estimate the shrinkage value that can be used to weight the regression coefficients in the final model.

Table 17.7   Volatility in Prognostic Coefficients Derived From 5,000 Bootstrap Samples

Parameter	Original Sample	Mean	Median	95% CI	p < .05, %
Intercept	3.401	3.435	3.416	3.05, 3.91	100
zACH	1.458	1.477	1.459	1.08, 1.96	100
zSES	1.203	1.216	1.206	0.87, 1.61	100
SESACH	0.581	0.588	0.581	0.24, 0.97	83.4

In essence, scholars in this area have suggested that a good estimate of the shrinkage in a model such as this could then be used to weight the final prediction equation with shrunken coefficients (shrunken coefficients = sb_1 to sb_k, where s is the shrinkage estimate and b_1 to b_k are the regression coefficients for the k predictors.

Although this might not create improved and flawless prediction equations, it will provide an estimate of how volatile the prediction equation is likely to be in a single external sample. For example, as you can see in Table 17.7, there is some significant volatility in the prediction equations derived from bootstrap resampling of the initial sample.

As you can see from Table 17.7, even within a sample as large as this one ($N = 1,337$), there seems to be substantial volatility around the regression coefficients (and even the intercept), with effect sizes generally doubling between the lower boundary of the CI and upper boundary (e.g., $b_1 = 1.08$ corresponds to an odds ratio [OR] of 2.94, whereas $b_1 = 1.95$ corresponds to an OR of 7.10).

Internal Validation

Using each of these 5,000 regression equations could provide insight into how well we can expect prediction equations to work within a single sample. Therefore, using recommendations by authors cited above, we can use each of these regression equations to make prognostications within the original data set, calculating the various indicators above (e.g., Brier score, discrimination, or PI). The first 10 examples of this calculation are presented in Table 17.8. As you can see, there is significant volatility in the regression equations, although the indicators of quality are relatively constant in their conclusion. The Brier score tends to be around 0.07, which is not bad, calibration (PI, c, or s, depending on the author) varies around 1.00, and discrimination is relatively strong. In fact, averaging the calibration scores in this example of 10 bootstrapped samples, we get an average of 1.0057, which we could then use to modify the prognostic equation for use in the external validation data.

Unfortunately, these indicators of shrinkage or optimism (Steyerberg et al., 2001b) do not seem to really capture the true volatility or shrinkage observed above with the external validation. As you see in the prior section, the external validation was reasonably good but was not as strong as this averaged calibration value, nor other estimators used previously.

Table 17.8 Volatility in Internal Validation

Regression Equation	Brier Score	Calibration (PI or s)	Discrimination (Area Under ROC)
$\hat{Y} = 3.611 + 1.446(\text{zACH}) + 1.558(\text{zSES}) + 0.784(\text{SESACH})$	0.0694	0.902	0.824
$\hat{Y} = 3.367 + 1.407(\text{zACH}) + 1.079(\text{zSES}) + 0.611(\text{SESACH})$	0.0689	1.128	0.826
$\hat{Y} = 3.283 + 1.185(\text{zACH}) + 1.127(\text{zSES}) + 0.365(\text{SESACH})$	0.0692	1.007	0.828
$\hat{Y} = 3.464 + 1.396(\text{zACH}) + 1.357(\text{zSES}) + 0.516(\text{SESACH})$	0.0691	0.890	0.827
$\hat{Y} = 3.357 + 1.375(\text{zACH}) + 1.315(\text{zSES}) + 0.824(\text{SESACH})$	0.0694	1.109	0.824
$\hat{Y} = 3.186 + 1.351(\text{zACH}) + 1.341(\text{zSES}) + 0.657(\text{SESACH})$	0.0692	1.003	0.826
$\hat{Y} = 3.671 + 1.783(\text{zACH}) + 1.435(\text{zSES}) + 0.841(\text{SESACH})$	0.0690	0.893	0.826
$\hat{Y} = 3.221 + 1.257(\text{zACH}) + 1.115(\text{zSES}) + 0.509(\text{SESACH})$	0.0689	1.104	0.828
$\hat{Y} = 3.815 + 1.771(\text{zACH}) + 1.445(\text{zSES}) + 0.800(\text{SESACH})$	0.0689	0.874	0.826
$\hat{Y} = 3.365 + 1.461(\text{zACH}) + 1.282(\text{zSES}) + 0.724(\text{SESACH})$	0.0690	1.043	0.826

Another methodology one could use would be to explore the boundary conditions (the more extreme iterations of the prediction equations) to explore the possible range of prediction. For example, the upper range of the 95% CIs for intercept and zACH slope is 3.91 and 1.96, respectively. Choosing bootstrap number 378, there is an equation with relatively extreme characteristics (top row of Table 17.9).

Alternatively, I could choose bootstrap number 2,697, which had a coefficient for the interaction term at the upper boundary of the CI (0.97; second row of Table 17.9).

I could then contrast the results of these equations in the external validation sample with equations at the other end of the CIs. For example, the lower boundary of the CI for the intercept is 3.05, which is found in bootstrap sample number 2,672 (which also contains coefficients for zSES very near the lower boundary of the 95% CI; see the third row of Table 17.9). An example of an equation in which the interaction term is near the lower 95% CI boundary is bootstrap number 4,278 (fourth row of Table 17.9).

As you can see in Table 17.9, when each of these extreme versions of the prognostic equation is tested within the independent external validation sample of the same size, all four of

Table 17.9   Volatility in External Validation

Regression Equation	Brier Score	Calibration (PI or s)	Discrimination (Area Under ROC)	Model χ²
Equations near the upper limits of 95% CIs for parameters				
$\hat{Y} = 3.911 + 1.967(zACH) + 1.340(zSES) + 0.787(SESACH)$	0.0474	0.941	0.865	144.54
$\hat{Y} = 3.594 + 1.719(zACH) + 1.421(zSES) + 0.970(SESACH)$	0.0482	1.048	0.861	139.09
Equations near the lower limits of 95% CIs for parameters				
$\hat{Y} = 3.050 + 1.228(zACH) + 0.859(zSES) + 0.329(SESACH)$	0.0469	1.355	0.871	148.97
$\hat{Y} = 3.144 + 1.153(zACH) + 0.813(zSES) + 0.240(SESACH)$	0.0468	1.363	0.872	149.48
Using original sample (from Table 17.4)				
$\hat{Y} = 3.401 + 1.458(zACH) + 1.203(zSES) + 0.581(SESACH)$	0.0471	1.158	0.869	147.64

these extreme examples of the prediction equation are about as good as the simple equation we first arrived at. All are fairly reasonable according to the Brier score, all have about the same discrimination index (area under the ROC curve), and all show a modest amount of shrinkage from the original model chi-square of 159.495. This example allows several conclusions.

First, prediction equations can be highly volatile, even when samples are relatively large and events (or the ratio of predictors to events) are reasonably strong. It is difficult to see how any researcher could publish one equation based on one sample and have the actual parameters in the equation replicate closely in an independent sample, particularly if there were more variables in the equation (or more complex effects).

Second, no matter how extensively a researcher tries to create a "shrunken" or more generalizable equation that is corrected for overfitting, it is probably not reasonable to assume that shrinkage and overfitting have been accounted for. Given the simple, limited examples above, attempts at correcting for overfitting seemed to universally fall short of the actual observed shrinkage.

Third, and what is somewhat unexpected given the above points, is that in general, most of the equations showed reasonably good prognostic capabilities. Despite relatively robust variability across the various bootstrap examples, the metrics for evaluating prognostic equations stayed relatively consistent. This may be due to the size of the sample, the small number of predictors, or the strength of the predictive power of the variables, and your experiences may vary.

SUMMARY

Linear modeling techniques can be an effective tool for creating prediction/prognostic equations in a context where there is adequate measurement, there are large-enough samples, assumptions are met, and care is taken to evaluate the regression equations for generalizability (shrinkage). Unfortunately, researchers seem overly optimistic in the generalizability of their prediction or prognostic equations even when using shrinkage estimation, which is problematic because the shrinkage estimation in this chapter was not accurate.

If you are attempting to create a prediction/prognostic equation that will be useful and generalizable, your best bet is to use the following principles developed during much of the book:

- Use large samples that are representative of the population of interest;
- Use clean data for optimal generalizability;
- Use complex terms (interactions or curvilinear effects) if they improve prediction substantially;
- Carefully evaluate the linear models through multiple means, including perhaps such techniques as bootstrap resampling for internal validation; and
- Be particularly careful to use external validation as a further reality check.

REFERENCES

Anderson, N. H., & Shanteau, J. (1977). Weak inference with linear models. *Psychological Bulletin, 84*(6), 1155–1170.

Cohen, J., Cohen, P., West, S., & Aiken, L. S. (2003). *Applied multiple regression/correlation analysis for the behavioral sciences.* Mahwah, NJ: Lawrence Erlbaum.

DeGroot, A. D. (1969). *Methodology: Foundations of inference and research in the behavioral sciences.* The Hague, The Netherlands: Mouton.

Harrell, F. E., Jr., Lee, K. L., & Mark, D. B. (1996). Multivariable prognostic models: Issues in developing models, evaluating assumptions and adequacy, and measuring and reducing errors. *Statistics in Medicine, 15*(4), 361–387. doi: 10.1002/(SICI)1097-0258(19960229)15:4<361::AID-SIM168>3.0.CO;2-4

Hoerl, A. E., & Kennard, R. W. (1970). Ridge regression: Biased estimation for nonorthogonal problems. *Technometrics, 12*(1), 55–67. doi: 10.1080/00401706.1970.10488634

Kaplan, A. (1964). *The conduct of inquiry: Methodology for behavioral science.* San Francisco, CA: Chandler.

Park, C., & Dudycha, A. (1974). A cross-validation approach to sample size determination. *Journal of the American Statistical Association, 69*(345), 214–218. doi: 10.2307/2285528

Pedhazur, E. J. (1997). *Multiple regression in behavioral research: Explanation and prediction.* Fort Worth, TX: Harcourt Brace College Publishers.

Royston, P., Moons, K. G., Altman, D. G., & Vergouwe, Y. (2009). Prognosis and prognostic research: Developing a prognostic model. *BMJ, 338*, b604. doi: dx.doi.org/10.1136/bmj.b604

Schafer, W. D. (1991). Reporting hierarchical regression results. *Measurement and Evaluation in Counseling and Development, 24*(3), 98–100.

Schafer, W. D. (1992). Reporting nonhierarchical regression results. *Measurement and Evaluation in Counseling and Development, 24*(2), 146–149.

Scriven, M. (1959). Explanation and prediction in evolutionary theory. *Science, 130*(3374), 477–482. doi: 10.1126/science.130.3374.477

Steyerberg, E. W., Borsboom, G. J., van Houwelingen, H. C., Eijkemans, M. J., & Habbema, J. D. (2004). Validation and updating of predictive logistic regression models: A study on sample size and shrinkage. *Statistics in Medicine, 23*(16), 2567–2586. doi: 10.1002/sim.1844

Steyerberg, E. W., Eijkemans, M., & Habbema, J. D. F. (2001a). Application of shrinkage techniques in logistic regression analysis: A case study. *Statistica Neerlandica, 55*(1), 76–88. doi: 10.1111/1467-9574.00157

Steyerberg, E. W., Harrell, F. E., Jr., Borsboom, G. J. J. M., Eijkemans, M. J. C., Vergouwe, Y., & Habbema, J. D. F. (2001b). Internal validation of predictive models: Efficiency of some procedures for logistic regression analysis. *Journal of Clinical Epidemiology, 54*(8), 774–781. doi: 10.1016/S0895-4356(01)00341-9

Tabachnick, B. G., & Fidell, L. S. (2001). *Using multivariate statistics* (4th ed.). New York, NY: Harper Collins.

Thompson, B. (1989). Why won't stepwise methods die? *Measurement and Evaluation in Counseling and Development, 21*(4), 146–148.

Thompson, B. (1995). Stepwise regression and stepwise discriminant analysis need not apply. *Educational and Psychological Measurement, 55*(4), 525–534. doi: 10.1177/001316449505500400

Tibshirani, R. (1996). Regression shrinkage and selection via the lasso. *Journal of the Royal Statistical Society, Series B (Methodological), 58*(1), 267–288.

van Houwelingen, H. C., & Sauerbrei, W. (2013). Cross-validation, shrinkage and variable selection in linear regression revisited. *Open Journal of Statistics, 3*, 79–102. doi: 10.4236/ojs.2013.32011

18

MODELING IN LARGE, COMPLEX SAMPLES

The Importance of Using Appropriate Weights and Design Effect Compensation

Advance Organizer

Large, governmental or international data sets (of the type we have used in many of the chapters in this book) are important resources for researchers across many disciplines and sciences. They present researchers with the opportunity to examine trends and hypotheses within nationally (or internationally) representative data sets that are difficult to acquire without the resources of a large research institution or governmental agency.

However, there are challenges to using these types of data sets. For example, individual researchers must take the data as given—in other words, we have no control over the types of questions asked, how they are asked, to whom they are asked, and when they are asked. The variables are often not ideally suited to answering the particular questions you, as an individual researcher, might wish to ask.

Despite their potential shortcomings, these valuable resources are often freely available to researchers (at least in public release formats that have had potentially identifying information removed). There is, however, one cost worth discussing: the expectation that researchers will utilize best practices in using these samples. Specifically, researchers must take the time to understand the sampling methodology used and appropriately utilize weighting and design effects, which can be potentially confusing and intimidating to a novice. There is mixed evidence on researchers' utilization of appropriate methodology (e.g., Johnson & Elliott, 1998), which highlights the need for more conversation around this important issue.

The goal of this brief chapter[1] is to introduce some of the issues around using complex samples and explore the possible consequences (e.g., Type I errors) of failure to appropriately model the complex sampling methodology.

In this chapter, we will cover

---

1 This chapter was originally published as Osborne, J. W. (2011). Best practices in using large, complex samples: The importance of using appropriate weights and design effect compensation. *Practical Assessment, Research & Evaluation, 16*(12).

- Issues around complex sampling common in large data sets
- Whether results are negatively impacted by failure to weight data from complex samples
- Whether scaled weights and design effect compensation can work as effectively as software that provides complex weighting

What Types of Studies Use Complex Sampling?

Many of the most interesting social science and health sciences databases available to researchers use complex sampling. Some examples include databases from the National Center for Education Statistics (NCES; the National Education Longitudinal Study of 1988 [NELS88] or the Third International Mathematics and Science Study [TIMSS]),[2] the US Centers for Disease Control and Prevention (CDC; the National Health Interview Survey [NHIS] and the National Health and Nutrition Examination Survey [NHANES]),[3] and the Bureau of Justice Statistics (BJS; the National Crime Victimization Survey [NCVS]).[4] Almost any survey seeking a representative sample from a large population will probably have a complex multistage probability sampling methodology, because it is relatively efficient and allows for estimation of representative samples.

Why Does Complex Sampling Matter?

In most of the examples cited above, the samples are not simple random samples; rather, they are complex samples with multiple goals. For example, in NELS88, students in certain underrepresented racial groups and in private schools were *oversampled* (i.e., more respondents selected than would typically be the case for a representative sample), meaning that the sample is not, in its raw form, necessarily representative (Ingels, 1994; Johnson & Elliott, 1998). Furthermore, in any survey such as the ones discussed above, there is a certain amount of nonresponse that may or may not be random, making unweighted samples potentially still less representative.

Finally, in multistage probability sampling, in contrast with simple random sampling, complex sampling often utilizes cluster sampling (especially where personal interviews are required), in which clusters of individuals within primary sampling units are selected for convenience. For example, in the Education Longitudinal Study of 2002, researchers sampled approximately 20,000 students from 752 schools, rather than simply conducting random sampling from the approximately 27,000 schools that met criteria within the United States (Bozick, Lauff, & Wirt, 2007). Thus, students within clusters are more similar than students randomly sampled from the population

---

2 Available from the NCES website (http://nces.ed.gov/) or the ICPSR website (http://www.icpsr.umich .edu).

3 Available from the CDC website (http://www.cdc.gov/nchs/index.htm).

4 Available from the BJS website (http://bjs.ojp.usdoj.gov/index.cfm?ty=dctp&tid=3).

as a whole. This effectively reduces the information contained in each degree of freedom. These effects of sampling, called "design effects" (Kish [1965] is often credited with introducing this concept), must also be accounted for or the researcher risks not only misestimating effects but also making Type I errors, because this common modern sampling strategy can lead to violation of traditional assumptions of independence of observations. Specifically, without correcting for design effects, standard errors (*SEs*) are often underestimated, leading to significance tests that are inappropriately sensitive (Johnson & Elliott, 1998; Lehtonen & Pahkinen, 2004).

Note that complex, multistage probability sampling is *not* the same as multilevel (i.e., nested or hierarchical) analyses or data. Both are important, modern techniques that correctly deal with different violations of assumptions or issues. Multistage probability sampling has to do with the sampling methodology employed, which violates our assumption that samples are drawn randomly from populations of interest.[5] In particular, using this methodology violates our assumption that each data point represents an equal amount of information; in the case of cluster sampling, it can also violate assumptions of independence of observations. Some subpopulations of interest might be oversampled, and others might be undersampled. From a conceptual point of view, each data point thus represents a different portion of the overall population, which can be corrected by methods discussed below.

Multilevel or nested data are data with variables measured at different levels of organization (e.g., students within classrooms within schools, or employees within corporations within sectors). This violates assumptions of independence of observations (for a brief primer on this topic, see Osborne, 2000, 2008) that can lead to misestimation of parameters and misspecification of analysis models if not taken into account.

Note also that the two are not mutually exclusive. For example, many of the data sets discussed above are both nested data *and* complex samples. For example, in the NCES data, researchers often want to model teacher- or school-level effects on student performance, which creates a multilevel analysis in the context of a complex, multistage probability sample. But researchers can also encounter nested data sets that are not produced using probability samples, and probability samples that are not nested.

In summary, there are two issues introduced by complex sampling: a sample that employs advanced sampling techniques (or has nonresponse or missing data that need to be accounted for), causing the sample to potentially deviate from being representative of the population of interest, and a sample that violates assumptions of independence of observations, potentially leading to significant misestimation of significance levels in inferential statistical tests.

What Are Best Practices in Accounting for Complex Sampling?

In most samples of this nature, the data provider includes information in the data set (and in the user documentation) to facilitate appropriate use of the data. For example,

---

5   In fact, I think most sampling methodologies violate this assumption. An inspection of top journals in any field reveals few studies that could be correctly classified as simple random samples from a population of interest.

the data may include weights for each individual, information about design effects (DEFFs) for the overall sample and different subpopulations, and information on which primary sampling unit and cluster each individual belongs to.

More information on these topics is available in most user manuals for those interested in the technical details of how each of these pieces of information is calculated and used.[6]

Most modern statistical packages can easily apply sample weights to a data set. Applying the appropriate weight creates a sample that is representative of the population of interest (e.g., eighth graders in the United States who remained in school through twelfth grade, to continue the previous example from NELS88). The problem is that application of weights dramatically increases the sample size to approximately the size of the population. For example, in NELS88, a sample of approximately 25,000 becomes the population of more than 3,000,000 students, dramatically (and illegitimately) inflating the degrees of freedom used in inferential statistics. Prior to statistical software packages commonly supporting this, you could seek to accomplish the same goal by scaling the weights so that the weighted sample has the same weighted number of participants as the original, unweighted sample. I did this in some of my early research (Osborne, 1995, 1997) thanks to the mentoring of Robert Nichols, who was also involved in some early NCES national studies. However, scaling the weights does not take into account the DEFFs, which should further reduce the degrees of freedom available for the statistical tests.

Not all statistical software provides for accurate modeling of complex samples (e.g., an add-on module is required with SPSS; in SAS, STATA, and SUDAAN, complex sampling appears to be incorporated, and there is also freely available software such as AM[7] that correctly deals with this issue).[8] For those without access to software that models complex samples accurately (again, as was the case long ago when I first started working with large data sets), one way to approximate best practices in complex sampling would be to further scale the weights to take into account DEFFs (e.g., if the DEFF = 1.80 for whatever sample or subsample a researcher is interested in studying, that researcher would divide all weights by 1.80).

However, the most desirable way of dealing with this issue is by using software that has the capability to directly model the weight, primary sampling unit, and cluster directly, which best accounts for the effects of the complex sampling (e.g., Bozick et al., 2007; Ingels, 1994; Johnson & Elliott, 1998). In most cases, a simple set of commands

---

6   In many data sets, there are multiple options for weights. For example, in NELS88, a survey of eighth-grade students who were then followed for many years, there is a weight only for individuals interested in using the first (BY) data collection. There is a similar weight for each other data collection point (F1, F2, F3, etc.). Yet not all students present in BY are also present in F1 and F2; if I want to perform an analysis following students from eighth grade to tenth and twelfth grade, there is also a weight (called a panel weight) for longitudinal analyses. This highlights the importance of being thoroughly familiar with the details of the user manual before using data from one of these studies.

7   AM is available from the American Institutes for Research (http://am.air.org/).

8   I was unable to determine whether R statistical software incorporates complex sample handling, but I encourage readers to explore R as an option because it often has advanced techniques incorporated prior to commercial programs. R is freely available on Unix, Windows, and Macintosh platforms (http://cran.r-project.org/).

informs the statistical software what weight you desire to use, what variable contains the Primary Sampling Unit (PSU) information, and what variable contains the cluster information, and the analyses are adjusted from that point on, automatically.

Does It Really Make a Difference in the Results?

Some authors have argued that, particularly for complex analyses such as multiple regression, it is acceptable to use unweighted data (e.g., Johnson & Elliott, 1998). This advice is in direct opposition to the sampling and methodology experts who create many of these data sets and sampling frames and is also in opposition to what makes conceptual sense. Thus, to explore whether this really does have the potential to make a substantial difference in the results of an analysis, I performed several example analyses below under four different conditions that might reflect various strategies researchers would take to using this sort of data: (a) unweighted (taking the sample as is); (b) weighted only (population estimate); (c) weighted, using weights scaled to maintain original sample size and scale weights to account for DEFF (best approximation); and (d) using appropriate complex sampling analyses via AM software, which is designed to accurately account for complex sampling in analyses.

To examine the effects of utilization of best practices in modeling complex samples, the original tenth-grade (G10COHRT = 1) cohort from the Education Longitudinal Study of 2002 (along with the first follow-up) public release data were analyzed. Only students who were part of the original cohort (G10COHRT = 1) and who had weight over 0.00 on F1PNLWT (the weight for using both tenth- and twelfth-grade data collection time points) were retained so that the identical sample is utilized throughout all analyses.

Conditions Used

Unweighted

In this condition, the original sample (meeting condition G10COHRT = 1 and F1PNLWT > 0.00) was retained with no weighting or accommodation for complex sampling. This resulted in a sample of $N = 14,654$.

Weighted

In this condition, F1PNLWT was applied to the sample of 14,654 who met the inclusion criteria for the study. Application of F1PNLWT inflated the sample size to 3,388,462. This condition is a likely outcome when researchers with only passing familiarity with the nuances of weighting complex samples attempt to use a complex sample.

Scaled Weights

In this condition, F1PNLWT was divided by 231.232 (the ratio of the inflated sample size with weights applied and the unweighted sample: 3,388,462/14,654), bringing the sample size back to approximately the original sample size but retaining the representativeness of the population. Further, the weights were scaled by the design effect (1.88 for examples using only males, yielding a final sample of 3,923 males, or 2.33 for examples using all subjects, yielding a final sample[9] of 6,289) to approximate use of best practices. This condition is a likely outcome when a researcher is sophisticated enough to understand the importance of correcting for these issues but does not have access to software that appropriately models the complex sampling (or is using advanced analytical techniques such as structural equation modeling that does not incorporate complex sampling methodology at this time).

Appropriately Modeled

In this case, AM software was utilized to appropriately model the weight, PSU, and cluster information provided in the data to account for all issues mentioned above. This is considered the "gold standard" for the purposes of this analysis.

Comparison of Unweighted Versus Weighted Analyses

Four different analyses were compared to explore the potential effects of failing to use best practices in modeling complex samples.

Large Effect in Ordinary Least Squares Regression

In this example, the twelfth-grade mathematics Item Response Theory (IRT) achievement score (F1TXM1IR) is predicted from base year reading IRT achievement score (BYTXRIRR) controlling for socioeconomic status (F1SES2). The results of this analysis across all four conditions are presented in Table 18.1.

As Table 18.1 shows, with a strong effect (e.g., $\beta > 0.60$), there is not a substantial difference in the effect regardless of whether the complex sampling design is accounted for. However, note that the *SEs* vary dramatically across condition, with the weighted only condition being misestimated by a factor of 16 times or more. Note also that the scaled-weights condition closely approximates the appropriately modeled condition. However, as following analyses will show, this is possibly the exception, rather than the rule.

Modest Effect in Binary Logistic Regression

To test the effects of condition on a more modest effect, African American males were selected for a logistic regression predicting dropout (F1DOSTAT; 0 = never, 1 = dropped out) from the importance of having children, controlling for standardized reading test scores in tenth grade. The results of these analyses are presented in Table 18.2.

The results indicate that the conclusions across all four analyses are similar—that as the importance of having children increases, the odds of dropping out decreases among African American males. However, there are several important differences across the

---

9 These DEFF estimates are usually easily found in the user manuals for these data sets. Researchers need to decide what aspects of the sample they are interested in using and utilize the appropriate DEFF estimate for that aspect of the sample, as I did. Not all DEFFs are the same for all subgroups.

Table 18.1   Large Effect: Ordinary Least Squares Regression Predicting F1 Math
Achievement From BY Reading ACH

Analysis	Group	b	SE	t (df)	p	Beta
SPSS—no weighting	White male	1.009	0.019	14.42 (3,858)	< .0001	0.647
	African American male	0.959	0.040	23.91 (807)	< .0001	0.638
SPSS— weight only	White male	1.027	0.001	872.25 (927,909)	< .0001	0.658
	African American male	0.951	0.003	379.40 (201,334)	< .0001	0.642
SPSS— weights scaled for N, DEFF	White male	1.027	0.025	41.806 (2132)	< .0001	0.658
	African American male	0.951	0.052	18.138 (460)	< .0001	0.642
AM weight, PSU, Strata modeled	White male	1.027	0.023	45.35 (362)	< .0001	
	African American male	0.951	0.049	19.41 (232)	< .0001	

NOTE: Males only; BYTXRIRR predicting F1TXM1IR controlling for F1SES2; identical sample. In all analyses, G10COHRT = 1, F1PNLWT > 0. The lower right cell is empty because AM does not provide standardized regression coefficients.

Table 18.2   Modest Effect: Logistic Regression Predicting Dropout From Importance of
Having Children

Analysis	b	SE	Wald	p	Exp(b)
SPSS—no weighting	−0.344	0.146	5.59	< .018	0.709
SPSS—weight only	−0.346	0.008	1,805.85	< .0001	0.708
SPSS—weights scaled for N, DEFF	−0.344	0.170	4.154	< .042	0.709
AM weight, PSU, Strata modeled	−0.346	0.177	3.806	< .052	0.708

NOTE: African American males only; F1DOSTAT never versus DO only; controlling for BYTXRSTD. AM did not give odds ratios, but the odds ratio is easily calculated.

conditions. First, the SE of b varies dramatically across the four analyses, again misestimating the SE by large magnitudes. Second, this analysis is an example of a potential Type I error: using the original sample with no weights or nonscaled weights produces a clear rejection of the null hypothesis, whereas the appropriately weighted analysis might not if one uses a rigid $p < .05$ cutoff criterion for rejection of the null hypothesis.[10]

---

10 I do not espouse rigid cutoffs in quantitative analysis, as you have seen in discussions of null hypothesis statistical testing (NHST) earlier in this book. Note here that a probability of <.052 is not much different from a probability of <.049, but very different decisions would be made as a result of a rigid and blind adherence to the $p < .05$ criterion.

Null Effect in Analysis of Variance

To test the effects of condition on an analysis in which the null hypothesis should be retained (no effect), an analysis of variance was performed examining sex differences (F1SEX) in the importance of strong friendships (F1S40D). Using the "gold standard" of modeling the complex sample effects via AM, as Table 18.3 indicates, there should be no differences across groups.

The results in Table 18.3 are a good example of the risks associated with failing to appropriately model or approximate complex sampling weights and DEFFs. A researcher using only the original weights would conclude there are sex differences in the importance of strong friendships among high school students when in fact there are probably not. Again, *SEs* are substantially misestimated (again by a factor of 20 or so). Finally, there is again similarity between the third (scaled weights) and fourth conditions (AM analysis), indicating that the approximation in this case yields similar results to the AM analysis.

Table 18.3 Null Effect: Sex Differences in Importance of Strong Friendships (F1S40D)

Analysis	Group	Mean	SE Mean	t (df)	p
SPSS—no weighting	Male	2.827	0.0050	−1.67 (14,539)	< .095
	Female	2.838	0.0048		
SPSS—weight only	Male	2.822	0.0003	−25.53 (3,360,675)	< .0001
	Female	2.833	0.0003		
SPSS—weights scaled for N, DEFF	Male	2.822	0.0077	−1.100 (6,236)	< .27
	Female	2.833	0.0075		
AM weight, PSU, Strata modeled	Male	2.822	0.0060	−1.366 (386)	< .17
	Female	2.833	0.0060		

NOTE: DEFF average of 2.33 used for these analyses.

Table 18.4 Null Effect: Predicting Student Grade Point Average From School Poverty, Controlling for Race, School Sector

Analysis	b	SE	t (df)	p
SPSS—no weighting	−0.21	0.069	−2.98 (5,916)	< .003
SPSS—weight only	−0.01	0.005	−2.09 (1,124,550)	< .04
SPSS—weights scaled for N, DEFF	−0.01	0.11	−0.09 (2,078)	< .93
AM weight, PSU, Strata modeled	−0.01	0.17	−0.058 (228)	< .95

Null Effect in Ordinary Least Squares Regression

In the final example, a multiple regression analysis predicted cumulative ninth-through twelfth-grade grade point average (GPA; F1RGPP2) from school poverty (percentage of students with free or reduced lunch; BY10FLP), controlling for dummy-coded race (based on F1RACE) and whether the school was public or private (BYSCTRL).

As Table 18.4 shows, in this case, there is a stark contrast between appropriately modeled complex sampling and less ideal analyses. In this example, researchers using the unweighted sample or a weighted sample would make a Type I error, rejecting the null hypothesis and concluding there is a significant (albeit weak) relationship between school poverty and student GPA once other background variables were covaried. The last two conditions (scaled weights and AM modeling) produced similar and contrary results—that there is no relationship between these two variables when the sampling frame is approximated or modeled appropriately.

SUMMARY

Although this might seem an esoteric topic to many researchers in the social or health sciences, there is a wealth of compelling data freely available to researchers, and some researchers have found evidence that researchers do not always model the sampling frame appropriately (Johnson & Elliott, 1998). In brief, most modern statistical software can take complex sampling into account, either through using weights scaled for N and DEFF, or through using information such as primary and secondary sampling units (often called clusters) directly in the software. There is also, as mentioned above, free software that correctly models complex samples, although it does have a small learning curve. Thus, there is little excuse for failing to take sampling into account when using these data sets.

In three of the four examples included above, there is a risk of potentially serious error if a researcher fails to take the sampling effects into account. In two of the four analyses, researchers would clearly make a Type I error, whereas it is less clear but still troubling in the logistic regression example.

Furthermore, most of the analyses highlight how unweighted samples can misestimate not only parameter estimates but also *SEs*. This is because the unweighted sample is *not* representative of the population as a whole and contains many eccentricities such as oversampling of populations of interest and perhaps nonrandom dropout patterns. Weighting provides a better parameter estimate, but unless further measures are taken, serious errors can occur in hypothesis testing and drawing of conclusions. Thus, although it requires extra effort to appropriately model the complex samples in these data sets, it is a necessary step to have confidence in the results arising from the analyses.

ENRICHMENT

This chapter is intended to be more informational; at the time of publication, no exercises were available. Check the book website to see if some have been developed.

REFERENCES

Bozick, R., Lauff, E., & Wirt, J. (2007). *Education Longitudinal Study of 2002 (ELS: 2002): A first look at the initial postsecondary experiences of the sophomore class of 2002 (NCES 2008-308)*. Washington, DC: Institute of Education Sciences, National Center for Education Statistics, US Department of Education.

Ingels, S. (1994). *National Education Longitudinal Study of 1988: Second follow-up: Student component data file user's manual*. Washington, DC: Office of Educational Research and Improvement, National Center for Education Statistics, US Department of Education.

Johnson, D. R., & Elliott, L. A. (1998). Sampling design effects: Do they affect the analyses of data from the national survey of families and households? *Journal of Marriage and Family*, *60*(4), 993–1001. doi: 10.2307/353640

Kish, L. (1965). Selection techniques for rare traits. In J. V. Neel, M. W. Shaw, and W. J. Schull (Eds.), *Genetics and the epidemiology of chronic diseases*. Washington, DC: US Department of Health, Education and Welfare, US Public Health Service.

Lehtonen, R., & Pahkinen, E. (2004). *Practical methods for design and analysis of complex surveys*. Chichester, UK: John Wiley & Sons.

Osborne, J. W. (1995). Academics, self-esteem, and race: A look at the assumptions underlying the disidentification hypothesis. *Personality and Social Psychology Bulletin*, *21*(5), 449–455.

Osborne, J. W. (1997). Race and academic disidentification. *Journal of Educational Psychology*, *89*(4), 728–735.

Osborne, J. W. (2000). Advantages of hierarchical linear modeling. *Practical Assessment, Research & Evaluation*, *7*(1).

Osborne, J. W. (2008). A brief introduction to hierarchical linear modeling. In J. W. Osborne (Ed.), *Best practices in quantitative methods*. Thousand Oaks, CA: SAGE.

APPENDIX A

A Brief User's Guide to z-Scores

As promised in Chapter 2, I include for your convenience a table of *z*-scores and a small amount of information about how to use and interpret them. But first, we must be sure we understand what a standard normal distribution is.

The Normal (Gaussian) Distribution

Normally distributed variables are common in various disciplines and not so common in others. Normal distribution of error terms is a common assumption in some statistical analyses (e.g., ordinary least squares regression). This distribution is a hallmark of many statistics classes and is easily identifiable by virtue of its shape, which is somewhat similar to a bell cross-section (hence, the common label "bell-shaped distribution").

Many distributions are similarly "bell shaped," including, as an example, some variants of Student's *t* distribution and some logistic distributions. Technically, when we talk about a "normally distributed" variable, we are referring to a variable that matches a specific distribution, which we can then use for important inferences. There are technically an infinite number of distributions with a mean of μ and standard deviation (*SD*) of σ that match this shape. The simplest case of the normal distribution is the *standard normal distribution*, which has $\mu = 0$ and $\sigma = 1$. The probability density function of the standard normal distribution is presented in Equation A.1.

$$\Phi(x) = \frac{e^{\frac{x^2}{2}}}{\sqrt{2\pi}} \tag{A.1}$$

This equation can be generalized to any distribution with mean = μ and *SD* of σ. Because it is simple to convert any of these distributions to the standard normal distribution (also referred to as *z*-scores by me in prior chapters and by many other authors, as presented in Equation A.2), this is unnecessary for our purposes.

$$z = \frac{x - \bar{x}}{sd} \tag{A.2}$$

The normal distribution is most often seen when the process underlying the development of the scores has a significant random component. Sir Francis Galton designed a machine, often called a "Galton Board" or "quincunx," in which balls are dropped

Figure A.1   The Standard Normal Distribution

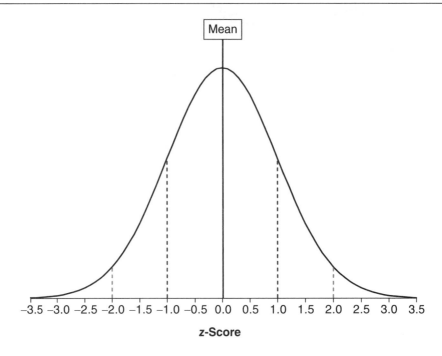

into a series of pegs. Balls hitting pegs have an equal chance of going left or right; if enough balls are dropped into a machine like this, the distribution will approach normality. Other classic examples include the number of times one side of a coin is facing up given a particular number of coin flips. Other physical, physiological, and cognitive measurements (e.g., blood pressure, the height of males or females in a country, or scores on a challenging test of ability) will be approximately normally distributed by virtue of being determined by a large number of variables (e.g., genetics, developmental history, current environment, psychological variables, etc.).

Why Is the Normal Distribution Such a Big Deal?

Because the normal distribution has a fixed shape with known quantities, we know the area under the curve at any given point. This allows us to understand the probability of observing a score at any given distance from the mean, but only when the shape of the distribution is normal. As you can see in Figure A.1, most scores are clustered around the mean; thus, in a distribution of this type, the probability of being relatively far from the mean decreases rapidly. These probabilities are usually captured in a z-table, as you can see in Table A.1.

As you can see in Figure A.1 and Table A.1, 68.26% of the values in a normal distribution are within 1 *SD* of the mean, and 95.45% of the values are within 2 *SD* of the mean. Once values get beyond 2 *SD* from the mean, probabilities of seeing those values drop toward zero rather quickly. I often use $z = \pm 3$ as a starting point for data

Table A.1   *z*-Scores and Probabilities

z-Score	*p (x < z)*	*p (x > |z|)*	*p (x < |z|)*
−4	.00003	.00003	.49997
−3.9	.00005	.00005	.49995
−3.8	.00007	.00007	.49993
−3.7	.00011	.00011	.49989
−3.6	.00016	.00016	.49984
−3.5	.00023	.00023	.49977
−3.4	.00034	.00034	.49966
−3.3	.00048	.00048	.49952
−3.2	.00069	.00069	.49931
−3.1	.00097	.00097	.49903
−3	.00135	.00135	.49865
−2.9	.00187	.00187	.49813
−2.8	.00256	.00256	.49744
−2.7	.00347	.00347	.49653
−2.6	.00466	.00466	.49534
−2.5	.00621	.00621	.49379
−2.4	.00820	.00820	.49180
−2.3	.01072	.01072	.48928
−2.2	.01390	.01390	.48610
−2.1	.01786	.01786	.48214
−2	.02275	.02275	.47725
−1.9	.02872	.02872	.47128
−1.8	.03593	.03593	.46407
−1.7	.04457	.04457	.45543
−1.6	.05480	.05480	.44520
−1.5	.06681	.06681	.43319
−1.4	.08076	.08076	.41924

(Continued)

Table A.1   (Continued)

z-Score	$p\ (x < z)$	$p\ (x > |z|)$	$p\ (x < |z|)$
−1.3	.09680	.09680	.40320
−1.2	.11507	.11507	.38493
−1.1	.13567	.13567	.36433
−1	.15866	.15866	.34134
−0.9	.18406	.18406	.31594
−0.8	.21186	.21186	.28814
−0.7	.24196	.24196	.25804
−0.6	.27425	.27425	.22575
−0.5	.30854	.30854	.19146
−0.4	.34458	.34458	.15542
−0.3	.38209	.38209	.11791
−0.2	.42074	.42074	.07926
−0.1	.46017	.46017	.03983
0	.50000	.50000	.00000
0.1	.53983	.46017	.03983
0.2	.57926	.42074	.07926
0.3	.61791	.38209	.11791
0.4	.65542	.34458	.15542
0.5	.69146	.30854	.19146
0.6	.72575	.27425	.22575
0.7	.75804	.24196	.25804
0.8	.78814	.21186	.28814
0.9	.81594	.18406	.31594
1	.84134	.15866	.34134
1.1	.86433	.13567	.36433
1.2	.88493	.11507	.38493
1.3	.90320	.09680	.40320
1.4	.91924	.08076	.41924

z-Score	p (x < z)	p (x > |z|)	p (x < |z|)
1.5	.93319	.06681	.43319
1.6	.94520	.05480	.44520
1.7	.95543	.04457	.45543
1.8	.96407	.03593	.46407
1.9	.97128	.02872	.47128
2	.97725	.02275	.47725
2.1	.98214	.01786	.48214
2.2	.98610	.01390	.48610
2.3	.98928	.01072	.48928
2.4	.99180	.00820	.49180
2.5	.99379	.00621	.49379
2.6	.99534	.00466	.49534
2.7	.99653	.00347	.49653
2.8	.99744	.00256	.49744
2.9	.99813	.00187	.49813
3	.99865	.00135	.49865
3.1	.99903	.00097	.49903
3.2	.99931	.00069	.49931
3.3	.99952	.00048	.49952
3.4	.99966	.00034	.49966
3.5	.99977	.00023	.49977
3.6	.99984	.00016	.49984
3.7	.99989	.00011	.49989
3.8	.99993	.00007	.49993
3.9	.99995	.00005	.49995
4	.99997	.00003	.49997

cleaning because 99.73% of cases are within 3 SD of the mean. In other words, there is only a 0.27% chance that a case will have a value more than 3 SD from the mean and will have arisen from the same process that produced that distribution. This means that it is highly likely a score that far from the mean is either an error or arose by another process (e.g., not being a legitimate member of the distribution). If I want to be even more conservative, I will use $z = \pm 4$ as a cutoff value, because there is only a 0.006% chance that a score that extreme will arise from the original processes that created the distribution.

You can use Table A.1 for reference or you can explore other values using Excel and the =NORMSDIST(z) function.

AUTHOR INDEX

SUBJECT INDEX